PATTERN RECOGNITION FOR RELIABILITY ASSESSMENT OF WATER DISTRIBUTION NETWORKS

T0304236

PATTERN RECOGNITION FOR RELIABILITY ASSESSMENT OF WATER DISTRIBUTION NETWORKS

DISSERTATION

Submitted in fulfilment of the requirements of
the Board for Doctorates of Delft University of Technology
and of
the Academic Board of the UNESCO-IHE Institute for Water Education
for the Degree of DOCTOR
to be defended in public,
on Monday, February 13, 2012, at 10:00 o'clock
in Delft, The Netherlands

by

Nemanja TRIFUNOVIĆ

Master of Science in Civil Engineering, University of Belgrade, Yugoslavia
born in Zagreb, Yugoslavia

This dissertation has been approved by the supervisor:
Prof. dr. K. Vairavamoorthy

Composition of Doctoral Committee:

Chairman	Rector Magnificus Delft University of Technology
Vice-Chairman	Rector UNESCO-IHE
Prof. dr. K. Vairavamoorthy,	UNESCO-IHE/Delft University of Technology, supervisor
Prof. dr. ir. L. Rietveld,	Delft University of Technology
Prof. dr. D. Solomatine,	UNESCO-IHE/Delft University of Technology
Prof. dr. D. Savić,	University of Exeter, UK
Prof. dr. M. Ivetić,	University of Belgrade, Serbia
Prof. dr. M. Kennedy,	UNESCO-IHE/Delft University of Technology, reserve

CRC Press/Balkema is an imprint of the Taylor & Francis Group, an informa business

© 2012, Nemanja Trifunović

Published by:
CRC Press/Balkema
PO Box 447, 2300 AK Leiden, the Netherlands
e-mail: Pub.NL@taylorandfrancis.com
www.crcpress.com - www.taylorandfrancis.co.uk - www.ba.balkema.nl

ISBN 978-0-415-62116-8 (Taylor & Francis Group)

This research is about water distribution network resilience. Resilience was crucial to bring it to the end.

My gratitude goes to my supervisor, Prof. Kala Vairavamoorthy, who has been persistently emphasising the relevance of this step in my career. His constant encouragements were necessary boosts for the sacrifice I was to go through besides my regular work and private life, both being very demanding in the last five years.

Furthermore, I would like to thank Dr. Assela Pathirana for his intellectual support in the programming side of my work and critical evaluation of my concepts. I was always of opinion that one learns more from those who oppose his/her ideas that from those who share them.

Coincidentally, both Kala and Assela have roots in Sri Lanka, the country until recently torn by ethnic conflicts. I mention this because I am originating from Yugoslavia, the country that did not survive such a conflict. As a consequence, I was building my life in The Netherlands during the most obvious period for doing a PhD. Only later, working closely with these two very sharp brains, I realised that the pointless drama of civil wars may have brought us together in spite of cultural differences caused by our origin.

Next, my thanks go to Mr. Jan-Herman Koster who understood that a PhD can hardly be completed working exclusively outside office hours. The support he gave as my department head in the last two years of the study, and specifically since September 2010, was essential to bring it to the end. Special thanks also go to those departmental colleagues who were picking bits of my regular work in order to open more space for the research. Being part of such a nice and diverse group of people has grossly enriched my life and is still bringing lots of pleasure in my work, in general.

And finally: to my dear Gordana, Stefan and Jana. They have always been my paramount focus and inspiration. Without them, this work would have possibly been finished earlier but the life would have little meaning.

Making persistent effort without any result can be frustrating.
Losing belief and patience can be devastating.
On a dead-end road, the only way is the way back.
After the point of no return, the only way is towards the end.

Contents

Summary

The reliability of water distribution networks has been increasingly recognised as one of the top challenges of water supply companies. Three questions have been particularly relevant, namely: (1) what parameters are the most accurate descriptors of the reliability, (2) what can be considered as an acceptable reliability level, and (3) what is the most appropriate reliability assessment method? Despite lots of efforts, it is still not easy to answer these questions because the reliability assessment is influenced by numerous factors, such as uncertain nature of demand variables, overall condition of the network, the pressure-flow relationship, different standards with respect to water consumption, etc. In fact, the literature reveals that there is no universally acceptable definition and measure for the reliability of water distribution systems.

The study presented in this manuscript aims to investigate the patterns that can be possibly used for reliability assessment of water distribution networks, focusing (1) to the node connectivity, (2) hydraulic performance and energy balance, and (3) economics of network construction, operation and maintenance assessed from the perspective of reliability. Furthermore, a number of reliability measures to evaluate network resilience has been developed and assessed to arrive at more accurate and complete diagnostics of network performance in regular and irregular scenarios. These measures have been proposed as a part of the methodology for snap-shot assessment of network reliability based on its configuration and hydraulic performance.

Practical outcome of the research is the decision support tool for reliability-based design of water distribution networks. This computer package named NEDRA (**NE**twork **D**esign and **R**eliability **A**ssessment) consists of the modules for network generation, filtering, initialisation, optimisation, diagnostics and cost calculation, which can be used for sensitivity analyses of single network layout or assessments of multiple layouts.

The key conclusion of the study is that none of the analysed aspects influencing network resilience develops clear singular patterns. Nevertheless, the proposed network buffer index (NBI) and the corresponding hydraulic reliability diagram (HRD) give sufficient snap-shot assessment; the diagram, as visual representation of the network resilience, clarifies the index composition and displays possible weak points in the network that can be hidden behind the averaged values of other reliability measures used in practice.

Regarding the measures of regular hydraulic performance and their relation to the network resilience, the two other indices proposed: the network power index (NPI) and the pressure buffer index (PBI), correlate to the average value of the available demand fraction (ADF) fairly well, and yield less conservative values compared to the similar factors from the literature. Nevertheless, they are not equally sensitive towards topographic conditions (NPI is more than PBI) and overall level of the head/pressure in the network (PBI is more than NPI). For those reasons, none of the two could have been considered as universally applicable measure. The third index, the network residence time (NRT) appeared to be less suitable as the reliability measure, which also applied for the indices analysing purely the node connectivity i.e. the network shape.

Despite numerous simulations done, the conclusions from the study are still based on the analyses of relatively small number of samples, looking from statistical perspective. Those should be therefore verified by generating even more layouts of various characteristics.

NEDRA package, generating 13,000 layouts for the design of a 50-node synthetic case network, looks capable of doing such task after introducing minor improvements. This decision support tool has been providing full assessment of numerous design alternatives, which can be used while drawing conclusions about the best alternative from the perspective of targeted reliability and accompanying costs.

Samenvatting

De leveringszekerheid van waterdistributienetwerken is steeds meer erkend als een top uitdaging voor watervoorziening bedrijven. Drie vragen zijn bijzonder relevant, namelijk: (1) welke parameters zijn de meest nauwkeurige beschrijvingen van de leveringszekerheid, (2) wat kan worden beschouwd als een acceptabel leveringszekerheid niveau en (3) wat is de meest geschikte leveringszekerheid beoordelingsmethodiek? Ondanks veel inspanningen is het nog steeds niet gemakkelijk deze vragen te beantwoorden omdat de beoordeling van leveringszekerheid wordt beïnvloed door talloze factoren, zoals de onzekere aard van vraag variabelen, de algemene toestand van het netwerk, de druk/volumestroom relatie, verschillende normen met betrekking tot waterverbruik, enz. In feite, blijkt uit de literatuur dat er geen universeel aanvaardbare definitie en de maatregel voor de leveringszekerheid van de waterdistributiesystemen is.

De studie, gepresenteerd in dit manuscript, heeft als doelstelling de patronen te onderzoeken die eventueel kunnen worden gebruikt voor de beoordeling van de leveringszekerheid van waterdistributienetwerken, gericht op (1) de netwerk connectiviteit, (2) een relatie tussen de hydraulische prestaties en energie balans, en (3) de economie van de netwerk opbouw en reconstructie, en exploitatie en onderhoud, allebei beoordeeld vanuit het perspectief van leveringszekerheid. Bovendien werden een aantal maatregelen voor de te evalueren veerkracht in het netwerk ontwikkeld en geëvalueerd om te komen tot meer nauwkeurige en volledige diagnostische gegevens van netwerkprestaties in regelmatige en onregelmatige scenario's. Deze maatregelen zijn voorgesteld als een onderdeel van de methodologie voor een snapshot beoordeling van de leveringszekerheid van het netwerk op basis van de configuratie en hydraulische prestaties.

Praktische uitslag van het onderzoek is het besluit ondersteuningsprogramma voor leveringszekerheid gebaseerde ontwerp van distributienetwerken. Dit softwarepakket met de naam NEDRA ('**NE**twork **D**esign and **R**eliability **A**ssessment' = netwerkontwerp en leveringszekerheid beoordeling) bestaat uit de modules voor netwerk generatie, filteren, initialisatie, optimalisering, diagnostiek en berekening van de kosten, die kan worden gebruikt voor analyses van de gevoeligheid van één netwerk lay-out of evaluaties van meerdere lay-outs.

De belangrijkste conclusie van de studie is dat geen van de geanalyseerde aspecten die netwerk veerkracht beïnvloeden, duidelijk enkelvoudige patronen ontwikkelt. Niettemin geven de voorgestelde netwerk buffer index (NBI) en het bijbehorende hydraulische betrouwbaarheidsdiagram (HRD) voldoende snapshot beoordeling; het diagram, als visuele representatie van de veerkracht van het netwerk, verduidelijkt de samenstelling van NBI en mogelijke zwakke punten van het netwerk worden weergegeven, die kunnen worden verborgen achter de gemiddelde waarden van andere leveringszekerheid maatregelen die in de praktijk zijn gebruikt.

Met betrekking tot de maatregelen van reguliere hydraulische prestaties en hun relatie tot de veerkracht van het netwerk, zijn twee andere indexen voorgesteld: de netwerk macht index ('Network Power Index' - NPI) en de druk buffer index ('Pressure Buffer Index' - PBI), zij correleren aan de gemiddelde waarde van de beschikbare vraag breuk ('Available Demand Fraction' - ADF) vrij goed, en bieden minder conservatieve waarden vergeleken met soortgelijke factoren uit de literatuur. Ze zijn echter niet even gevoelig naar topografische voorwaarden (NPI is meer dan PBI) en het algemene niveau van de hoofd/druk in het netwerk (PBI is meer dan NPI). Om deze redenen kon geen van de twee beschouwd worden als universeel toepasbare maatregel. De derde index, de netwerk verblijftijd ('Network Residence Time' - NRT) bleek minder geschikt als betrouwbaarheidsmaatregel, die ook was toegepast voor de indexen die puur de connectiviteit i.e. de vorm van het netwerk analyseerden.

Ondanks dat talrijke simulaties zijn gedaan, zijn de conclusies van de studie toch gebaseerd op de analyses van een relatief klein aantal voorbeelden, vanuit statistisch perspectief. Deze zouden dan ook moeten worden geverifieerd door het genereren van steeds meer lay-outs van verschillende kenmerken. Na het genereren van 13.000 lay-outs voor het ontwerp van een 50-knooppunt netwerk, blijkt het NEDRA pakket in staat dit te doen, na de introductie van kleine verbeteringen. Dit ondersteuningsprogramma heeft voorzien in volledige beoordeling van talrijke ontwerp alternatieven, welke gebruikt kunnen worden tijdens het trekken van conclusies over het beste alternatief vanuit het perspectief van gerichte leveringszekerheid en de bijbehorende kosten.

List of Figures

List of Tables

Abbreviations

ADF   - Available demand fraction
ANGel   - Artificial network generator
AST   - Appended spanning trees
BFS   - Breadth First Search (algorithm)
BSD   - Berkeley Software Distribution (permissive free software licences)
CSG   - Case study generator
DD   - Demand driven (calculation)
DST   - Decision support tool
EO   - Evolving objects (GA optimiser)
FC   - First cost (of an asset)
GA   - Genetic algorithms
HPP   - Homogeneous Poisson process
HRD   - Hydraulic reliability diagram
MA   - Mechanical availability
MAP   - Minimum acceptable pressure
MDS   - Modular design system (tool)
MTTF   - Mean time to failure
MTTR   - Mean time to repair
MU   - Mechanical unavailability
NBI   - Network buffer index
NCI   - Network connectivity index
NCF   - Network connectivity factor
NDT   - Network diagnostics tool
NEDRA   - Network design and reliability assessment (tool)
NGI   - Network grid index
NGT   - Network generation tool
NHPP   - Non-homogeneous Poisson process
NSI   - Network shape index
EO   - Evolving objects (optimisation software)
PDD   - Pressure-driven demand
PRNG   - Pseudo-random number generator
PW   - Present worth (method)
RG   - Random graph
SWMM   - Storm water management model (software)
VCS   - Virtual case study
VIBe   - Virtual infrastructure benchmarking (tool)
WDN   - Water distribution network

CHAPTER 1

Introduction

_____

1.1 WATER DISTRIBUTION MODELS

Water is life. Water distribution networks live as humans: they are born, grow, get old, may suffer from 'stress', 'high cholesterol', 'blood pressure', 'haemorrhage', 'heart attack' or else. However, they rarely 'die' although they are nearly dead when they have been poorly designed and/or constructed, or little has been done about their operation and maintenance. On contrary, some of their users <u>can</u> die as a result of such an abysmal situation.

A lot has been written about the relevance of potable water for public health. Researchers in the field of water distribution are continuously concerned with performance improvement of distribution networks, analysing specifically the 'diseases' related to water demand and leakage reduction, corrosion growth, water hammer impacts, pump failures or else. Drawing parallels with medicine may therefore not be so ridiculous; it is almost that the average life expectancy of water users could be brought into a proportion with the average lifetime of the network components. Some futuristic research topic could aim at possible correlations between the condition of distribution networks and medical records of the population supplied from them. It is not quite clear how feasible such a research could be, but it is very clear that computer models would be playing essential role in it.

Hardly any field of civil engineering has benefited so early from the development of PC computer technology, as it did the field of water distribution. Hydraulic modelling software launched massively in the developed countries in early eighties, has been speedily introduced all over the world. Water quality modelling applications that followed with the delay of some 10-15 years are nowadays equally available in practice.

Such breakthrough allowed a single water distribution expert to analyse dozens of design and operational scenarios for the same time that would be required by a dozen of experts to analyse a single one in the era of manual calculations and hydraulic tables and diagrams, being daily practice just a couple of decades ago. The trend accelerated significantly by the end of the millennium, and the challenges of water distribution in the 21st century go even beyond optimisation of design and operation of selected layouts. New possibilities are opened to look deeper into the mechanisms of corrosion, sediment transport and other phenomena that are affecting maintenance practices and eventually play role in the overall ageing of the system.

Based on what has been achieved only within the last decade, it is fair to believe that a model that could simulate full behaviour of distribution network throughout a longer period of time than just a few days is not necessarily a dream. Apart from readily available results showing the network hydraulic performance, such model could suggest how the network should be managed throughout its entire design period, namely:
1. What are the most effective maintenance practices?
2. How to deal with renovation and expansion of the network?
3. How to improve the network reliability?

The prerequisites for having such a model, very likely sorted according to the degree of research complexity, are:
- To have complete picture of complex mechanisms that take place in pipes, namely the growth patterns of various deposits, corrosion, biofilms, etc.

- To establish clear relation between these processes and appropriate maintenance methods.
- To have a system of monitoring providing information that can allow reasonable calibration of the model.
- To have good information about economic consequences of certain decisions.
- To develop and/or acquire powerful hardware and programming tools.

Back in the reality, the vast majority of problematic water distribution networks are located in developing countries, whereas the vast majority of good networks exist in developed countries. The problem is in the fact that modern computer software alone cannot help to erase this difference; it is just a tool to work with. Applying it on a network case where good quality information is missing is just like attempting to drive an expensive limousine on a road where maximum possible speed is 30 km/h. In spite of futuristic predictions, the world is also to achieve the Millennium Development Goals, much of them dealing with safe drinking water.

1.2   LIMITS AND RISKS

Any system is composed of components that interact in an equilibrium established within certain limits. Once this equilibrium becomes disturbed, the system tries to restore it. If this is impossible, a calamity is going to occur. Sometime after the calamity, the equilibrium will be restored. Without external intervention, this will be achieved at lower level i.e. at lower limits. In extreme cases, this cannot be achieved, even with external intervention, because the system has collapsed.

Systems operate at certain level of risks. It may be valid to say that *risks partly originate from the lack of awareness about the limits.* For example, a pretty inexperienced driver can drive his/her very bad car, passing by an accident caused by very experienced driver in his/her brand new car. What initially appears to be more reliable transportation system will suffer calamity if the limits of its components: the car, the driver and/or the road in this case, have been underestimated; the latter person was simply driving 'too risky'.

How to quantify risks? These will be normally associated with probability that something bad is going to happen. The less aware of the limits we are, the higher the risk will be. However, we are not necessarily concerned with high risks of catching cold, as we are afraid of low risks to contract some fatal disease. 'High' and 'low' quite often coincide with 'very bad' and 'not that bad'. Hence, the risk is high when something 'unacceptably bad' might happen; no matter if the calculated chance is 10% or 0.0001%. This means that the *level of acceptable risk is proportional to the magnitude of calamity.*

Proper assessment of limits helps to reduce the level of risks. This is easy to say, but where the real limits are is often a difficult question. Science has equipped us with methods and tools that allow fairly accurate estimate of limits in many cases. Nevertheless, some systems are quite complex and it is not always possible to judge interactions between numerous components. Not only that we do not posses sufficient knowledge about their limits but those also change in time. Being unsure about it, engineers will usually try to protect the system from calamities by introducing higher safety factors in their designs. Effectively, the result of this is a system that most of the time operates below its limits, i.e. an inefficient system. Contrary to this, what one would expect from any system is an optimal performance i.e. a performance close to, but not beyond its limits.

Obviously, good knowledge of the system is prerequisite while defining its limits. A certain degree of calamities is inevitable in this learning process. Inflicting a bearable damage can be helpful while assessing and adapting the limits; the equilibrium will be restored after the calamity and conclusions can be drawn accordingly. Moreover, when properly defined system objectives have matched the risks, the effect of calamities will be reduced. Such risks can then be better controlled. Only by accepting controlled risks, we are able to learn more about the limits and how to expand them. By doing this, the growth of the system will be sustainable and its lifetime prolonged.

1.3   RELIABILITY

Systems are deemed reliable if they can withstand predicted level of calamity. Each calamity creates an impact. For instance, in the field of water distribution a pipe burst or electricity failure creates a drop of pressure and consequently a loss of demand. This calamity/impact is more or less probable but can also be more or less intense, and can cover a larger- or smaller area.

Talking about the low- or high reliability, one would have to define it in situations that are sometimes hardly comparable. For instance, a burst of pipe can affect relatively small area severely, or a larger area moderately. Which of these two scenarios depict more/less reliable network? Equally, would a system with new pipes, whose failure can create severe impact for consumers, be considered as more reliable than the system of old pipes, whose failure results in moderate impact?

In general, failures inflict consequences based on wide- or limited spatial coverage of the impact, and its high- or low intensity. Combined with high/low probability of the failure, this leads to four typical cases with growing concerns for consumers:
-   Case 0 (no concern): low spatial coverage with low intensity of the impact.
-   Case 1 (low concern): wide spatial coverage but low intensity of the impact.
-   Case 2 (moderate i.e. limited concern): limited spatial coverage but high intensity of the impact.
-   Case 3 (potentially high concern): wide spatial coverage with high intensity but low probability of impact.

Table 1.1 Reliability bandwidth

Reliability	Low > High							
Impact probability	High	Low	High	High	High	Low	Low	Low
Impact intensity	High	High	Low	High	Low	High	Low	Low
Impact coverage	High	High	High	Low	Low	Low	High	Low

In any calculation of water distribution network reliability, the meaning of low- or high index usually taking a value between 0 and 1 could possibly be associated with a bandwidth as shown in Table 1.1. The final verdict will however depend on the definition of the 'High' and the 'Low', which can be quite different in different countries. According to the studies done in The Netherlands in the late eighties, the average frequency of interruptions affecting the consumers was remarkably low; the chance that no water would run after turning on the tap was once in fourteen years. In the similar period, in 1993, the frequency of interruptions in the water supply system of Sana'a, the capital of the Republic of Yemen, was once in every two days. Some 20 years later, not much has changed in either of the places!

1.4   AIM OF THE STUDY

Water distribution network reliability has become an important issue emerging from increased perception of water as economic goods, which forces water supply companies to start managing their systems in more efficient manner. As a result, the customers' expectations have grown substantially and water companies are confronted with situations where they have to guarantee service levels with higher degree of confidence. To achieve this objective, water experts have at their disposal a number of methods and a range of powerful network modelling software that can run without much of limitations. Despite all these advances, no universally accepted approach in defining and assessing water distribution reliability exists in this moment. It is, above all, the complexity caused by interactions between numerous factors that influence the service level, technical as well as managerial, which make it difficult to address the reliability issues.

The study presented in this manuscript aims to investigate the patterns that can be possibly used for reliability assessment of water distribution networks, focusing (1) to the node connectivity, (2) hydraulic performance and energy balance, and (3) economics of network construction, operation and maintenance assessed from the perspective of reliability. The conducted research has been driven by the fact that the reliability is affected by particular choices in all of these aspects.

Furthermore, a number of reliability measures to evaluate network resilience has been developed and assessed to arrive at more accurate and complete diagnostics of network performance in regular and irregular scenarios. These measures have been proposed as a part of the methodology for snap-shot assessment of reliability based on the network configuration and hydraulic performance. Moreover, the network diagnostics has been developed to compare the costs of reliability increase with the consequent lowering of the calamity impact i.e. the costs of the damage. The optimal reliability is therefore the one achieved with the most effective investment into reconstruction and/or operation of the network to minimize the failures and customer complaints.

Practical outcome of the research is the decision support tool for reliability-based design of water distribution networks. This computer package named NEDRA (**NE**twork **D**esign and **R**eliability **A**ssessment) consists of the modules for network generation, filtering, initialisation, optimisation, diagnostics and cost calculation, which can be used for sensitivity analyses of single network layout or assessments of multiple layouts.

1.5   STRUCTURE OF THE THESIS

The manuscript consists of total ten chapters, seven of which are presenting the results of particular research segments:
-    Chapter 3, which deals with pressure-driven demand calculations applied to arrive at demand calculations under stress conditions.
-    Chapter 4, which explains the algorithm of network generation tool used for statistical analyses of network patterns, and eventually as module of NEDRA decision support tool.

- Chapter 5, which introduces the principles of graphical representation of network reliability and compares the proposed measures i.e. the resilience indices with similar measures known in literature.
- Chapter 6, which focuses to the assessment of network resilience based on different nodal connectivity and pipe diameters.
- Chapter 7, which investigates energy balance during regular operation as the measure of network reliability/resilience.
- Chapter 8, which discusses the preliminary assessment of network reliability based on various combinations of hydraulic operation against the different construction-, and operation and maintenance costs.
- Chapter 9, which illustrates the use of NEDRA decision support tool for reliability-based design of water distribution networks.

Next to this introductory chapter, Chapter 2 discusses the theoretical and conceptual framework of water distribution network reliability, including the discussion on the research gaps identified at the beginning of the research, and the objectives and methodology developed to tackle some of these. The final chapter (10) gives the summary of conclusions and recommendations for further research.

1.6   ACKNOWLEDGEMENTS

Particular results presented in this manuscript have been produced with contributions (in alphabetical order) of:
1. Mr. Bharat Maharjan from Nepal,
2. Mr. Lytone Kanowa from Zambia,
3. Mr. Santosh Chobhe from India, and
4. Mr. Sung Jae Bang from South Korea.

All four gentlemen are alumni of UNESCO-IHE who have graduated in the MSc programme in Municipal Water and Infrastructure, defending their MSc theses on the topics I developed within the framework of this research, and guided as their mentor. In the process that is following the submission of this manuscript, all of them will be recognised as the co-authors of corresponding journal/conference papers.

Finally, I would like to thank Mr. Kees van der Drift, the Head of Water Distribution Research and Development of WATERNET Amsterdam, for providing me with the EPANET model of the part of Amsterdam water distribution network used as a case study in Chapter 5.

CHAPTER 2

Theoretical and Conceptual Framework

2.1   BACKGROUND

Even in developing countries, where reliable distribution network is still not a high priority, the customer awareness regarding the service levels has risen significantly in present days. Although many of the systems there suffer from water shortage at the source, poor state of assets and lack of funds caused by low tariffs and/or poor revenue collection, a reliability analysis makes sense in order to explore optimal operation under given constraints. As mentioned by Vairavamoorthy et al. (2001), many water companies operating such systems will by default be unable to offer 24-7 supply but could, by applying a kind of 'constrained optimisation', achieve a 'regular irregularity' i.e. a fair share of limited water quantities for all consumers and at predictable intervals.

Nowadays, water experts feel overall more confident to address the reliability concerns having at their disposal powerful network modelling software. Some of these programmes are distributed through the Internet at nominal costs or even free of charge; EPANET software developed by the US Environmental Protection Agency is a well-known representative of this group (Rossman, 2000). Other, commercially available software include direct interaction with data available in geographical information systems and are able to process the results and calibrate the model in more sophisticated manner. On top of it, the latest generation of computer programmes, used in water distribution, includes optimisation modules based on methods applying genetic algorithms. Walski et al. (2003) offer a comprehensive overview of the state-of-the-art water distribution network modelling practices.

2.2   RELIABILITY ASSESSMENT OF WATER DISTRIBUTION NETWORKS

Already in the last two decades, the reliability of water distribution networks has been recognised as one of the top water supply challenges. Three questions have been particularly relevant, namely: (1) what parameters are the most accurate descriptors of the reliability, (2) what can be considered as an acceptable reliability level, and (3) what is the most appropriate reliability assessment method? Despite lots of efforts, it is still not easy to answer these questions because the reliability assessment relates to numerous factors, such as uncertain nature of demand variables, overall condition of the network, the pressure-flow relationship, different standards with respect to water consumption, etc. In fact, the literature reveals that there is no universally acceptable definition and measure for the reliability of water distribution systems.

Several researchers have tried to define the network reliability; Table 2.1 contains a few frequently cited quotes in the literature. In most of these, the reliability is related to the (nodal) pressures and flows. As such, the reliability reflects the ability of distribution network to satisfy consumer demands i.e. a certain level of service subject to the pressure and demand requirements under both normal and abnormal conditions. This conclusion brings the definition of reliability back to its roots, stating that quantified reliability, in general, is a measure of system performance. Having said that in case of water distribution network implies its hydraulic performance, which is in the literature defined as the *hydraulic reliability.*

Table 2.1 Definitions of water distribution system reliability

Source	Definitions
Cullinane et al. (1992)	'Reliability is the ability of the system to provide service with an acceptable level of interruption in spite of abnormal conditions of water distribution system to meet the demand that are placed on it.'
Gouter (1995)	'Reliability is the ability of a water distribution system to meet the demands that are placed on it where such demands are specified in terms of the flows to be supplied (total volume and flow rate) and the range of pressures at which the flows must be provided.'
Xu & Goulter (1999)	'Reliability is the ability of the network to provide an adequate supply to the consumers, under both regular and irregular operating conditions.'
Tanyimboh et al. (2001)	'Reliability is the time-averaged value of the flow supplied to the flow required.'
Lansey (2002)	'Reliability is the probability that a system performs its mission under a specified set of constraints for a given period of time in a specified environment.'

Low pressure is unanimously considered as the primary indicator of poor hydraulic performance/level of service, leading to almost certain demand reduction. This can be a result of inadequate operation, for instance an insufficient pumping caused by electricity failure. Furthermore, in many distribution systems, low pressures occur as a result of scarce water source. Common denominator for both of these cases is that the consumer demand has exceeded the supply. On a longer term, this can also happen from ageing of the system and/or the growth of population. If, on the other hand, a certain level of service is not satisfied due to poor condition of the network resulting in frequent failures of its components, such situation will be described by so-called *mechanical reliability*. Component failures in distribution systems involve pipe bursts, blockage of valves, failure of pumping stations, etc., which all reduce the delivery capacity of the network such that it is no longer able to meet the required service level. Pipes as major network components are subject to structural deterioration because of physical, environmental and operational stresses leading to a failure, as shown in Table 2.2 (Source: NGSMI, 2002).

Mathematically, the mechanical reliability $R(t)$ of a component is defined as the probability that the component experiences no failures during an interval from time 0 to time t. In other words, the reliability is the probability that the time to the failure T exceeds t. The formula for $R(t)$ is:

$$R(t) = \int_{t}^{\infty} f(t)dt \qquad\qquad 2.1$$

where $R(t)$ is the reliability factor, having the values between 0 and 1, and $f(t)$ is the *probability density function* of the time to the failure, which can be developed from the failure records.

This concept of reliability is meant for so called non-repairable components, in which the component has to be replaced after it fails. Nevertheless, most of the components in water distribution systems are generally repairable and can be put back into operation. It seems therefore more appropriate to use the concept of *component availability* as a surrogate.

Table 2.2 Factors that contribute to water system deterioration (Source: NGSMI, 2002)

Physical factors	Impact /Cause
Pipe material	Pipes made from different materials fail in different ways.*
Pipe wall thickness	Corrosion will penetrate thinner walled pipe more quickly.
Pipe age	Effects of pipe degradation become more apparent over time.
Pipe vintage	Pipes made at a particular time and place may be more vulnerable to failure.
Pipe diameter	Small diameter pipes are more susceptible to beam failure.
Type of joints	Some types of joints have experienced premature failure.
Thrust restraint	Inadequate restraint can increase longitudinal stresses.
Pipe lining and coating	Lined and coated pipes are less susceptible to corrosion.
Dissimilar metals	Dissimilar metals are susceptible to galvanic corrosion.
Pipe installation	Poor installation practices can damage pipes, making them vulnerable to failure.
Pipe manufacture	Defects in pipe walls produced by manufacturing errors can make pipes vulnerable to failure.
Environmental factors	**Impact /Cause**
Pipe bedding	Improper bedding may result in premature pipe failure.
Trench backfill	Some backfill materials are corrosive or frost susceptible.
Soil type	Some soils are corrosive; some soils experience significant volume changes in response to moisture changes, resulting in changes to pipe loading. Presence of hydrocarbons and solvents in soil may result in some pipe deterioration.
Groundwater	Some groundwater is aggressive toward certain pipe materials.
Climate	Climate influences frost penetration and soil moisture.
Pipe location	Migration of road salt into soil can increase the rate of corrosion.
Disturbances	Underground disturbances in the immediate vicinity of an existing pipe can lead to actual damage or changes in the support and loading structure on the pipe.
Stray electrical currents	Stray currents cause electrolytic corrosion.
Seismic activity	Seismic activity Seismic activity can increase stresses on pipe and cause pressure surges.
Operational factors	**Impact /Cause**
Internal water pressure, transient pressure	Changes to internal water pressure will change stresses acting on the pipe.
Leakage	Leakage erodes pipe bedding and increases soil moisture in the pipe zone.
Water quality	Some water is aggressive, promoting corrosion.
Flow velocity	Rate of internal corrosion is greater in unlined dead-ended mains.
Backflow potential	Cross connections with systems that do not contain potable water can contaminate water distribution system.
O & M practices	Poor practices can compromise structural integrity and water quality.

* More details can be found in Brandon (1984)

Whereas the reliability is the probability that the component experiences no failures during the time interval 0 - t, the availability of a component is the probability that the component is functional at time t, assuming that the component is as good as when it was new at time 0. For example, after a segment of broken pipe is replaced, this pipe will function again as one of the system components. Evaluating the pipe itself, it could be considered as a non-repairable component if it has to be replaced by a new one, while in the evaluation of the pipe as a component of the water distribution system, it is repairable i.e. available component.

The availability of a component can also be expressed as a percentage of the time during which the component is in operational state. The remaining is the time when the component is not functional, most frequently due to a failure or some maintenance. There are two basic categories of maintenance events:
-   *Corrective maintenance*, which includes repair after a breakdown or unscheduled maintenance resulting from the equipment failure. This additional information should be available from the maintenance records, depending on the pipe material and its condition; the records can be summarised in the form of a diagram as shown in Figure 2.1 (source: Vreeburg and van den Hoven, 1994).
-   *Preventive maintenance*, which includes regular activities to prevent breakdowns before they occur.

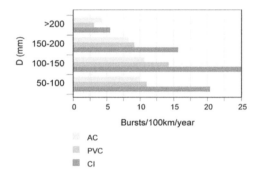

Figure 2.1 Pipe bursts frequency; example from The Netherlands (Vreeburg and van den Hoven, 1994)

On annual basis, the component availability can be calculated as:

$$A = \frac{8760 - CMT - PMT}{8760} \qquad\qquad 2.2$$

where A stands for component availability, CMT is annual corrective maintenance time in hours, and PMT is annual preventive maintenance time in hours.

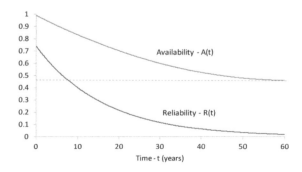

Figure 2.2 Reliability and availability trends (adapted from Tung, 1996)

The final outcome of the reliability assessment will be a coefficient that, for given level of demand, takes the values between 0 and 1. As a consequence, the availability index of a distribution system will always be higher than its reliability index, as Figure 2.2 shows (adapted from Tung, 1996). Parallel to the mechanical availability, some researchers express the hydraulic reliability as *the hydraulic availability*, pointing to the availability of (sufficient) demands and pressures in the system.

Next to the concept of availability, several other measures have been proposed as reliability surrogates at the early stage of computer modelling applications in water distribution, but none of them sufficiently complete to be universally accepted for reliability analysis. For instance, the concepts of *nodal connectivity* and *reachability* were first applied by Wagner et al. (1988a) using the algorithms of Satyanarayana and Wood (1982). The nodal connectivity is defined by the probability that every demand node in the network is connected to at least one source, while the nodal reachability is the probability that a specified demand node is connected to at least one source. The corresponding reliability indices consider the pipe connection and failure rate without running a hydraulic simulation. In reality, a particular node may not receive any water, due to low supplying heads and/or high energy losses, although a fully operational path may exist to the water source. For that reason, both methods lack practical validity and can only be used for initial screening of network geometry. Furthermore, a concept of *network redundancy* was discussed by Morgan and Goulter (1985) and Goulter (1987), which evaluates network reliability by looking for a presence of independent, hydraulically adequate paths between the source and demand node, which may be used in case of a failure of the standard supply path. Finally, Goulter et al. (2000), proposed reliability index referred to as *network vulnerability*, which is based on economic consequences of the shortages, related to the frequency, duration, and severity of failure events.

2.3   CLASSIFICATION OF METHODS FOR RELIABILITY ASSESSMENT

Ostfeld (2004) has classified the approaches to the reliability assessment of water distribution systems in three groups: (1) analytical- (connectivity/topological), (2) simulation- (hydraulic) and (3) heuristic (entropy) approaches; these arer summarised in Table 2.3.

Analytical approaches deal with the layout of water distribution network, which is associated with the probability that a given network keeps physically connected, given its component reliabilities. These are the approaches linked to the above-mentioned concepts of connectivity and reachability that are not based on hydraulic simulations.

Simulation approaches deal with the hydraulic reliability and availability. Thus, they analyse the hydraulic performance of the network, i.e. a conveyance of desired quantities and qualities of water at required pressure to the appropriate locations at any given time. Therefore these approaches rely heavily on hydraulic models and require very good information about the network layout and operation, including the records related to the component failures. Owing to the availability of powerful computers and software, these are the most widely explored approaches, nowadays.

Heuristic approaches reflect the redundancy using entropy measure, as a surrogate for network reliability. Tanyimboh and Templeman (2000) suggested algorithms for maximizing the entropy of flows for single-source networks and they summarized the existing approaches to develop the relationships between the reliability and entropy. Here, the key question is what a given level of entropy means in terms of reliability for a given distribution system.

Table 2.3 Approaches for reliability assessment of water distribution systems (Ostfeld, 2004)

Authors	Reliability Measure	Methodology	Applications
ANALYTICAL APPROACHES			
Goulter (1987)	General overview/trends	Overview	Overview
Jacobs and Goulter (1988)	Enumeration of all possible combinations of working/non-working system components	State enumeration, filtering and heuristic procedures	Small illustrative example
Wagner et al. (1988)	Connectivity and Reachability	Graph theory algorithms (Rosenthal, 1977; Satyanarayana and Wood, 1982)	Small illustrative example
Jacobs and Goulter (1989)	Redundancy measures arising from system layout	Integer goal programming	AnyTown USA (Walsky et al., 1987)
Shamsi (1990)	Nodal Pair Reliability (NPR): the probability of two nodes being connected	Minimal path sets/minimum cut-sets.	Small illustrative looped network
Quimpo and Shamsi (1991)	Idem	Idem	City of Norwich, state of New-York Hydraulic
SIMULATION APPROACHES			
Su et al. (1987)	Probability of satisfying nodal demands and pressure heads for various possible pipe failures	Minimum cut-set	Small illustrative looped network
Wagner et al. (1987b)	List of Even-Related, Node-Related, Link-Related and System-Related	Stochastic (Monte Carlo) simulation	AnyTown USA (Walsky et al., 1987)
Bao and Mays (1990)	Probability of satisfying nodal demands at required pressure heads	Stochastic (Monte Carlo) simulation	Small illustrative looped network
Cullinane et al. (1992)	Hydraulic availability: the proportion of time the system satisfactory fulfils its function – minimum pressure at consumer nodes	Hydraulic simulation linked with non-linear optimisation	Small illustrative looped network
Fujiwara and Ganesharajan (1992)	Expected served demand	Markov chain approach	Small illustrative looped network
Xu and Goulter (1988)	Probability of meeting nodal demands at, or above, a minimum prescribed pressure	Two-stage stochastic assessment based on a linearised hydraulic model	Small illustrative example
Shinistine et al. (2001)	Probability of satisfying nodal demands and pressure heads for various possible pipe failures	Minimum cut-set (Su et al., 1987)	Two large-scale municipal water distribution networks, Tucson Arizona
Weintrob at al. (2001)	Required demands at acceptable pressures	Fast stochastic simulation (Lieber et al. 1999) + a linear optimisation model	EPANET Example 3 (Rossman, 2000)

Ostfeld (2001)	Probability of zero annual shortage	Hydraulic analysis + stochastic simulation	The regional water supply system of Nazareth, Israel
Ostfeld et al. (2002)	Probability of Fraction of Delivered Volume (FDV), Fraction of Delivered Demand (FDD) and Fraction of Delivered Quality (FDQ)	Stochastic (Monte Carlo) simulation	EPANET Examples 1 and 3 (Rossman, 2000)
HEURISTIC APPROACHES			
Awumah et al. (1990, 1991)	Entropy measures based on flow and consumption	Simple entropy reliability expression calculations	Illustrative examples
Awumah and Goulter (1992)	Idem	Entropy measures as constraints in optimal design of WDS; use of non-linear optimisation	Illustrative examples
Tanyimboh and Templeman (1993)	Idem	Tailored maximum entropy flow algorithm for single source networks	Small illustrative example
Tanyimboh and Templeman (2000)	Summary of previous work	Tailored maximum constrained approach	Small illustrative example

2.4   MODELLING FAILURES IN WATER DISTRIBUTION SYSTEMS

Generally, two distinct types of events can induce a water distribution system to a failure state. The first one is termed as the *hydraulic performance failure*, which is related to the situations where the demand imposed on a system exceeds the capacity of the system. The second one is related to a component failure, or the *mechanical failure*, which can lead (but not necessarily) to the hydraulic performance failure. It involves actual failures of the network reducing its conveying capacity during the failure but also after the failed component is isolated and undergoing a repair. This situation can be prevented through a selection of larger diameter pipes and larger capacities for other components of the network (Lansey et al., 2002; Trifunović, 2006). Many researchers have focused on the mechanical failures, assuming their probability as a crucial aspect of the network reliability assessment.

2.4.1 Pipe Failures

Given their huge number and variety of conditions (different material, size, age, type and frequency of maintenance), pipes are commonly analysed on mechanical failures. The objective of modelling pipe failure rate is to reproduce adequately the average tendency of the annual number of pipe breaks and to predict breakage rates in the future. According to the Watson et al. (2001), the modelling of pipe failures can be mainly grouped into three categories: (1) survival analyses, (2) aggregated (regression) models, and (3) probabilistic predictive models.

Survival analyses focus on the lifetime of a pipe and are primarily used for a long term financial planning. The pipe lifetime is treated as a random variable and a standard statistical distribution is then fitted to a collection of similar pipes. The pipe group can then be aged to assess what the likely future costs of replacement would be. Data stratification is required in this analysis (Pelletier at al. 2003). For example, it can be necessary to identify the time to failure between the installation and the first break, between the first and second break, etc. in order to distinguish between the subsequent break occurrences. This analysis relies heavily on a proper estimate of the lifetime and is highly dependent on the individual pipe characteristics.

In the *aggregated (regression) models*, pipes with the same intrinsic properties are grouped together and then use linear regression to establish a relationship between the age of the pipe and the number of failures. Shamir and Howard (1979) proposed an exponential model at which the pipe failure is increased with time (also quoted by Engelhardt et al., 2000):

$$\lambda(t) = \lambda(t_o)e^{A(t-t_o)}$$
2.3

where, $\lambda(t)$ is the average annual number of failures per unit length of the pipe surveyed at year t, t_0 is the base year for analysis, and A is the growth rate coefficient between year t_0 and t. This approach does not provide any information about the variability that may exist between individual pipes in general. A number of researchers have used the multiple regressions to improve the above equation to relate the environmental and intrinsic properties of the pipe. Su et al. (1987) have obtained a regression equation that correlates the failure rate λ and pipe diameter D using the data from the 1985 St. Louis Main Break Report (also quoted by Gargano and Pianese, 2000):

$$\lambda = \frac{0.6858}{D^{3.26}} + \frac{2.7158}{D^{1.3131}} + \frac{2.7685}{D^{3.5792}} + 0.042$$
2.4

where, D is the pipe diameter in inches and λ is the failure rate in breaks/mile/year.

Probabilistic predictive models try to predict the probability that a pipe will burst at a particular moment. This probability can then help to calculate the economic life of the pipe and therefore predict when it should be replaced. Andreo et al. (1987) proposed the use of the *Cox Proportional Hazard Model* to consider the hazard function to this kind of model. The basic form of this model is:

$$h(t : z) = h_o(t)e^{zb}$$
2.5

where, $h(t:z)$ is the failure rate at time t related to factor z , $h_0(t)$ baseline hazard function, z is a vector of explanatory variables (diameter, soil, etc.), and b is a vector of regression coefficients. These models are more sophisticated but typically lack statistical rigour in their information and suffer from a reliance on complete long-term failure records.

2.4.2 Lifetime Distribution Models

The failure rates can also be analysed using the *lifetime distribution models*. For a well-designed and well-installed water distribution network, the series of pipes can be considered repairable as the cost of failure is small in comparison to the replacement. The main reasons for this are that firstly, the pipes are mostly constructed under the ground and possible environmental effects of a failure are minimal. Secondly, storage tanks usually available within the network will keep a buffer volume for emergency situations. And thirdly, networks typically contain redundancy and flexibilities planned in their layout.

Many repairable systems, including pipes, typically have a 'bathtub' shaped intensity function. This is shown in Figure 2.3, where $\lambda(t)$ indicates the failure rate (adapted from Neubeck, 2004).

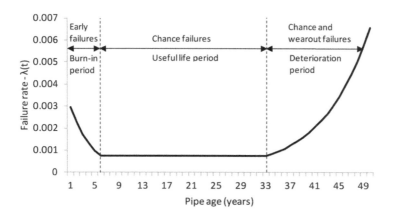

Figure 2.3 Pipe failure rate as a function of age (adapted from Neubeck, 2004)

Immediately after a pipe has been laid and put into operation, the failure rate can be high and because of poor transportation, stacking or workmanship during the installation. After early faults have settled down, the intensity of bursts will be decreasing and remain relatively constant for long periods of the pipe useful life. Following this period, the pipe starts deteriorating faster, and the intensity of bursts will increase again. These bursts are considered as wear-out failures.

Two approaches have been suggested in the literature to model the failure life time distribution: (1) Homogeneous Poisson Process (HPP) and (2) Non-homogeneous Poisson Process (NHPP). The HPP model neglects the time component of the failure and as such is mostly appropriate for renewable systems where repairs are executed regularly and the age of pipe is within the useful life period. According to Shinistane et al. (2002), the probability of a pipe failure using the *Poisson probability distribution* is:

$$p_j = 1 - e^{-\beta_j}, \text{ where } \beta_j = \lambda_j L_j \qquad\qquad 2.6$$

Exponent β_j is the expected number of failures per year for pipe j, and λ_j and L_j its failure rate and length, respectively. Goulter and Coals (1986) and Su et al. (1987) also used similar methods with the HPP model to determine the probability of failure of individual pipes.

On the other hand, the NHPP model considers the time component and is therefore suitable for the burn-in- and deterioration period during which the times between the failures are neither independent nor identically distributed. The NHPP model also assumes negligible repair times, meaning that the repair time will have no effect on the increase of the pipe failure rate. This is considered acceptable when comparing a pipes lifetime measured in years with repair times counted in hours (Watson et al., 2001).

Tobias and Trindade (1995) describe the NHPP model in two versions:

(1) Power relation model: $\lambda(t) = \dfrac{dM(t)}{dt} = abt^{b-1}$ 2.7

(2) Exponential model: $\lambda(t) = e^{c+bt}$ 2.8

In the above equations $\lambda(t)$ is the pipe failure rate at time t, $dM(t)$ is expected number of failures between time 0 and t, and a, b, and c are empirically determined parameters from the historical burst records.

Although the probabilistic aspect of pipe failures have been incorporated in many reliability studies (e.g., Cullinane et al., 1992; Gargano and Pianese, 2000; Shinstine et al., 2002), the underlying assumption was that the successive pipe failures can be modelled using HPP with the constant failure rate, λ. Hence, the Poisson probability distribution, given by Equation 2.6 has been commonly used in reliability model of repairable systems.

2.5 SIMULATION APPROACHES USING DEMAND-DRIVEN MODELS

In the assessment of water network reliability, the simulation approaches that are applied by manipulating the demand-driven models were firstly used, despite the inability of the demand-driven algorithm to calculate the demand reduction resulting from the pressure drop caused by failure. These deficiencies have been coped with by applying an alternative way of modelling of nodes, for instance as virtual tanks that should supply the demand.

The application is based on simulations of single pipe failures throughout the system. The effects of the failure of subsequent pipes are analysed by comparing the level of service after each failure to the one established by setting a threshold pressure under normal operating conditions. The hydraulic reliability and/or availability can therefore be determined based on the pressure- or demand deficit in the system. Two of such approaches are elaborated further in this section.

2.5.1 Reliability Approach Based on Pressure Drop Analysis

In its most rudimentary form, this approach does not quantify the reliability although it clearly gives an idea about the magnitude of the failure; the drop of the pressure will obviously be more significant in case of the burst of a major pipe rather than the peripheral one. On the other hand, such kind of analysis does not require any additional information next to standard input required for the demand-driven models, which makes it relatively simple to apply in situations where the network information is scarce.

A method that quantifies reliability by analysing the pressure deficit in the system has been proposed by Cullinane (1989). According to this method, the nodal reliability can be defined as a percentage of the time in which the pressure at the node is above the defined threshold. Cullinane states it as follows:

$$R_i = \sum_{k=1}^{m} \frac{r_{ki} t_k}{T}$$

2.9

where R_i = hydraulic reliability of node i; r_{ki} = hydraulic reliability of node i during time step k; t_k = duration of the time step; m = total number of the time steps; and T = length of the simulation period. $r_{ki} = 1$ for the nodal pressure p_{ki} equal or above the threshold pressure p_{min}, and $r_{ki} = 0$ in the remaining case of $p_{ki} < p_{min}$. For equal time intervals, $t_k = T/m$. The reliability of the entire system consisting of n nodes can be defined as the average of all nodal reliabilities:

$$R = \sum_{i=1}^{n} \frac{R_i}{n}$$

2.10

The above equations assume that the components and sub-components are fully functional, i.e. 100% available, which is rarely the case. Applying so called expected value of the nodal reliability includes impacts of the availability on the hydraulic performance. This value can be determined as follows:

$$RE_{ij} = A_j R_{ij} + U_j R_i$$

2.11

where RE_{ij} = expected value of the nodal reliability while considering pipe j; A_j = availability of pipe j i.e. the probability that this pipe is operational; U_j = unavailability of pipe j i.e. the probability that it is non-operational; R_{ij} = reliability of node i with pipe j available i.e. operational; and R_i = reliability of node i with pipe j unavailable i.e. non-operational.

Availability A_j is determined by Equation 2.2, while $U_j = 1 - A_j$. The values for R_{ij} and R_i are calculated by Equation 2.9, running the network simulation once with pipe j operational, and then again, by excluding it from the layout. With such correction of the nodal reliability, the overall system reliability can be calculated by Equation 2.11.

Table 2.4 Case network: node properties

Node	Elevation	Demand
	msl	l/s
n1	16.8	3
n2	22.2	3
n3	17.5	3
n4	20.2	3
n5	14.6	3
n6	14.3	4
P1	14.0	-
P2	15.0	-

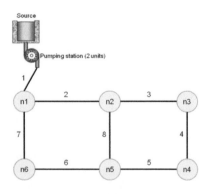

Figure 2.4 Sample case network (Trifunović and Umar, 2003)

The method is illustrated on a simple system of two loops, shown in Figure 2.4 (Trifunović and Umar, 2003). The data about the nodes and pipes are given in Tables 2.4 and 2.5, respectively. An arbitrary demand pattern with the morning- and afternoon peaks has been applied.

Table 2.5 Case network: pipe properties

Pipe	Diameter	Length	Roughness	Burst rate[1]	Burst frequency
	mm	m	mm	$km^{-1}yr^{-1}$	year/burst
1	200	300	0.5	0.71	4.69
2	200	300	0.5	0.71	4.69
3	200	300	0.5	0.71	4.69
4	150	400	0.5	1.04	2.40
5	100	300	0.5	1.04	3.21
6	100	300	0.5	1.04	3.21
7	100	400	0.5	1.04	2.40
8	150	400	0.5	1.04	2.40

1) Assumed based on Gupta & Bhave (1994)

The pumping station at the source consists of two units with following characteristics:
1. *P1* – duty head = 25 mwc, duty flow = 40 l/s (operated constantly).
2. *P2* – duty head = 40 mwc, duty flow = 40 l/s (operated during peak supply).

Operation of the system should have met the following requirements:
- The threshold pressure of 20 mwc.
- Additionally, node 6 must keep minimum 30 mwc during the maximum peak hour.

Assuming that only the corrective maintenance is applied, Table 2.6 displays the information about the pipe availabilities based on the information from Table 2.5 and applying Equation 2.6. The average repair time of two days has been adopted for each burst.

Table 2.6 Case network: pipe availability

Pipe	Burst rate	CMT	A_i	U_i
	yr^{-1}	days/year		
1	0.213	0.426	0.9988	0.0012
2	0.213	0.426	0.9988	0.0012
3	0.213	0.426	0.9988	0.0012
4	0.416	0.832	0.9977	0.0023
5	0.312	0.624	0.9983	0.0017
6	0.312	0.624	0.9983	0.0017
7	0.416	0.832	0.9977	0.0023
8	0.416	0.832	0.9977	0.0023

Nine simulation runs have been conducted using EPANET software, one with the complete system and eight by disconnecting each of the pipes, one at a time. Based on the pressure variation in the system, the nodal reliabilities were calculated by Equation 2.9; these results are shown in Table 2.7. The figures in bold demonstrate the principle: with all pipes available, the pressure in node *n4* is above the threshold during 14 out of 24 hours, whereas the calculation with pipe 3 closed shows the sufficient pressure in this node during seven hours, only. Hence, $R_{43} = 14/24 = 0.5833$, $R_4 = 7/24 = 0.2917$ and $RE_{ij} = 0.5833 \times 0.9988 +$

0.2917 x 0.0012 = 0.583. Finally, the overall reliability factor of the whole system, from Equation 2.11, is 0.833.

Table 2.7 Sample of the reliability calculation based on the method of Cullinane

Node		Pipe								RE$_{ij}$
		1	2	3	4	5	6	7	8	average
n1	R$_{1i}$	1	1	1	1	1	1	1	1	
	R$_1$	0	1	1	1	1	1	1	1	0.9999
	RE$_{1i}$	0.9988	1	1	1	1	1	1	1	
n2	R$_{2i}$	0.5417	0.5417	0.5417	0.5417	0.5417	0.5417	0.5417	0.5417	
	R$_2$	0	0	0.5417	0.5417	0.5417	0.5417	0.5417	0.5417	0.5415
	RE$_{2i}$	0.5410	0.5410	0.5417	0.5417	0.5417	0.5417	0.5417	0.5417	
n3	R$_{3i}$	0.8750	0.8750	0.8750	0.8750	0.8750	0.8750	0.8750	0.8750	
	R$_3$	0	0	0.4167	0.8750	0.8750	0.8750	0.8333	0.8750	0.8747
	RE$_{3i}$	0.8740	0.8740	0.8745	0.8750	0.8750	0.8750	0.8749	0.8750	
n4	R$_{4i}$	0.5833	0.5833	**0.5833**	0.8533	0.5833	0.5833	0.5833	0.5833	
	R$_4$	0	0	**0.2917**	0.5417	0.5833	0.6250	0.5833	0.5833	0.5831
	RE$_{4i}$	0.5827	0.5827	**0.5830**	0.5832	0.5833	0.5834	0.5833	0.5833	
n5	R$_{5i}$	1	1	1	1	1	1	1	1	
	R$_5$	0	0	0.8750	0.9167	1	1	0.9167	0.8333	0.9996
	RE$_{5i}$	0.9988	0.9988	0.9999	0.9998	1	1	0.9998	0.9996	
n6	R$_{6i}$	1	1	1	1	1	1	1	1	
	R$_6$	0	0	0.8750	0.9167	1	0.8333	0.6667	0.8333	0.9995
	RE$_{6i}$	0.9988	0.9988	0.9999	0.9998	1	0.9997	0.9992	0.9996	

2.5.2 Reliability Based on Demand Reduction Analysis

The second group of the methods that look into the network reliability by analysing single pipe failures are those based on the assessment of the demand reduction inflicted by the failure. The overall network reliability factor can be determined by a simplified formula:

$$R = 1 - \frac{Q - Q_f}{Q}$$

2.12

where Q_f represents the available demand in the system after the pipe failure, against the original demand Q. The approach suggested by Ozger and Mays (2003) takes into consideration the probabilities of the pipe failures; the system reliability is expressed in terms of *available demand fraction* (ADF) under the minimum required service pressure:

$$R = 1 - \frac{1}{m}\sum_{j=1}^{m}(1 - ADF_{net}^{j})P_j$$

2.13

where ADF_{net} is the network available demand fraction resulting from the failure of pipe j, P_j is the probability of the failure of that pipe, and m is the total number of pipes in the network. In the above equation, P_j is determined by using the Poission probability distribution described by Equation 2.6. In calculation of the system availability, this method considers the first and the second order failures:

$$A = ADF_{net}^{0} MA + \sum_{j=1}^{m} ADF_{net}^{j} u_j + \sum_{j=1}^{m-1} \sum_{k=j+1}^{m} ADF_{net}^{jk} u_{jk} \qquad 2.14$$

where ADF^{0} is the available demand fraction with fully functional network, ADF^{j} is the available demand fraction after the failure of pipe j, and ADF^{jk} is the available demand fraction after the simultaneous failure of pipes j and k. Furthermore, MA stands for the mechanical availability of the system that can be calculated as:

$$MA = \prod_{j=1}^{n} MA_j \qquad 2.15$$

which is the probability that all n pipes are operational. The mechanical availability of pipe j is defined as:

$$MA_j = \frac{MTTF_j}{MTTF_j + MTTR_j} \qquad 2.16$$

where $MTTR_j$ is the *mean time to repair* and $MTTF_j$ is the *mean time to failure* of this pipe, which is given by:

$$MTTF_j = \frac{1}{\lambda_j L_j} \qquad 2.17$$

Furthermore, in Equation 2.14, the probability of a failure of the jth pipe and all other components remaining fully functional is given as:

$$u_j = MA \frac{MU_j}{MA_j} \qquad 2.18$$

For simultaneous failure of pipes j and k:

$$u_{jk} = MA \frac{MU_j}{MA_j} \frac{MU_k}{MA_k} \qquad 2.19$$

In both equations, MU stands for the mechanical unavailability that equals $1-MA$.

Yoo et al. (2005) tested this method on the sample network taken from Khomsi et al. (1996), shown in Figure 2.5. The Hazen-Williams coefficient and the length of all pipes were set at 130 and 1000 m, respectively. The diameter of pipe 3 is 100 mm, pipes 7 and 8 = 150 mm, pipes 4 and 5 = 200 mm, pipes 1 and 2 = 250 mm and pipe 6 = 300 mm. A hypothetical $MTTF$ was specified for each pipe, in the range of 0.039 and 0.329 bursts per km per year, and the threshold pressure and $MTTR$ were set at 20 mwc and one day, respectively. Figure 2.5 clarifies the procedure to calculate ADF, which has six steps:

Step 1: A fully functional network with no pipe failures is analysed. All nodes satisfy the threshold (i.e. the minimum acceptable pressure) of 20 mwc (MAP). The total demand is 150.75 l/s.

Step 2: Simulation of the system without pipe 6 in operation shows all nodes to have the pressure below MAP, calculated in the demand-driven mode. Nodes 3 to 6 have negative pressures.

Step 3: All nodes with the pressure below MAP (in this case all in the network) are transformed into a system consisting of virtual reservoir with the head equal to the sum of nodal elevation and MAP. Each reservoir is connected with short pipe containing a non-return valve to prevent the backflow from the reservoir. To assess the actual demand, the original nodal demand has been set to zero.

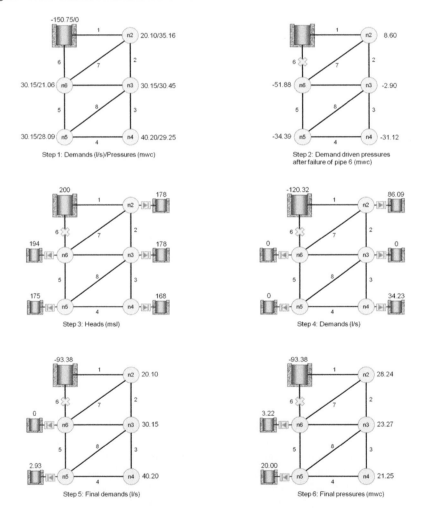

Figure 2.5 Steps for calculation of ADF (Yoo et al., 2005)

Step 4: After the modification of the network, the model simulation runs again. If one or more virtual reservoirs have received more water than the original demand of the corresponding node, those virtual reservoirs are removed from the network and the original nodal demands are restored. For example, the demand of virtual reservoir 7 (86.09 l/s) has exceed the original demand of node 2 (20.10 l/s) and this reservoir can be removed from the network in the next iteration. The reservoir in node 4 receives 34.23 l/s, which is less than the original demand of 40.20 l/s. For that reason, this reservoir stays attached to node 4 in the next iteration, as is the case with those in nodes 3, 5 and 6.

Step 5: With the original demand of node 2 restored, the simulation is run again. In nodes where the reservoirs receive less water than the original nodal demand, their supply shows the degree of the demand reduction. The iterative process continues until there is no change in the setup of the virtual reservoirs and ordinary nodes in two consecutive iterations. In this example, the total available demand stabilises at 93.38 l/s. The overall available demand fraction resulting from the burst of pipe 6 (ADF^6) is therefore 61.94% (93.38/150.75). The original demand of nodes 2, 3, and 4 is fully satisfied, whilst the demand of node 5 will be partially satisfied ($ADF_5 = 9.72$ %). Node 6 will receive no water as a result of the pipe burst ($ADF_6 = 0$ %).

Step 6: In the final step, the actual nodal pressures are calculated based on the true demand.

The above procedure is rather complex if it is to be applied for extended period simulations and any larger network, where each pipe is to be failed in the availability/ reliability analysis. Using the programme for ADF calculation, developed by Yoo et al. (2005), the results of the test network calculated for the applied demand pattern are shown in Figure 2.6, with the average availability factor of 0.8789, higher than the average reliability factor of 0.7742. This is normal knowing that the availability assumes possibility of repair of the failed system component. Also logical is that the higher demand in the system inflicts the lower values of the availability and reliability factors, during particular hours of the day.

Applying the same procedure for any other demand scenario, makes possible to compile the reliability and availability curves for 24-hour period, as is shown in Figure 2.6.

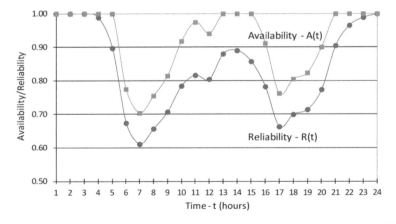

Figure 2.6 Hydraulic reliability and availability for network from Figure 2.5 (Yoo et al., 2005)

2.6   MAIN GAPS IN NETWORK RELIABILITY ANALYSES

The main gaps in water distribution system reliability analyses can be clustered per (A) definitions, (B) methods and (C) tools, which are elaborated in the following paragraphs.

2.6.1 Definitions (A)

The major definitions are not seen as gaps in analysing the network reliability, as such. However, they affect the uniformity of approaches to the reliability assessment. The key questions to be considered with that respect are:

(A.1) *What service level can be considered as acceptable?*
Universal definition on this, which could be taken as a starting point for any reliability assessment, does not exist. Consumers' satisfaction is the only real parameter for general comparison of the service levels in various distribution systems. Nevertheless, that one rests on a wide range of 'what is needed' and 'what is available and/or affordable'.

(A.2) *When is a water distribution system sufficiently reliable?*
Here as well, a universal conclusion is difficult to reach, although most would agree that in case the issue of acceptable service level has been resolved, the system would be sufficiently reliable if it can maintain that service level defined for regular and irregular demand scenarios. Customised service levels based on various degrees of irregularity are acceptable provided that the water shortages are reasonable in quantities and duration.

(A.3) *What network parameters are the closets indicators of its reliability?*
Nodal pressures and demands are the most convenient i.e. accessible parameters for network reliability assessment. It is indeed that sufficiently high pressures will almost always mean sufficient quantities of water in the system. Nevertheless, the pressure-demand dependency is a 'chicken and egg story': the lower pressures affect the demand but the increased demand will also affect the pressures! Ultimately, consumers are those who decide when to open the tap, with or without sufficient pressure. Examples of trying to save excessive water quantities once they have observed a drop in the pressure, anticipating forthcoming shortages, are well known in water distribution practice.

Furthermore, more research could be done to assess the reliability by analysing economic parameters. The most economic designs commonly start from some optimisation of the investment and operational costs taking into consideration regular supply conditions. More light should therefore be thrown on the costs of failures and shortages in relation to the investment into the reliability improvement (larger diameter/parallel pipes, more pumps, storage volume, etc.). Apart from guaranteeing the adopted service levels, the reliability is also to be achieved at minimum costs for water companies and consumers. A robust tool that can compare investments into the reliability improvement and clearly describe their implications for the maintenance of service levels in irregular conditions would certainly be helpful in decision making process.

Finally, the literature review showed little research done on the role of water quality parameters in the reliability assessment; this aspect is missing even in the basic definitions of

the reliability where the quantity comes clearly before the quality. In theory, there are two ways in which the water quality parameters could be included in the reliability assessment:
1. by analysing a probability that once it has been tapped, the water will (not) satisfy prescribed quality standards; the reasons for this can be various: poor water treatment, pipe sediments and corrosion, intrusion of pollution through pipe bursts, etc.;
2. by analysing effects of particular water quality on ageing of pipes and consequently having an idea about the link between the measures taken to maintain a certain water quality and frequency of bursts, all with implications for the technical/economic life time of the pipe.

Practically, both of these are very difficult problems to solve; complex mechanisms that explain interactions between various water composition and pipe materials are the main reason for it.

(A.4) *What are the most appropriate reliability measures/indices?*
The network reliability is in the first place about the probability of getting 'a glass of potable water' any time the tap is open, and less about the probability of having a component failure in the system. Although these two events are closely related, possible shortage of supply is not always a result of mechanical failures in the system. Therefore, the concept of hydraulic availability describes the picture better than the mechanical availability, which describes one part of the problem, only.

As demonstrated by the two examples in the previous section, many reliability calculations often end-up with a factor, which takes values between 0 and 1. The following, more or less general conclusions, can be drawn with respect to such result:
- Expressing the network reliability with single index gives an advantage of easy check of its sensitivity towards possible reconstructions attempted in the system or changes of various network parameters. It is possible to assess the improvement i.e. the increase of the index resulting from the investment and operational costs implying from certain scenario.
- Nevertheless, while they give fairly straightforward conclusions in terms of 'higher/lower i.e. better/worse than....' when comparing several alternatives, the index still offers a bit unclear description of the magnitude of the problem when evaluated independently. It is therefore not clear in advance which values/ranges can be considered as high/medium/low reliability, in other words acceptable or unacceptable. This suggests the need for a research that would deal more in-depth with sensitivity analysis of the reliability indexes in absolute terms.
- Once the hydraulic reliability has been improved to a certain degree, the question is if and how this translates into the improvement of mechanical reliability, for instance the pipe failure frequency, and are the costs involved really justified.

Ideally, any hydraulic reliability index should be convertible into a probability of getting a sufficient quantity of potable water any time the tap is open. The surrogate indexes described in this chapter are missing this conversion, next to the fact that they seem to be less applicable in practice than the concept of availability indices.

2.6.2 Methods (B)

The literature review reveals large number of reliability approaches, the simulation approaches being the most frequently developed and tested. These approaches also look the most transparent and applicable in reality, although they rely heavily on the quality of data used for input and calibration of the hydraulic- and failure models. The approaches presented in the other two groups, analytical and heuristic, are either too simplified, or too complex, i.e. contain some parameters that are difficult to survey, monitor or interpret in reality. Nevertheless, the analytical concepts of connectivity, reachability and redundancy look interesting and would require further research. It is very clear that in cases of pipe failures, the water changes its path in the system and the network resistance will play essential role in redistribution of pressures and flows. The questions to address the gaps in analysing the network reliability are formulated in the following five paragraphs.

(B.1) *Are the simulation approaches offering complete picture about network reliability?*
The conclusions from the literature review and two applications illustrated in the previous section can be summarised as follows:
- The simulation approaches take the major aspects of reliability into consideration and as such are sufficiently practical, provided that reliable data for the model building and calibration are available.
- In some aspects, namely the determination of nodal reliability, and consequently the overall reliability as an average value, the methods simplify the reality. Averaged figures potentially hide failures that can by no means be tolerated, due to their bad consequences for the system operation. If these are just a few, but very weak points in otherwise reliable system, the picture about its overall reliability may be distorted. Introducing some kind of weighting of nodal reliabilities could be considered to rectify this problem.
- Categorising the pipe availability per diameter may not be entirely accurate as it neglects implications of the pipe location in the system. The same diameter pipe can have more chance to burst if located in a high pressure area (e.g. closer to the source), or is exposed to more aggressive soil conditions, excessive surface loading, etc. than if it is located in the area where all these risks are smaller or can be avoided. Moreover, the pipe age and material will also have different impact on the frequency of bursts of the same pipe diameters. Including at least these two variables should therefore help to diversify this information.
- Most of the methods focus to pipe failures exclusively, neglecting operations of other components: pumps, storage and valves. Typically for many water distribution systems in developing countries, electricity failures are often the major cause of water shortage, and the hydraulic reliability in such cases will depend largely on available buffer volume in the reservoirs. Depending on the moment of the electricity failure, the chance that these reservoirs will not function properly, if normally refilled by pumping, also exists. More research would therefore be needed on these issues.

(B.2) *Is it possible to incorporate economic aspects into reliability assessment?*
The presented simulation approaches predominantly deal with hydraulic aspects and much less with economic aspects of reliability. From the perspective of water companies, a crucial decision is where their money should be invested: into cleaning, repair, replacements, additional pipes, pumps, storage volume, etc. What that means for the improvement of the service level, and reduction of failure rates, losses of water/revenue/consumer confidence, etc. resulting from the calamities, is important question. It is often a case that the network

optimised at certain minimum pressure and minimum investment costs will result in a layout that performs well under regular conditions but is not necessarily sufficiently reliable.

(B.3) *Is it possible to incorporate the impact of water quality into reliability assessment?*
In the wider context of reliability, a distribution network would not be reliable if the full demand has been satisfied but with water of deteriorated quality, carrying pathogens or unacceptable colour, taste or odour, even if not hazardous. This aspect is particularly relevant in view of the fact that the improvement of hydraulic reliability by enlarging the pipes enhances water stagnation, which is a source of a range of water quality problems. Correlating the threshold values of velocities and chlorine residuals to the system pressures, in regular and irregular supply scenarios, would likely add value to the overall reliability analysis.

(B.4) *Are the pipe failure models sufficiently describing the reality?*
Water quality and its impact on corrosion is the factor affecting the pipe failure rates. Corrosion growth results from very complex mechanism and is a long-lasting event that is difficult to model with confidence. Nevertheless, verifying more complex hypothesis and models seems increasingly possible, owing to improved monitoring and recording of the field data. A field study that correlates service conditions (component materials, installation, operation, maintenance, environment, etc.) to the failure rates could make very useful contribution to the modelling of the frequency of failure events.

(B.5) *Is a snapshot method for preliminary assessment of reliability possible?*
The literature review did not result in any quick method for preliminary assessment of reliability that could call for deeper analyses of reliability, similarly as it is done in medical investigations. Such approach could correlate the basic system parameters: the network layout, topography and supplying scheme and would develop an index hinting, already during regular operation, if the system is sufficiently reliable. Two examples elaborate this concept:
- From the perspective of demand reduction, a network with higher pressures is likely to be more hydraulically reliable than the network with low pressures, despite the negative effects on leakage. A snapshot assessment of the hydraulic performance during regular operation could indicate threshold pressures that shall not drop below the value that could significantly affect the demand, once the pipe failure has occurred.
- Similarly, a network with larger pipe diameters, more (connected) pipes, and newer i.e. smoother pipes is potentially more reliable than an insufficiently developed network of old pipes. One way to survey the reliability in both of these cases could be to compare their resistances for various demand scenarios and suggest a critical resistance that will almost surely lead to a drop of pressure below the same threshold value as in the first example.

A feeling is that there is more room to investigate the correlations between reliability measures and major performance parameters than is presented in the literature.

2.6.3 Tools (C)

The simulation approaches for reliability assessment rely predominantly on the steady-state network hydraulic calculations. The pressure-driven models conceptually offer more accurate representation of the low-pressure conditions but most of the commercial software still operates based on the demand-driven principles as a default calculation mode, offering faster and more stable simulations of real-life networks. The questions that can be raised in relation to the tools for reliability assessment are:

(C.1) *Are the existing numerical approaches appropriate for reliability analyses?*
Applying hydraulic solver based on demand-driven approach makes the simulation approaches practical i.e. applicable on a system of any size and complexity, provided that the reliable data for the model building and the burst frequency can be extracted from the water distribution company records. That is why the adaptation of this approach into a 'quasi pressure-driven' approach, such as the one described in the works of Ozger and Mays (2003), and Yoo et al. (2005) seems to be a good compromise between the reality and robustness of the numerical algorithm. Nevertheless, the concept of emitter coefficients available in EPANET software becomes more frequently used for pressure-driven demand calculations, being pretty robust while simpler to code.

The commercial introduction of genetic algorithms (GA) opens interesting opportunities for optimisation of network performance based on economic grounds that also include reliability aspects. For example, particular design layouts can potentially be optimised to minimise the reduction of demand during the failures in the system. Such a software tool would certainly be instrumental in developing the methods suggested under (B.2).

(C.2) *What is the real nature of pressure-demand dependency and its implication for demand patterns?*
In all kinds of pressure-driven demand considerations, the real nature of the dependency between these two parameters is difficult to generalise. It is likely that a specific way of water use and type of water appliances and outlets typically installed in one area contributes to this relation. Secondly, the diurnal demand pattern applied in regular situations will almost surely be distorted in the times of failures. It is very difficult to predict how much of this will be caused by lowering of the pressures and how much by changing usual habits in water use, which makes running of extended period simulations for reliability analyses rather problematic. Consequently, the results like the one shown in Figure 2.6, would need to be more rigorously scrutinised. In both of these cases, a thorough collection and analyses of the relevant field data could be helpful to discover the pressure-demand patterns.

(C.3) *Is accurate calibration of the models studying network reliability possible?*
One of the difficulties in the reliability analyses is that the model is difficult to calibrate using the field data. Water companies will hardly inflict a failure in order to be able to test its consequences on the level of service; any failure will happen spontaneously i.e. mostly unpredictably. Unfortunately, those networks that are mostly suffering from general negligence are the same networks where poor data collection is typical problem. A thorough monitoring of the system performance during any calamity is vital source of information for computer modelling of network reliability.

(C.4) *What steps of the reliability assessment process could become standard features of commercial network modelling software?*
In the absence of generally accepted reliability method, most of the applications from the literature show that the simulation approaches are computationally intensive and automation of the calculation process is a must, especially if networks above a few hundred nodes would be considered. With the current speed of state-of-the-art PCs this is however not seen as a serious constraint. Nevertheless, the commercial water distribution software hardly posses

standard features that could be helpful to assess the network reliability. A few possible add-on's, in the order of complexity, could be:

- Simulation run that models single failures of consecutive pipes and calculates the deficit of pressure in the system.
- The similar simulation that calculates the demand reduction based on the adopted (quasi) pressure-driven demand method.
- Various optimisation modules that consider optimal design, operation and maintenance measures from the perspective of the least cost that could also include reliability aspects, such as the loss of revenue due to water shortage.

2.7   RESEARCH OBJECTIVES AND SCOPE

The general objective of the research presented in this manuscript has been to define a reliability measure that is more tangible for day-to-day management and design/renovation/expansion of water distribution networks, which could then be assessed i.e. diagnosed having a user-friendly tool at disposal. Looking at the above list of the main gaps, the specific research objectives have been dealing with the issues raised in the paragraphs (A.3), (A.4), (B.2), (B.5), (C.1) and (C.4). The research has not tackled any aspect of failure probability or mechanical reliability/availability and has been strictly focused to the network hydraulic reliability. In that sense, using the term 'reliability' further in this manuscript, actually relates to the network resilience.

2.7.1 Key Research Questions

In order to arrive at what has been defined as 'more tangible' reliability assessment, the following research questions have been considered in this study:
1. What kind of network buffer can be best correlated to its reliability? Namely:
    1.1. Is the available demand fraction (ADF) true descriptor of network reliability?
    1.2. Are the demand-driven based reliability measures, sufficiently accurate?
2. What are the implications of the choice of particular network layout for the reliability? Namely:
    2.1. How the change in node connectivity affects the levels of service?
    2.2. What is the effect of selected supply schemes, with or without balancing tanks in the system?
3. To which extent can the monitoring of the basic hydraulic parameters help in predicting the network reliability? Namely:
    3.1. Do threshold values of flows and velocities for given layout exist, above which the negative effect of the failure significantly increases?
    3.2. What is the link between the pressure levels and the effects of potential failure in the system?
    3.3. Is there a link between the reliability and energy balance in the network?
4. How the reliability assessment affects the most economic design? Namely:
    4.1. What are the differences between the most economic design and the most reliable design?
    4.2. Under what conditions is the increase of investment costs more affective for improvement of the reliability than the increase of operation and maintenance costs?

It has been assumed that the answers to the above questions can throw more light to the choice between the range of operational and investment measures to improve the network reliability. Secondly, in terms of the tool development, the conclusions should help while analysing to which extent the irregular supply conditions can be modelled by expanding the existing algorithms or actually transforming them into more effective ones. This issue is specifically relevant having in mind the huge models nowadays built with the information derived from GIS databases, which require algorithms that shall be robust enough and will not increase the calculation times significantly.

2.7.2 Research Hypotheses

Based on the above research questions, the following research hypotheses have been tested in this research:

1. For a given topography, network layout and demand scenario, there is a unique reliability/resilience footprint that can be described as a function of ADF. This footprint reflects the network buffer.
2. Reliability measures derived using demand-driven hydraulic models are less accurate than those derived by the use of pressure-driven demand models.
3. Increasing the connectivity between the pipes improves network reliability in general. For a given supply scheme, there is an optimal network geometry that can be described by a 'shape index', which can be correlated to other reliability measures.
4. It is possible to make a quick reliability snapshot of a network by looking at typical hydraulic indicators. There is a clear implication from the interrelation between the pressures, flows/velocities and network resistance, namely:
    4.1. The networks with generally higher pressures, despite potential for increased leakage, have more of a buffer to maintain the minimum service level during a single event of the component failure.
    4.2. The reliability is disproportional to the increase of the system resistance i.e. hydraulic loss.
5. The most economic design shall always be cheaper than the most reliable design. There is a threshold velocity and/or network resistance that can be taken as a border between the increase of investment- or operational costs, resulting in the most effective reliability improvement.

In summary, it can be assumed that for chosen supply scheme, network configuration, demand scenario, and adopted preventive and reactive maintenance, it is possible to predict an event of single component failure that will not affect the guaranteed minimum service under irregular supply conditions.

2.8   RESEARCH METHODOLOGY

This research, compiled of the elements related to the hydraulic performance of water distribution systems, economics of operation, maintenance and renovation of these systems, and computer programming, is presented as an omnibus of the topics formulated from the research hypotheses. Brought together, the outcome of the analyses of those hypotheses has served to design a decision support tool (DST) for reliability-based design of water distribution networks.

Two main outputs of the study have been used to support the DST: (a) a set of measures for quick assessment of the network reliability using the performance indicators under regular operation, and (b) an algorithm that incorporates the developed method for reliability assessment and full diagnostics of single or multiple network layouts. The computer programme has been developed with the help of EPANET Toolkit Functions available with the main programme, also distributed by US Environmental Protection Agency (EPA). Owing to its robustness, simplicity and free distribution via the Internet, EPANET software has become very popular amongst the researchers and practitioners worldwide.

The first phase of the study upon the completion of the literature survey has focused to the development of emitter based pressure-demand driven algorithm needed for calculation of the available demand fraction. Furthermore, a choice has been made to analyse the patterns by drawing statistical correlations on larger samples of networks. A network generation tool has therefore been necessary in the second step, in order to create consistent samples, yet sufficiently variable to arrive at solid conclusions about the patterns. This tool, based on the principles of graph theory, has been developed for arbitrary spatial distribution of (demand) nodes, using their coordinates in the generation process.

To provide reference layouts in the pattern analysis, the network generation tool has been further expanded by GA-optimiser, building in the feature of the least-cost diameter optimisation. In parallel, a number of indices has been developed that should quantify network shape, its buffer and resilience. All generated networks, optimised or not, have been further processed using the developed network diagnostics tool, which includes all proposed reliability measures combined with a few others from the literature. Next to the reliability assessment, the tool has been intended for assessment of network geometry, hydraulic performance and annual costs of the assets, including operation and maintenance. To be able to do it, the diagnostics tool has been designed to allow gradual or random increase/reduction of the most typical network parameters to be able to monitor their possible correlation with the reliability measures. With this, sufficient volume of the results of the network diagnostics tool could have been statistically correlated.

In the final stage, the DST has been completed by adding the network filtering and initialisation modules, allowing selection of networks that fit pre-selected routes of the pipes. Hence, an integrated tool has been created that can generate, filter, optimise, and diagnose large number of layouts in relatively short period of time.

Nearly 20,000 network layouts have been processed using the tool, varying between three and 5044 pipes. The largest network analysed has been a real-case network supplying part of Amsterdam. Additional simulation runs have been conducted on several samples, by altering particular parameters of some layouts, namely the nodal demand multiplier and pipe diameters.

REFERENCES

1. Andreo, S., Marks, D., and Clark, R. (1987). *A New Methodology for Modelling Break Failure Patterns in Deteriorating Water Distribution Systems: Theory.* Journal of Advanced Water Resources, 10(March), 2-10.

2. Awumah, K., Goulter, I.C., Bhatt, S.K. (1990) *Assessment of reliability in water distribution networks using entropy-based measures.* Stochastic Hydrology and Hydraulics, 4(4), 325-336.

3. Awumah, K., Goulter, I.C., Bhatt, S.K. (1991) *Entropy-based redundancy measures in water-distribution networks.* Journal of Hydraulic Engineering, ASCE,117(5), 595-614.

4. Awumah, K., Goulter, I. (1992), *Maximizing entropy defined reliability of water distribution networks.* Engineering Optimization, 20(1), 57-80.

5. Bao, Y., Mays, L.W. (1990) *Model for water distribution system reliability.* Journal of Hydraulic Engineering, ASCE, 116(9), 1119-1137.

6. Brandon, T.W. (1984). *Water Distribution Systems.* The Institution of Water Engineers and Scientists.

7. Cullinane, M.J. (1989). *Determining availability and reliability of water distribution system,* in L.W. Mays, ed., Reliability analysis of water distribution system. ASCE, New York, N.Y., 190-224.

8. Cullinane, M.J., Lansey, K., and Mays, L.W. (1992). *Optimization-availability based design of water-distribution networks.* Journal of Hydraulic Engineering, ASCE, 118(3), 420-441.

9. Engelhardt, M.O., Skipworth, P.J., Savic, D.A., Saul, A.J., and Walters, G.A. (2000). *Rehabilitation Strategies for Water Distribution Networks: a Literature Review with a UK Perspective.* Urban Water, 2(2000), 153-170.

10. Fujiwara, O., Ganesharajah, T. (1993) *Reliability assessment of water supply systems with storage and distribution networks.* Journal of Water Resources Res 29(8), 2917-2924.

11. Gargano, R., and Pianese, D. (2000). *Reliability as a Tool for Hydraulic Network Planning.* Journal of Hydraulic Engineering, ASCE, 126(5), 354-364.

12. Germanopoulos, G. (1985). *A Technical Note on The Inclusion of Pressure Dependant Demand and Leakage Terms in Water Supply Network Models,* Civil Engineering Systems, 2(3) 171-179.

13. Goulter, I.C. (1995). *Analytical and Simulation Models for Reliability Analysis in Water Distribution System,* in E. Caberra and A. Vela, eds., Improving Efficiency and Reliability in Water Distribution System. Kluwer Academic Publisher, Dordrecht, The Netherlands.

14. Goulter, I.C., Walski, T.M., Mays, L.W., Sakarya, A.B.A., Bouchart, F., Tung, Y.K. (2000). *Reliability Analysis for Design,* in L.W. Mays, ed., Water Distribution Systems Handbook. The McGraw-Hill Companies, Inc, New York, N.Y.,18.1-18.48.

15. Goulter, I.C., and Coals, A.V. (1986). *Quantitative Approaches to Reliability Assessment in Pipe Networks.* Journal of Transportation Engineering, ASCE, 112(3), 287-301.

16. Grayman, W.M., Rossman, L.A., Arnold, C., Deininger, R.A., Smith, C., Smith, J.F., and Schnipke, R. (2000). *Water Quality Modeling of Distribution System Storage Facilities,* Denver, Colorado, AWWA and AwwaRF.

17. Gupta, R., and Bhave, P. (1994). *Reliability Analysis of Water Distribution Systems.* Journal of Hydraulic Engineering, ASCE, 120 : 447-460.

18. Khomsi,D., Walters,G.A., Thorley, A.R.D., and Ouazar, D. (1996). *Reliability Tester for Water Distribution networks.* Journal of Computing in Civil Engineering, ASCE, 10(1), 10-19.

19. Lansey, K., Mays, L.W., and Tung,Y.K. (2002). *Chapter 10: Reliability and Availability Analysis of Water Distribution,* in Urban Water Supply Handbook, Mays,L.W., Editos-in Chief. Mc-Graw-Hill, New York, NY.

20. Morgan, D.R., Goulter, I.C. (1985) *Optimal Urban Water Distribution Design,* Journal of Water Resources Research 21(5), 642-652.

21. Neubeck, K. (2004). *Practical Reliability Analysis.* Pearson Education Inc., Ohio USA

22. *NGSMI National Guide to Sustainable Municipal Infrastructure* (2002). Alternative Funding Mechanisms, Ottawa, Ontario Canada.

23. Ostfeld, A., Kogan, D., and Shamir, U. (2001). *Reliability Simulation of Water Distribution Systems-single and Multiquality.* Urban Water, Elsevier Science Ltd, 129, 1-9.
24. Ostfeld,A. (2004). *Reliability Analysis of Water Distribution Systems,* Journal of Hydroinformatics, IWA, 6(4), 281-294.
25. Ozger, S.S., and Mays, L.W. (2003). *A Semi-pressure-driven Approach to Reliability Assessment of Water Distribution Networks.* PhD Thesis, Arizona State University, Arizona USA
26. Pelletier, G., Mailhot, A. and Villeneuve, J.P. (2003). *Modeling Water Pipe Breaks-Three Case Studies.* Journal of Water Resources Planning and Management, ASCE, 129(2), 115-123.
27. Rossman, L.A. (2000). *EPANET 2 Users Manual.* U.S. Environmental Protection Agency's, National Risk Management Research Laboratory, Cincinnati, OH 45268, USA.
28. Satyanarayana, A. and Wood, R.K. (1982), *Polygon-to-Chain Reductions and Network Reliability,* CSA Illumina.(http://md1.csa.com/)
29. Shamir, U. and Howard, C.D.D. (1979), *An Analytical Approach to Scheduling Pipe Replacement.* Journal of AWWA, Vol.71, No.5, pp. 248-258.
30. Shinstine,D.S., Ahmed,I., and Lansey,K. (2002). *Reliability/availability analysis of municipal water distribution networks: Case studies.* Journal of Water Resources Planning and Management, ASCE,128(2),140-151.
31. Su, Y.C., Mays, L.W., Duan, N. and Lansey, K. (1987). *Reliability-Based Optimization Model for Water Distribution Systems.* Journal of Hydraulic Engineering Division, ASCE, 114(12), 1539-1556.
32. Tanyimboh, T.T., Templeman, A.B. (1993) *Maximum entropy flows for single-source networks.* Engineering Optimization, 22(1), 49-63.
33. Tanyimboh, T.T., Templeman, A.B. (2000). *A Quantified Assessment of the Relationship Between the Reliability and Entropy of Water Distribution Systems.* Engineering Optimization, 33(2), 179-199.
34. Tanyimboh, T.T., Tabesh, M., and Burrows, R. (2001). *Appraisal of source head methods for calculating reliability of water distribution networks.* Journal of Water Resources Planning and Management, ASCE, 127(4), 206-213.
35. Tobias, P.A., and Trindade, D.C. (1995). *Applied reliability.* Second Edition, Chapman & Hall/CRC, New York, NY.
36. Trifunović, N. (2006). *Introduction to Urban Water Distribution.* Taylor & Francis, UK
37. Trifunović, N., Umar, D. (2003). *Reliability Assessment of the Bekasi Distribution Network by the Method of Cullinane.* International Conference on Computing and Control for the Water Industry, London, UK.
38. Tung, Y.K. (1996). *Uncertainty Analysis in Water Resources Engineering.* Stochastic Hydraulics '96 Editor: Goulter, I. and Tickle, K., A.A. Balkema Publishers, NL, 29-46.
39. Vairavamoorthy, K., Akinpelu, E., and Lin, Z. (2001). *Design of Sustainable Water Distribution Systems in Developing Countries.* Proceedings of EWRI - World Water & Environmental Resource Congress.
40. Vreeburg, J.H.G., Hoven, T.J.J. van den. (1994). *Maintenance and Rehabilitation of Distribution Networks in The Netherlands.* KIWA, Nieuwegein.
41. Wagner, M., Shamir, U., and Marks, D.H. (1998a). *Water Distribution Reliability: Analytical Methods.* Journal of Water Resources Planning and Management, ASCE, 114(3), 253-275.
42. Walski, T.M. Chase, D.V., Savić D.A., Grayman, W., Beckwith, S., Koelle, E. (2003). *Advanced Water Distribution Modeling and Management.* Haestad Methods, Haestad Press, USA.
43. Watson, T., Christian, C., Mason, A., and Smith, M. (2001). *Maintenance of Water Distribution Systems.* Proc., The 36th Annual Conference of the Operational Research Society of New Zealand, University of Canterbury, 57-66.
44. Wood, D.J., Charles, C.O.A. 1972. *Hydraulic Network Analysis Using Linear Theory.* Journal of Hydraulic Division of ASCE, 98(HY76): p.1157-1170.
45. Xu, C., and Goulter, I.C. (1999). *Reliability-based Optimal Design of Water Distritution Networks.* Journal of Water Resources Planning and Management, ASCE, 125(6), 352-362.

46. Yoo, T.J., Trifunović, N., Vairavamoorthy, K. (2005). *Reliability Assessment of the Nonsan Distribution Network by the Method of Ozger*. The 31st WEDC International Conference on Maximizing the Benefits from Water and Environmental Sanitation, Kampala, Uganda.

CHAPTER 3

Emitter Based Algorithm for Pressure-Driven Demand Calculations of Water Distribution Networks[1]

Pressure-driven demand (PDD) models have become essential tools for hydraulic analyses of water distribution networks under stress conditions. They are commonly based on the concept of emitter coefficients initially developed within EPANET software of US EPA (Rossman, 2000). Thanks to substantially improved computational speed in the past decade, fairly large networks can be processed by PDD models, nowadays. Nevertheless, the PDD models are still unable to simulate some specific situations appearing in reality. The research presented in this chapter assesses the PDD algorithm for calculations of networks laid in topography with extreme altitude range. The negative emitter demands resulting from high altitudes are taken as an indicator of total demand loss and are eliminated after removing the emitter and setting the node demand to zero. In addition, a situation is analysed where the loss of demand can also be dependent from altitude range preventing conveyance through non-demand nodes of extremely high elevation, which are also generating negative pressures. The approach has been demonstrated on a few simple scenarios. The results point the deficiencies of neglecting high elevated nodes without the demand and the introduction of lower PDD threshold limit for (temporary) disconnection of the pipes connected to the negative-pressure nodes.

---

[1] Extended abstract submitted by Trifunović, N. and Vairavamoorthy, K., under the title *Simplified Emitter Based Approach for Pressure-Driven Demand Calculations of Networks with Extreme Topography*, for the 14th Water Distribution Systems Analysis Conference in Adelaide, Australia, 24-27 September 2012.

3.1   INTRODUCTION

Water distribution modelling practice still relies mostly on the demand-driven (DD) calculations of steady and uniform flows in pressurised networks. For the scenarios describing regular conditions with sufficient pressures, this approach is accurate enough while providing very fast and robust algorithm. In situations when the pressure in the network drops, either due to a pipe- or pump failure, or as a result of 'regular' intermittent supply caused by inadequate source capacity, the low pressures can affect the demand. Demand-driven models are unable to capture this reduction and the only indicators of the failure they are able to produce are negative pressures. When this is about to happen, the hydraulic simulation should switch to more computationally intensive pressure-driven demand mode (PDD).

PDD models have become essential tools for hydraulic analysis of water distribution networks under stress conditions as well as they are applied for modelling of leakages. No universally accepted method exists here, due to inability of mathematical equations to precisely describe the hydraulic complexity of irregular supply, on one hand, and practically impossible monitoring of the data that would enable full model calibration, on the other hand. In simplified approach, the PDD models can be based on the principle of emitter coefficient available in EPANET software (Rossman, 2000). Common approach assumes the definition of pressure threshold as an indicator of sufficient service level, which is then used to switch between the DD- and PDD mode.

3.2   PRESSURE-DRIVEN DEMAND CONCEPT

The concept of pressure-driven demand can be compared to the discharge through an orifice, as shown in Figure 3.1 (Trifunović, 2006).

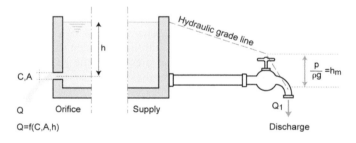

Figure 3.1 Analogy between discharge through orifice and pressure-driven demand (Trifunović, 2006)

The pressure, $p/\rho g$, above the water tap eventually becomes destroyed and can therefore be considered as the minor loss, h_m. Equation 3.1, describes the flow relation in both cases.

$$Q = CA\sqrt{2gh} \quad \Leftrightarrow \quad h_m = \xi \frac{Q^2}{A^2 2g} \quad \Rightarrow \quad Q = \frac{1}{\sqrt{\xi}} A\sqrt{2gh_m} \qquad 3.1$$

A is the surface area of the orifice/tap opening, and C is the shape factor of the orifice, which corresponds to the minor loss factor ξ.

Low heads or high losses calculated in the DD mode can result in negative pressures, as shown in Figure 3.2. This is a clear indication of irregular supply conditions that makes proper interpretation of the results difficult, because the pipe flows are based on the nodal demands that are false.

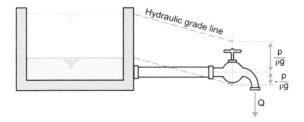

Figure 3.2 Negative pressures as a result of DD calculation (Trifunović, 2006)

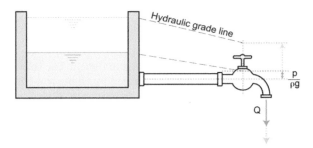

Figure 3.3 Pressures as the result of PDD calculation (Trifunović, 2006)

In the PDD mode, the nodal discharges will be reduced, causing a slower drop of the reservoir levels, as Figure 3.3 shows. This is the result of reduced pipe flows causing smaller friction losses.

Tanyimboh et al. (2001) describe the PDD relationship as:

$$H_i = H_i^{min} + K_i Q_i^n \qquad\qquad 3.2$$

where H_i represents the actual head at demand node i, H_i^{min} is the minimum head below which the service becomes terminated, K_i is the resistance coefficient for node i, Q_i is the nodal discharge, and n is the exponent that theoretically (and usually in practice) takes value of 2.0 (Gupta and Bhave, 1996). Equation 3.2 actually follows the similar format as the one of Equation 3.1. To determine the unknown value of Q_i for any given nodal head, Equation 3.2 should be rearranged as:

$$Q_i = \left(\frac{H_i - H_i^{min}}{K_i} \right)^{1/n} \qquad\qquad 3.3$$

When Q_i equals the required demand, Q_{req}, the value for H_i should equal the desired head, H_{des}, in the node. It is the head that should be available if the demand at that node is to be satisfied in full. Hence:

$$Q_i^{req} = \left(\frac{H_i^{des} - H_i^{min}}{K_i} \right)^{1/n} \Rightarrow \frac{1}{K_i^{1/n}} = \frac{Q_i^{req}}{\left(H_i^{des} - H_i^{min} \right)^{1/n}}$$

3.4

Finally, substituting K_i in Equation 3.3 yields:

$$Q_i^{avl} = Q_i^{req} \left(\frac{H_i^{avl} - H_i^{min}}{H_i^{des} - H_i^{min}} \right)^{1/n}$$

3.5

where Q_i^{avl} is the discharge available for the head available at the node (H_i^{avl}). Equation 3.5 considers three possible situations:

1. $H_i^{avl} \leq H_i^{min} \Rightarrow Q_i^{avl} = 0$
2. $H_i^{min} < H_i^{avl} < H_i^{des} \Rightarrow 0 < Q_i^{avl} < Q_i^{req}$
3. $H_i^{avl} \geq H_i^{des} \Rightarrow Q_i^{avl} = Q_i^{req}$

and as such it is used in balancing of the flows in the pipes connected to node i. The solution algorithms for solving the system of head equations have been described by Gupta and Bhave (1996). Eventually, the head-driven simulation is able to determine the nodes with insufficient supply. Apart from more complex and longer simulation, the key problem here is the correct definition of the values for H_i^{min} and H_i^{des} head, i.e. their correlation to the nodal resistance K_i, which describes the nature of the PDD relationship that is essentially empirical. Figure 3.4 shows the approximate PDD relation from the Dutch experience (KIWA, 1993). A linear relation is suggested until a certain threshold, which is typically a pressure around 20 mwc. In this way of presentation, the critical pressure $p_i^{crit}/\rho g = H_i^{des} - H_i^{min}$.

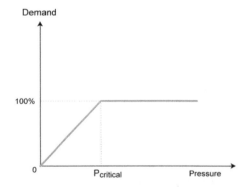

Figure 3.4 Pressure-related demand relation (KIWA, 1993)

The concept of EPANET emitter coefficients uses similar relationship as in Equation 3.2. An emitter is modeled as a setup of a dummy pipe connecting the actual node with a dummy reservoir whose initial head equals the nodal elevation, z. Hence, $H_i^{min} = z_i$ and:

$$Q_i = \frac{1}{K_i^{1/n}}(H_i - z_i)^{1/n} \hspace{6cm} 3.6$$

Strictly speaking, the K-value in Equation 3.6 stands for the resistance of the dummy pipe, but actually it has the same meaning as in Equations 3.2 to 3.4. Finally,

$$Q_i = k_i \left(\frac{p_i}{\rho g}\right)^\alpha \;\; ; \;\; \frac{p_i}{\rho g} = H_i - z_i \;\; ; \;\; \alpha = 1/n \;\; ; \;\; k_i = \frac{1}{K_i^\alpha} \hspace{3cm} 3.7$$

where k_i stands for the emitter coefficient in node i and α is general emitter exponent.

Emitter coefficients were first introduced in EPANET to simulate operation of fire hydrants. By specifying the emitter coefficient, the demand node would turn into an emitter node. In essence, this is a node in which the demand shall be adjusted based on the actual pressure in the system, following Equation 3.7. The default value for exponent α in EPANET is 0.5, which can be adjusted if necessary. Using the emitter approach gives clear advantages while exploring the effects of pressure management on the leakage reduction in the system. Furthermore, by connecting an emitter node to the normal consumption node (with a dummy pipe of low resistance), the amount of water loss shall be clearly visible, although the size of the model increases significantly. In the analyses of failures, EPANET has to be upgraded with the value of pressure threshold which will be used to switch between the DD- and PDD mode of calculation. Turning the DD hydraulic calculations into a PDD mode, the nodal demands being an input in the DD mode then become dependent on the nodal pressure that is an output parameter, as long its value is below the threshold value. A PDD add-on has been developed by Pathirana (2010) who defines an emitter cut-off point (*ECUP*) which is the PDD threshold pressure.

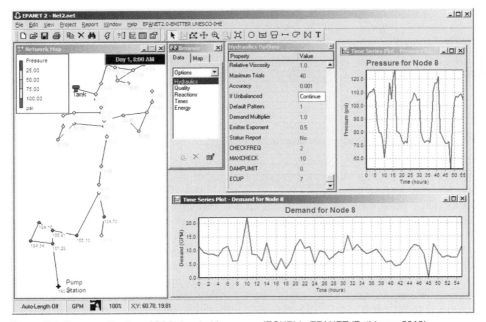

Figure 3.5 User specification of PDD threshold pressure (ECUP) in EPANET (Pathirana, 2010)

A layout of this version of EPANET programme, called EPANET-EMITTER is shown in Figure 3.5. In the iteration process, for:
1. $p_i/\rho g > ECUP$, $Q_{i,PDD} = Q_{i,DD}$,
2. $0 < p_i/\rho g < ECUP$, $Q_{i,PDD} = k_i(p_i/\rho g)^\alpha$
3. $p_i/\rho g < 0$, $Q_{i,PDD} = 0$

Before the iteration process starts, the values for emitter coefficients in each node will be estimated based on Equation 3.8.

$$k_i = \frac{Q_{i,DD}}{ECUP^\alpha} \qquad\qquad 3.8$$

3.3 EMITTER PERFORMANCE UNDER EXTREME TOPOGRAPHIC CONDITIONS

As shown above, in cases the nodal pressure tends to become negative in the PDD mode of calculation, the nodal demand should be set to zero, which is the situation that can occur in cases of extreme topography i.e. where negative pressures are mostly inflicted by high elevations. Setting the demand to zero only in the nodes with negative pressures assumes that the effect of high elevation is restricted only to that particular node, whilst based on the situation in the network, this effect can propagate i.e. affect the demand in surrounding nodes, too. On the other hand, without additional intervention on the nodal demand, emitters in the nodes with negative pressure will cause a negative demand i.e. a dummy supply, as shown in Figure 3.6.

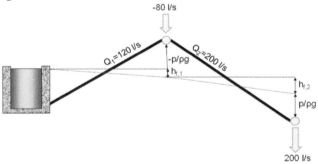

Figure 3.6 Negative demand resulting from negative pressures in PDD models using emitters

Piller and Van Zyl (2009) summarise the deficiencies of emitters used for PDD algorithm, reported by other researchers, namely Todini (2003, 2006), Cheung et al. (2005) and Wu et al. (2006). They point three inconveniences of the emitter use: the negative demands, the reduced convergence of the solving method, and the upper bound of consumption that in reality is not only dependant on the available head but also on physical characteristics such as the number of connected premises. They further propose the algorithm for solving a problem of network section connected to a source via a high-altitude node, causing negative pressure. These situations in reality inflict reduction of pipe conveying capacity because of the air entrainment through air valves and taps. The algorithm based on the energy minimisation concept that avoids definition of PDD relationship, based on their work published in 2007, is however tested on rather small case network restricted to a single node/route connection.

The PDD algorithm of Pathirana (2010) deals with negative nodal pressures in a straightforward way, by setting the nodal demand to zero. Nevertheless, high altitude nodes without the demand, and with negative pressure, will keep the same conveyance of the connecting pipes, regardless their elevation i.e. the (negative) pressure value. This resembles hydraulic conditions of a siphon, which in reality is possible, but very likely causes the reduction of capacity, and in more extreme conditions can disable the route for conveyance, entirely. To which extent the water distribution becomes affected will certainly depend on the nodal elevations, next to the supplying heads and pipe resistance.

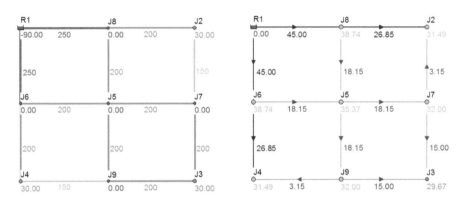

Figure 3.7 Simple network (L_j = 1500 m, z_i = 5 msl) – pipes: D (mm), Q (l/s), nodes: Q (l/s), p/pg (mwc)

These phenomena have been illustrated on a simple network shown in Figure 3.7. The figure shows the network under normal operation. For the sake of simplicity, all the pipe lengths have been set at 1500 m, the nodal elevations at 5 msl, and the k-values at 0.5 mm. The water level in the reservoir has been kept at 50 msl.

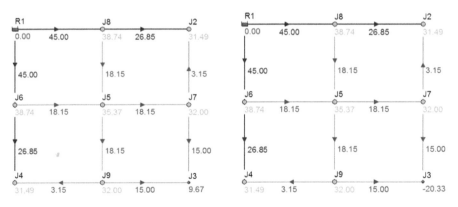

Figure 3.8 DD simulation (left: z_{J3} = 25 msl, right: z_{J3} = 55 msl) – pipes: Q (l/s), nodes: p/pg (mwc)

Figure 3.8 shows the results of demand-driven simulations where the elevation of node $J3$ has been increased first at 25 msl (left) and then at 55 msl (right). Consequently, only the pressure in that node changed, while all other nodes and pipes kept the same pressures and flows, which is the well-known anomaly of the demand-driven mode of calculation and obviously inaccurate representation of the reality.

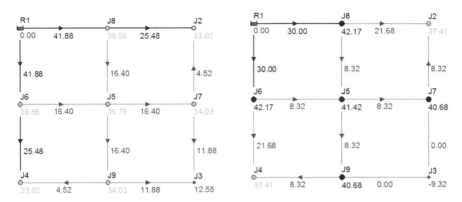

Figure 3.9 PDD by Pathirana (2010) of the nets from Figure 3.8, – pipes: Q (l/s), nodes: p/ρg (mwc)

The same networks as in Figure 3.8 have been recalculated by the PDD adaptation of EPANET developed by Pathirana (2010), assuming the PDD threshold of 20 mwc and the emitter exponent of 0.5; the results can be seen in Figure 3.9. Here, the PDD calculation gives more logical results. The drop of pressure in the node *J3* to 12.55 mwc resulting from the elevation of 25 mwc has caused the drop of the demand in that node from 30 to 23.76 l/s. The pressure in *J3* is however higher than if calculated in the DD mode, which is also logical; the demand drop has also caused the reduction of friction losses in the network. In the extreme case of elevation z_{J3} = 55 msl, which is higher than the head of the reservoir, the demand in this node has been reduced to zero. The negative pressure of -9.32 msl, reflecting the high elevation of the node, has been higher than in case of the DD calculation, again for the same reason of reduced friction losses. Consequently, the pressures in the network with higher loss of demand (on the right) will be generally higher.

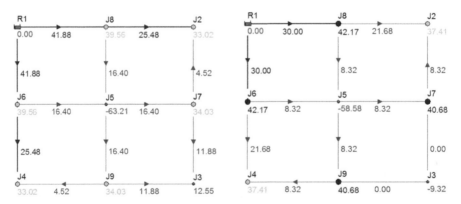

Figure 3.10 PDD by Pathirana (2010), increased z of *J5* – pipes: Q (l/s), nodes: p/ρg (mwc)

Repeating the same calculation by increasing the elevation of non-demand node *J5* from 5 to 105 msl will give the results as in Figure 3.10. It shows that the PDD calculation affects only the demand nodes i.e. the flow patterns will not be affected by the nodal elevations of non-demand nodes. Equally, the supply of node *J3* will not be necessarily affected by the increase of the elevations in additional two nodes, *J2* and *J4*, as long the pressure in node *J3* is sufficient; this has been illustrated in Figure 3.11.

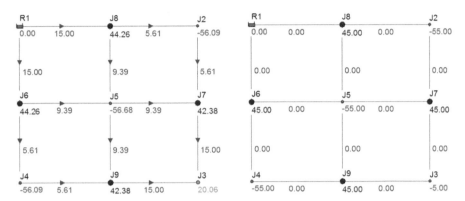

Figure 3.11 PDD by Pathirana (2010), increased z of J2, J4 and J5 – pipes: Q (l/s), nodes: p/pg (mwc)

No matter how little possible the situations in Figure 3.10 and 3.11 are in reality, the PDD model shows the results that would hardly be possible, too. However, the hydraulic picture in Figure 3.11 on the right is correct. All three demand nodes, *J2* to *J4*, are elevated higher than the reservoir, and the total demand therefore equals zero; all the node pressures respond to the reservoir head of 50 msl. The full supply has been provided in node *J3* in Figure 3.11 on the left, despite the fact that all the routes from the reservoir towards this node go via extremely high elevated nodes. Because the entire demand of nodes *J2* and *J4* has been lost, low friction losses enable the pressure in node *J3* to build above the PDD threshold of 20 mwc, meaning that the entire demand of 30 l/s in this node will be supplied; possible in the model, but questionable in reality.

The hydraulic picture in Figure 3.10 does not appear to be affected by the highly elevated node *J5*, compared to that shown in Figure 3.9, except for the pressure in that node. That the flows of 16.40 l/s and 8.32 l/s can be transported via the node located 55 metres above the source, by gravity, is of course impossible. More accurate simulation would possibly be to simply disconnect i.e. close the pipes connected to node J5, which is shown in Figure 3.12.

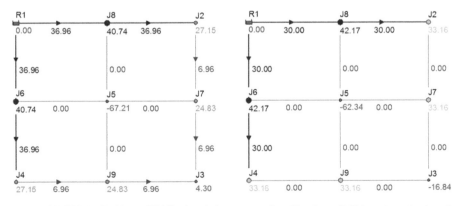

Figure 3.12 PDD by Pathirana (2010), closed pipes connecting J5 – pipes: Q (l/s), nodes: p/pg (mwc)

The figure shows more radical loss of demand which partly originates from the simplified i.e. radical approach of closing the entire pipe. It is reasonable to believe in reality, that the head of the source would enable partial supply of the consumers connected to the pipe closer to the nodes opposite of *J5*. Despite this inaccuracy, the results in Figure 3.12 look more logical than those in Figure 3.10. Due to unavailability of the routes passing *J5*, the total supply of *J3* elevated at 25 msl has reduced to 13.92 l/s at the pressure of 4.30 mwc. In case of the elevation of 55 msl, *J3* will receive no supply whatsoever, which has been also shown in Figure 3.10 on the right, but at generally higher pressures caused by different flow and friction loss distribution. Needless to mention, the results in Figure 3.10 depict a distorted picture leading to lower demand losses in any possible pipe failure analysis. One or another way, the network reconfiguration may be necessary to arrive at more accurate results.

Knowing that the pressure in node *J3* is going to drop below the set PDD threshold of 20 mwc, after increasing its elevation, the drop of demand can also be simulated in the DD mode by putting an emitter in this node and removing the node demand of 30 l/s. The emitter value will be calculated from Equation 3.8 as:

$$k_{J3} = \frac{Q_{J3,DD}}{ECUP^\alpha} = \frac{30}{20^{0.5}} = 6.7082 \qquad\qquad 3.9$$

Using this value, the situation from Figure 3.9 has been simulated in the DD mode and the results are shown in Figure 3.13; the emitter in node *J3* has been indicated by a rhomb. It shows that the results for the node elevation of 25 msl (on the left) are identical as those calculated by the PDD adapted EPANET by Pathirana (2010). In case of the node elevation of 55 msl in *J3*, the hydraulic picture is similar as in Figure 3.6 and differs from that in Figure 3.9 on the right. EPANET has balanced the system hydraulically at the pressure of -5.87 mwc in *J3*, and the emitter is consequently supplying the network with the dummy flow of 16.26 l/s; this is the figure that inflicts the reduction of total flow next to the total loss of demand. Combined with the supply from the reservoir, the total demand will be 60 l/s.

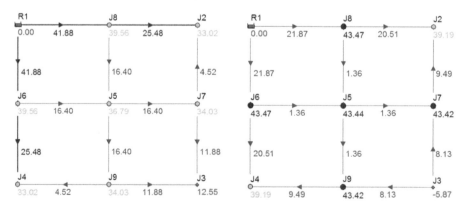

Figure 3.13 DD simulation of the nets from Figure 3.9, emitter in *J3*, – pipes: Q (l/s), nodes: p/pg (mwc)

The emitter supply can be eliminated in two ways shown in Figure 14. By adding the negative emitter pressure to the source head, in a number of iterations, the negative pressure and dummy supply shall gradually disappear, as it is shown after the final iteration in the network on the left. Eventually, the surplus of head of exactly 9.32 mwc, is to be deducted from all the pressures. More usual and simple approach is to remove the emitter coefficient in node *J3* and set its demand to zero, as it has been the case in the network on the right. Both procedures will be of course more difficult in complex networks with multiple sources of supply and nodes with negative pressures.

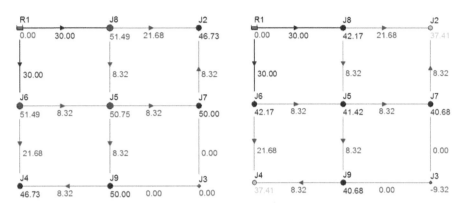

Figure 3.14 DD simulation, elimination of emitter supply in *J3*, – pipes: Q (l/s), nodes: p/ρg (mwc)

3.4   EMITTER BASED PDD ALGORITHM

The PDD algorithm based on the EPANET emitters can be relatively easily developed using the EPANET toolkit functions. Essentially, it is a computational loop consisting of three steps:
1. Removing the node baseline demand and introducing emitter coefficients in all the nodes where the pressure is lower than the PDD threshold.
2. Reinstating the node baseline demand and removing the emitter coefficients in all the nodes where the pressure is equal or higher than the PDD threshold.
3. Running a snap-shot DD hydraulic simulation.

Preliminary hydraulic simulation will be executed and the emitter coefficients calculated for all the nodes, as in Equation 3.9, before the loop has been executed. Furthermore, the loop execution will stop when acceptable difference in the results in two consecutive iterations has been reached.

Based on the selected exponent, the introduced emitters will start reducing the node demand which has as a consequence the pressure increase in the rest of the network. In the event that an emitter operates in the node with negative pressure, resulting from extreme topographic conditions, such emitter will generate negative demand i.e. a dummy inflow that reduces the source supply. This peculiarity can be mitigated in two ways:
1. by removing the emitter while retaining the baseline demand equal to zero;
2. by disconnecting i.e. closing all the pipes connected to that node; this action can be expanded to all the pipes that have at least one connecting node with negative pressure, which will then also include non-demand nodes with negative pressure.

Both of these approaches will have advantages and disadvantages. The first one is more straightforward but neglects reduced conveyance through high-altitude nodes. The latter one takes this conveyance into consideration but results in reconfiguration of the network layout, potentially breaking its integrity. Furthermore, the convergence could be an issue in both cases.

The combination of the two amendments to the basic emitter approach has been proposed to explore the benefit from well-proven robustness of EPANET hydraulic engine, leading to a relatively simple code. The hydraulic simulations presented further in this chapter have been executed in five PDD variants:

1. The basic mode, which allows negative demands in nodes with emitters (coded further as 'pdd1').
2. The mode which disconnects the pipes connected to all the nodes with negative pressures ('pdd2').
3. The mode in which all the dummy negative nodal demands have been set to zero and the corresponding emitters removed ('pdd3').
4. The mode coded as 'pdd4' is a combination of the 2nd and the 3rd approach. To which extent a siphon is really possible will depend on the topography and the head at the source(s). To leave this possibility open, the PDD threshold has been redefined into a PDD range, having also the lower value: a (negative) pressure threshold depicting severe effects of the failure. There will be four ways of node demand adjustment, based on the calculated pressures. For defined threshold pressure range between PDD_{min} and PDD_{max}:
   4.1. if $p_i/\rho g \geq PDD_{max}$, $Q_{i,PDD} = Q_{i,DD}$ (full demand is supplied);
   4.2. if $0 \leq p_i/\rho g < PDD_{max}$, $Q_{i,PDD} = k_i(p_i/\rho g)^{\alpha}$ (partial demand is supplied);
   4.3. if $PDD_{min} \leq p_i/\rho g < 0$, $Q_{i,PDD} = 0$ (no demand, conveyance possible);
   4.4. if $p_i/\rho g < PDD_{min}$, $Q_{i,PDD} = 0$ (pipes connected to node i closed).
5. The PDD software developed by Pathirana (2010), as a benchmark ('pdd5').

In situations when $PDD_{min} = 0$, the fourth approach will become identical to the second one. In fact, that also shows that the third approach is a subset of the second one. Hence, the conditions of the fourth approach can also be written as:

1. if $Q_{i,PDD} > 0$ & $p_i/\rho g \geq PDD_{max}$, $Q_{i,PDD} = Q_{i,DD}$ (full demand is supplied);
2. if $Q_{i,PDD} > 0$ & $p_i/\rho g < PDD_{max}$, $Q_{i,PDD} = k_i(p_i/\rho g)^{\alpha}$ (partial demand is supplied);
3. if $Q_{i,PDD} \leq 0$, $Q_{i,PDD} = 0$ (no demand, conveyance possible);
4. if $p_i/\rho g < PDD_{min}$, $Q_{i,PDD} = 0$ (pipes connected to node i closed).

These conditions have been adopted enabling better convergence of the reconfigured test networks.

3.5   TEST CASE

The first simulations have been conducted for the same scenarios as in the Figures 3.7 to 3.11. The results are shown in Tables 3.1 to 3.7. The results for five PDD variants are split into the tables comparing the node demands and pressures and those comparing the pipe flows and friction losses. The pressure threshold has been $PDD_{max} = 20$ mwc in all the cases, while the lower constraint used for 'pdd4' variant has been $PDD_{min} = -10$ mwc. To demonstrate the sensitivity of this variant, the calculations of the scenarios shown in Tables 3.4 to 3.6 have been repeated with the value of $PDD_{min} = -80$ mwc. Those results have been given in Tables 3.8 to 3.10.

Table 3.1a Test1: Flat terrain (as in Figure 3.7), comparison for node demands and pressures

Node ID	Z (msl)	Qdd (l/s)	Pdd (mwc)	Qpdd1 (l/s)	ppdd1 (mwc)	Qpdd2 (l/s)	ppdd2 (mwc)	Qpdd3 (l/s)	ppdd3 (mwc)	Qpdd4 (l/s)	ppdd4 (mwc)	Qpdd5 (l/s)	ppdd5 (mwc)
J4	5	30.00	31.49	30.00	31.49	30.00	31.49	30.00	31.49	30.00	31.49	30.00	31.49
J3	5	30.00	29.67	30.00	29.67	30.00	29.67	30.00	29.67	30.00	29.67	30.00	29.67
J2	5	30.00	31.49	30.00	31.49	30.00	31.49	30.00	31.49	30.00	31.49	30.00	31.49
J5	5	0.00	35.37	0.00	35.37	0.00	35.37	0.00	35.37	0.00	35.37	0.01	35.37
J6	5	0.00	38.74	0.00	38.74	0.00	38.74	0.00	38.74	0.00	38.74	0.01	38.74
J7	5	0.00	32.00	0.00	32.00	0.00	32.00	0.00	32.00	0.00	32.00	0.01	32.00
J8	5	0.00	38.74	0.00	38.74	0.00	38.74	0.00	38.74	0.00	38.74	0.01	38.74
J9	5	0.00	32.00	0.00	32.00	0.00	32.00	0.00	32.00	0.00	32.00	0.01	32.00

Table 3.1b Test1: Flat terrain (Figure 3.7), comparison for pipe flows and friction losses

Pipe ID	From node	To node	D (mm)	Qdd (l/s)	hfdd (mwc)	Qpdd1 (l/s)	hfpdd1 (mwc)	Qpdd2 (l/s)	hfpdd2 (mwc)	Qpdd3 (l/s)	hfpdd3 (mwc)	Qpdd4 (l/s)	hfpdd4 (mwc)	Qpdd5 (l/s)	hfpdd5 (mwc)
P03a	R1	J6	250	45.00	6.26	45.00	6.26	45.00	6.26	45.00	6.26	45.00	6.26	45.00	6.27
P05a	J9	J4	150	3.15	0.51	3.15	0.51	3.15	0.51	3.15	0.51	3.15	0.51	3.15	0.51
P01a	R1	J8	250	45.00	6.26	45.00	6.26	45.00	6.26	45.00	6.26	45.00	6.26	45.00	6.27
P04a	J7	J2	150	3.15	0.51	3.15	0.51	3.15	0.51	3.15	0.51	3.15	0.51	3.15	0.51
P03b	J6	J4	200	26.85	7.25	26.85	7.25	26.85	7.25	26.85	7.25	26.85	7.25	26.85	7.25
P04b	J7	J3	200	15.00	2.32	15.00	2.32	15.00	2.32	15.00	2.32	15.00	2.32	15.00	2.33
P08a	J8	J5	200	18.15	3.37	18.15	3.37	18.15	3.37	18.15	3.37	18.15	3.37	18.15	3.38
P08b	J5	J9	200	18.15	3.37	18.15	3.37	18.15	3.37	18.15	3.37	18.15	3.37	18.15	3.38
P01b	J8	J2	200	26.85	7.25	26.85	7.25	26.85	7.25	26.85	7.25	26.85	7.25	26.85	7.25
P05b	J9	J3	200	15.00	2.32	15.00	2.32	15.00	2.32	15.00	2.32	15.00	2.32	15.00	2.33
P07a	J6	J5	200	18.15	3.37	18.15	3.37	18.15	3.37	18.15	3.37	18.15	3.37	18.15	3.38
P07b	J5	J7	200	18.15	3.37	18.15	3.37	18.15	3.37	18.15	3.37	18.15	3.37	18.15	3.38

Table 3.2a Test1: z_{J3} = 25 msl (Figure 3.8 & 3.9 left), node demands and pressures

Node ID	Z (msl)	Qdd (l/s)	Pdd (mwc)	Qpdd1 (l/s)	ppdd1 (mwc)	Qpdd2 (l/s)	ppdd2 (mwc)	Qpdd3 (l/s)	ppdd3 (mwc)	Qpdd4 (l/s)	ppdd4 (mwc)	Qpdd5 (l/s)	ppdd5 (mwc)
J4	5	30.00	31.49	30.00	33.02	30.00	33.02	30.00	33.02	30.00	33.02	30	33.02
J3	25	30.00	9.67	23.76	12.55	23.76	12.55	23.76	12.55	23.76	12.55	23.76	12.55
J2	5	30.00	31.49	30.00	33.02	30.00	33.02	30.00	33.02	30.00	33.02	30	33.02
J5	5	0.00	35.37	0.00	36.79	0.00	36.79	0.00	36.79	0.00	36.79	0.01	36.79
J6	5	0.00	38.74	0.00	39.56	0.00	39.56	0.00	39.56	0.00	39.56	0.01	39.56
J7	5	0.00	32.00	0.00	34.03	0.00	34.03	0.00	34.03	0.00	34.03	0.01	34.03
J8	5	0.00	38.74	0.00	39.56	0.00	39.56	0.00	39.56	0.00	39.56	0.01	39.56
J9	5	0.00	32.00	0.00	34.03	0.00	34.03	0.00	34.03	0.00	34.03	0.01	34.03

Table 3.2b Test1: z_{J3} = 25 msl (Figure 3.8 & 3.9 left), pipe flows and friction losses

Pipe ID	From node	To node	D (mm)	Qdd (l/s)	hfdd (mwc)	Qpdd1 (l/s)	hfpdd1 (mwc)	Qpdd2 (l/s)	hfpdd2 (mwc)	Qpdd3 (l/s)	hfpdd3 (mwc)	Qpdd4 (l/s)	hfpdd4 (mwc)	Qpdd5 (l/s)	hfpdd5 (mwc)
P03a	R1	J6	250	45.00	6.26	41.88	5.44	41.88	5.44	41.88	5.44	41.88	5.44	41.88	5.45
P05a	J9	J4	150	3.15	0.51	4.52	1.01	4.52	1.01	4.52	1.01	4.52	1.01	4.52	1.01
P01a	R1	J8	250	45.00	6.26	41.88	5.44	41.88	5.44	41.88	5.44	41.88	5.44	41.88	5.45
P04a	J7	J2	150	3.15	0.51	4.52	1.01	4.52	1.01	4.52	1.01	4.52	1.01	4.52	1.01
P03b	J6	J4	200	26.85	7.25	25.48	6.54	25.48	6.54	25.48	6.54	25.48	6.54	25.48	6.54
P04b	J7	J3	200	15.00	2.32	11.88	1.48	11.88	1.48	11.88	1.48	11.88	1.48	11.88	1.49
P08a	J8	J5	200	18.15	3.37	16.40	2.77	16.40	2.77	16.40	2.77	16.40	2.77	16.4	2.76
P08b	J5	J9	200	18.15	3.37	16.40	2.77	16.40	2.77	16.40	2.77	16.40	2.77	16.4	2.76
P01b	J8	J2	200	26.85	7.25	25.48	6.54	25.48	6.54	25.48	6.54	25.48	6.54	25.48	6.54
P05b	J9	J3	200	15.00	2.32	11.88	1.48	11.88	1.48	11.88	1.48	11.88	1.48	11.88	1.49
P07a	J6	J5	200	18.15	3.37	16.40	2.77	16.40	2.77	16.40	2.77	16.40	2.77	16.4	2.76
P07b	J5	J7	200	18.15	3.37	16.40	2.77	16.40	2.77	16.40	2.77	16.40	2.77	16.4	2.76

Table 3.3a Test1: z_{J3} = 55 msl (Figure 3.8 & 3.9 right), node demands and pressures

Node ID	Z (msl)	Qdd (l/s)	Pdd (mwc)	Qpdd1 (l/s)	ppdd1 (mwc)	Qpdd2 (l/s)	ppdd2 (mwc)	Qpdd3 (l/s)	ppdd3 (mwc)	Qpdd4 (l/s)	ppdd4 (mwc)	Qpdd5 (l/s)	ppdd5 (mwc)
J4	5	30.00	31.49	30.00	39.19	30.00	37.41	30.00	37.41	30.00	37.41	30.00	37.41
J3	55	30.00	-20.33	-16.25	-5.87	0.00	0.00	0.00	-9.32	0.00	-9.32	0.00	-9.32
J2	5	30.00	31.49	30.00	39.19	30.00	37.41	30.00	37.41	30.00	37.41	30.00	37.41
J5	5	0.00	35.37	0.00	43.44	0.00	41.42	0.00	41.42	0.00	41.42	0.01	41.42
J6	5	0.00	38.74	0.00	43.47	0.00	42.17	0.00	42.17	0.00	42.17	0.01	42.17
J7	5	0.00	32.00	0.00	43.42	0.00	40.68	0.00	40.68	0.00	40.68	0.01	40.68
J8	5	0.00	38.74	0.00	43.47	0.00	42.17	0.00	42.17	0.00	42.17	0.01	42.17
J9	5	0.00	32.00	0.00	43.42	0.00	40.68	0.00	40.68	0.00	40.68	0.01	40.68

Table 3.3b Test1: z_{J3} = 55 msl (Figure 3.8 & 3.9 right), pipe flows and friction losses

Pipe ID	From node	To node	D (mm)	Qdd (l/s)	hfdd (mwc)	Qpdd1 (l/s)	hfpdd1 (mwc)	Qpdd2 (l/s)	hfpdd2 (mwc)	Qpdd3 (l/s)	hfpdd3 (mwc)	Qpdd4 (l/s)	hfpdd4 (mwc)	Qpdd5 (l/s)	hfpdd5 (mwc)
P03a	R1	J6	250	45.00	6.26	21.87	1.53	30.00	2.83	30.00	2.83	30.00	2.83	30.00	2.84
P05a	J9	J4	150	3.15	0.51	9.49	4.22	8.32	3.27	8.32	3.27	8.32	3.27	8.32	3.27
P01a	R1	J8	250	45.00	6.26	21.87	1.53	30.00	2.83	30.00	2.83	30.00	2.83	30.00	2.84
P04a	J7	J2	150	3.15	0.51	9.49	4.22	8.32	3.27	8.32	3.27	8.32	3.27	8.32	3.27
P03b	J6	J4	200	26.85	7.25	20.51	4.28	21.68	4.77	21.68	4.77	21.68	4.77	21.68	4.77
P04b	J7	J3	200	15.00	2.32	-8.13	0.71	0.00	0.00	0.00	0.00	0.00	0.00	0.00	0.00
P08a	J8	J5	200	18.15	3.37	1.36	0.03	8.32	0.75	8.32	0.75	8.32	0.75	8.32	0.75
P08b	J5	J9	200	18.15	3.37	1.36	0.03	8.32	0.75	8.32	0.75	8.32	0.75	8.32	0.75
P01b	J8	J2	200	26.85	7.25	20.51	4.28	21.68	4.77	21.68	4.77	21.68	4.77	21.68	4.77
P05b	J9	J3	200	15.00	2.32	-8.13	0.71	0.00	0.00	0.00	0.00	0.00	0.00	0.00	0.00
P07a	J6	J5	200	18.15	3.37	1.36	0.03	8.32	0.75	8.32	0.75	8.32	0.75	8.32	0.75
P07b	J5	J7	200	18.15	3.37	1.36	0.03	8.32	0.75	8.32	0.75	8.32	0.75	8.32	0.75

Table 3.4a Test1: z_{J3} = 25 msl, z_{J5} = 105 msl (Figure 3.10 left), node demands and pressures

Node ID	Z (msl)	Qdd (l/s)	Pdd (mwc)	Qpdd1 (l/s)	ppdd1 (mwc)	Qpdd2 (l/s)	ppdd2 (mwc)	Qpdd3 (l/s)	ppdd3 (mwc)	Qpdd4 (l/s)	ppdd4 (mwc)	Qpdd5 (l/s)	ppdd5 (mwc)
J4	5	30.00	31.49	30.00	33.02	30.00	27.15	30.00	33.02	30.00	27.15	30.00	33.02
J3	25	30.00	9.67	23.76	12.55	13.92	4.30	23.76	12.55	13.92	4.30	23.76	12.55
J2	5	30.00	31.49	30.00	33.02	30.00	27.15	30.00	33.02	30.00	27.15	30.00	33.02
J5	105	0.00	-64.63	0.00	-63.21	0.00	-67.21	0.00	-63.21	0.00	-67.21	0.01	-63.21
J6	5	0.00	38.74	0.00	39.56	0.00	40.74	0.00	39.56	0.00	40.74	0.01	39.56
J7	5	0.00	32.00	0.00	34.03	0.00	24.83	0.00	34.03	0.00	24.83	0.01	34.03
J8	5	0.00	38.74	0.00	39.56	0.00	40.74	0.00	39.56	0.00	40.74	0.01	39.56
J9	5	0.00	32.00	0.00	34.03	0.00	24.83	0.00	34.03	0.00	24.83	0.01	34.03

Table 3.4b Test1: z_{J3} = 25 msl, z_{J5} = 105 msl (Figure 3.10 left), pipe flows and friction losses

Pipe ID	From node	To node	D (mm)	Qdd (l/s)	hfdd (mwc)	Qpdd1 (l/s)	hfpdd1 (mwc)	Qpdd2 (l/s)	hfpdd2 (mwc)	Qpdd3 (l/s)	hfpdd3 (mwc)	Qpdd4 (l/s)	hfpdd4 (mwc)	Qpdd5 (l/s)	hfpdd5 (mwc)
P03a	R1	J6	250	45.00	6.26	41.88	5.44	36.96	4.26	41.88	5.44	36.96	4.26	41.88	5.45
P05a	J9	J4	150	3.15	0.51	4.52	1.01	-6.96	2.31	4.52	1.01	-6.96	2.31	4.52	1.01
P01a	R1	J8	250	45.00	6.26	41.88	5.44	36.96	4.26	41.88	5.44	36.96	4.26	41.88	5.45
P04a	J7	J2	150	3.15	0.51	4.52	1.01	-6.96	2.31	4.52	1.01	-6.96	2.31	4.52	1.01
P03b	J6	J4	200	26.85	7.25	25.48	6.54	36.96	13.59	25.48	6.54	36.96	13.59	25.48	6.54
P04b	J7	J3	200	15.00	2.32	11.88	1.48	6.96	0.53	11.88	1.48	6.96	0.53	11.88	1.49
P08a	J8	J5	200	18.15	3.37	16.40	2.77	0.00	0.00	16.40	2.77	0.00	0.00	16.40	2.76
P08b	J5	J9	200	18.15	3.37	16.40	2.77	0.00	0.00	16.40	2.77	0.00	0.00	16.40	2.76
P01b	J8	J2	200	26.85	7.25	25.48	6.54	36.96	13.59	25.48	6.54	36.96	13.59	25.48	6.54
P05b	J9	J3	200	15.00	2.32	11.88	1.48	6.96	0.53	11.88	1.48	6.96	0.53	11.88	1.49
P07a	J6	J5	200	18.15	3.37	16.40	2.77	0.00	0.00	16.40	2.77	0.00	0.00	16.40	2.76
P07b	J5	J7	200	18.15	3.37	16.40	2.77	0.00	0.00	16.40	2.77	0.00	0.00	16.40	2.76

Table 3.5a Test1: z_{J3} = 55 msl, z_{J5} = 105 msl (Figure 3.10 right), node demands and pressures

Node ID	Z (msl)	Qdd (l/s)	Pdd (mwc)	Qpdd1 (l/s)	ppdd1 (mwc)	Qpdd2 (l/s)	ppdd2 (mwc)	Qpdd3 (l/s)	ppdd3 (mwc)	Qpdd4 (l/s)	ppdd4 (mwc)	Qpdd5 (l/s)	ppdd5 (mwc)
J4	5	30	31.49	30.00	39.19	30.00	33.16	30.00	37.41	30.00	33.16	30.00	37.41
J3	55	30	-20.33	-16.25	-5.87	0.00	0.00	0.00	-9.32	0.00	-16.84	0.00	-9.32
J2	5	30	31.49	30.00	39.19	30.00	33.16	30.00	37.41	30.00	33.16	30.00	37.41
J5	105	0	-64.63	0.00	-56.56	0.00	-62.34	0.00	-58.58	0.00	-62.34	0.01	-58.58
J6	5	0	38.74	0.00	43.47	0.00	42.17	0.00	42.17	0.00	42.17	0.01	42.17
J7	5	0	32.00	0.00	43.42	0.00	33.16	0.00	40.68	0.00	33.16	0.01	40.68
J8	5	0	38.74	0.00	43.47	0.00	42.17	0.00	42.17	0.00	42.17	0.01	42.17
J9	5	0	32.00	0.00	43.42	0.00	33.16	0.00	40.68	0.00	33.16	0.01	40.68

Table 3.5b Test1: z_{J3} = 55 msl, z_{J5} = 105 msl (Figure 3.10 right), pipe flows and friction losses

Pipe ID	From node	To node	D (mm)	Qdd (l/s)	hfdd (mwc)	Qpdd1 (l/s)	hfpdd1 (mwc)	Qpdd2 (l/s)	hfpdd2 (mwc)	Qpdd3 (l/s)	hfpdd3 (mwc)	Qpdd4 (l/s)	hfpdd4 (mwc)	Qpdd5 (l/s)	hfpdd5 (mwc)
P03a	R1	J6	250	45.00	6.26	21.87	1.53	30.00	2.83	30.00	2.83	30.00	2.83	30.00	2.84
P05a	J9	J4	150	3.15	0.51	9.49	4.22	0.00	0.00	8.32	3.27	0.00	0.00	8.32	3.27
P01a	R1	J8	250	45.00	6.26	21.87	1.53	30.00	2.83	30.00	2.83	30.00	2.83	30.00	2.84
P04a	J7	J2	150	3.15	0.51	9.49	4.22	0.00	0.00	8.32	3.27	0.00	0.00	8.32	3.27
P03b	J6	J4	200	26.85	7.25	20.51	4.28	30.00	9.01	21.68	4.77	30.00	9.01	21.68	4.77
P04b	J7	J3	200	15.00	2.32	-8.13	0.71	0.00	0.00	0.00	0.00	0.00	0.00	0.00	0.00
P08a	J8	J5	200	18.15	3.37	1.36	0.03	0.00	0.00	8.32	0.75	0.00	0.00	8.32	0.75
P08b	J5	J9	200	18.15	3.37	1.36	0.03	0.00	0.00	8.32	0.75	0.00	0.00	8.32	0.75
P01b	J8	J2	200	26.85	7.25	20.51	4.28	30.00	9.01	21.68	4.77	30.00	9.01	21.68	4.77
P05b	J9	J3	200	15.00	2.32	-8.13	0.71	0.00	0.00	0.00	0.00	0.00	0.00	0.00	0.00
P07a	J6	J5	200	18.15	3.37	1.36	0.03	0.00	0.00	8.32	0.75	0.00	0.00	8.32	0.75
P07b	J5	J7	200	18.15	3.37	1.36	0.03	0.00	0.00	8.32	0.75	0.00	0.00	8.32	0.75

Table 3.6a Test1: z_{J3} = 25 msl, z_{J2} = z_{J4} = z_{J5} = 105 msl (Figure 3.11 left), node demands and pressures

Node ID	Z (msl)	Qdd (l/s)	Pdd (mwc)	Qpdd1 (l/s)	ppdd1 (mwc)	Qpdd2 (l/s)	ppdd2 (mwc)	Qpdd3 (l/s)	ppdd3 (mwc)	Qpdd4 (l/s)	ppdd4 (mwc)	Qpdd5 (l/s)	ppdd5 (mwc)
J4	105	30	-68.5	-44.13	-43.27	0.00	0.00	0.00	-56.09	0.00	-67.50	0.00	-56.09
J3	25	30	9.67	30.00	25.32	0.00	0.00	30.00	20.06	0.00	0.00	30.00	20.06
J2	105	30	-68.5	-44.13	-43.27	0.00	0.00	0.00	-56.09	0.00	-67.50	0.00	-56.09
J5	105	0	-64.6	0.00	-52.34	0.00	-67.50	0.00	-56.68	0.00	-67.50	0.01	-56.68
J6	5	0	38.7	0.00	47.67	0.00	45.00	0.00	44.26	0.00	45.00	0.01	44.26
J7	5	0	32	0.00	47.64	0.00	20.00	0.00	42.38	0.00	20.00	0.01	42.38
J8	5	0	38.7	0.00	47.67	0.00	45.00	0.00	44.26	0.00	45.00	0.01	44.26
J9	5	0	32	0.00	47.64	0.00	20.00	0.00	42.38	0.00	20.00	0.01	42.38

Table 3.6b Test1: z_{J3} = 25 msl, z_{J2} = z_{J4} = z_{J5} = 105 msl (Figure 3.11 left), pipe flows and friction losses

Pipe ID	From node	To node	D (mm)	Qdd (l/s)	hfdd (mwc)	Qpdd1 (l/s)	hfpdd1 (mwc)	Qpdd2 (l/s)	hfpdd2 (mwc)	Qpdd3 (l/s)	hfpdd3 (mwc)	Qpdd4 (l/s)	hfpdd4 (mwc)	Qpdd5 (l/s)	hfpdd5 (mwc)
P03a	R1	J6	250	45.00	6.26	-29.13	2.67	0.00	0.00	15.00	0.74	0.00	0.00	15.00	0.74
P05a	J9	J4	150	3.15	0.51	-14.05	9.09	0.00	0.00	-5.61	1.53	0.00	0.00	-5.61	1.53
P01a	R1	J8	250	45.00	6.26	-29.13	2.67	0.00	0.00	15.00	0.74	0.00	0.00	15.00	0.74
P04a	J7	J2	150	3.15	0.51	-14.05	9.09	0.00	0.00	-5.61	1.53	0.00	0.00	-5.61	1.53
P03b	J6	J4	200	26.85	7.25	-30.08	9.06	0.00	0.00	5.61	0.35	0.00	0.00	5.61	0.35
P04b	J7	J3	200	15.00	2.32	15.00	2.32	0.00	0.00	15.00	2.32	0.00	0.00	15.00	2.33
P08a	J8	J5	200	18.15	3.37	0.95	0.01	0.00	0.00	9.39	0.94	0.00	0.00	9.39	0.95
P08b	J5	J9	200	18.15	3.37	0.95	0.01	0.00	0.00	9.39	0.94	0.00	0.00	9.39	0.95
P01b	J8	J2	200	26.85	7.25	-30.08	9.06	0.00	0.00	5.61	0.35	0.00	0.00	5.61	0.35
P05b	J9	J3	200	15.00	2.32	15.00	2.32	0.00	0.00	15.00	2.32	0.00	0.00	15.00	2.33
P07a	J6	J5	200	18.15	3.37	0.95	0.01	0.00	0.00	9.39	0.94	0.00	0.00	9.39	0.95
P07b	J5	J7	200	18.15	3.37	0.95	0.01	0.00	0.00	9.39	0.94	0.00	0.00	9.39	0.95

Table 3.7a Test1: z_{J3} = 55 msl, z_{J2} = z_{J4} = z_{J5} = 105 msl (Figure 3.11 right), node demands and pressures

Node ID	Z (msl)	Qdd (l/s)	Pdd (mwc)	Qpdd1 (l/s)	ppdd1 (mwc)	Qpdd2 (l/s)	ppdd2 (mwc)	Qpdd3 (l/s)	ppdd3 (mwc)	Qpdd4 (l/s)	ppdd4 (mwc)	Qpdd5 (l/s)	ppdd5 (mwc)
J4	105	30.00	-68.51	-42.98	-41.06	0.00	0.00	0.00	-55.00	0.00	-55.00	0.00	-55.00
J3	55	30.00	-20.33	7.63	1.29	0.00	0.00	0.00	-5.00	0.00	-5.00	0.00	-5.00
J2	105	30.00	-68.51	-42.98	-41.06	0.00	0.00	0.00	-55.00	0.00	-55.00	0.00	-55.00
J5	105	0.00	-64.63	0.00	-49.38	0.00	-52.50	0.00	-55.00	0.00	-55.00	0.01	-55.00
J6	5	0.00	38.74	0.00	49.77	0.00	45.00	0.00	45.00	0.00	45.00	0.01	45.00
J7	5	0.00	32.00	0.00	51.46	0.00	50.00	0.00	45.00	0.00	45.00	0.01	45.00
J8	5	0.00	38.74	0.00	49.77	0.00	45.00	0.00	45.00	0.00	45.00	0.01	45.00
J9	5	0.00	32.00	0.00	51.46	0.00	50.00	0.00	45.00	0.00	45.00	0.01	45.00

Table 3.7b Test1: z_{J3} = 55 msl, z_{J2} = z_{J4} = z_{J5} = 105 msl (Figure 3.11 right), pipe flows and friction losses

Pipe ID	From node	To node	D (mm)	Qdd (l/s)	hfdd (mwc)	Qpdd1 (l/s)	hfpdd1 (mwc)	Qpdd2 (l/s)	hfpdd2 (mwc)	Qpdd3 (l/s)	hfpdd3 (mwc)	Qpdd4 (l/s)	hfpdd4 (mwc)	Qpdd5 (l/s)	hfpdd5 (mwc)
P03a	R1	J6	250	45.00	6.26	-39.17	4.77	0.00	0.00	0.00	0.00	0.00	0.00	0.00	0.00
P05a	J9	J4	150	3.15	0.51	-12.71	7.48	0.00	0.00	0.00	0.00	0.00	0.00	0.00	0.00
P01a	R1	J8	250	45.00	6.26	-39.17	4.77	0.00	0.00	0.00	0.00	0.00	0.00	0.00	0.00
P04a	J7	J2	150	3.15	0.51	-12.71	7.48	0.00	0.00	0.00	0.00	0.00	0.00	0.00	0.00
P03b	J6	J4	200	26.85	7.25	-30.27	9.17	0.00	0.00	0.00	0.00	0.00	0.00	0.00	0.00
P04b	J7	J3	200	15.00	2.32	3.82	0.17	0.00	0.00	0.00	0.00	0.00	0.00	0.00	0.00
P08a	J8	J5	200	18.15	3.37	-8.90	0.85	0.00	0.00	0.00	0.00	0.00	0.00	0.00	0.00
P08b	J5	J9	200	18.15	3.37	-8.90	0.85	0.00	0.00	0.00	0.00	0.00	0.00	0.00	0.00
P01b	J8	J2	200	26.85	7.25	-30.27	9.17	0.00	0.00	0.00	0.00	0.00	0.00	0.00	0.00
P05b	J9	J3	200	15.00	2.32	3.82	0.17	0.00	0.00	0.00	0.00	0.00	0.00	0.00	0.00
P07a	J6	J5	200	18.15	3.37	-8.90	0.85	0.00	0.00	0.00	0.00	0.00	0.00	0.00	0.00
P07b	J5	J7	200	18.15	3.37	-8.90	0.85	0.00	0.00	0.00	0.00	0.00	0.00	0.00	0.00

Table 3.8a Test1: z_{J3} = 25 msl, z_{J5} = 105 msl, PDD_{min} = -80 mwc, node demands and pressures

Node ID	Z (msl)	Qdd (l/s)	Pdd (mwc)	Qpdd1 (l/s)	ppdd1 (mwc)	Qpdd2 (l/s)	ppdd2 (mwc)	Qpdd3 (l/s)	ppdd3 (mwc)	Qpdd4 (l/s)	ppdd4 (mwc)
J4	5	30.00	31.49	30.00	33.02	30.00	27.15	30.00	33.02	30.00	33.02
J3	25	30.00	9.67	23.76	12.55	13.92	4.30	23.76	12.55	23.76	12.55
J2	5	30.00	31.49	30.00	33.02	30.00	27.15	30.00	33.02	30.00	33.02
J5	105	0.00	-64.63	0.00	-63.21	0.00	-67.21	0.00	-63.21	0.00	-63.21
J6	5	0.00	38.74	0.00	39.56	0.00	40.74	0.00	39.56	0.00	39.56
J7	5	0.00	32.00	0.00	34.03	0.00	24.83	0.00	34.03	0.00	34.03
J8	5	0.00	38.74	0.00	39.56	0.00	40.74	0.00	39.56	0.00	39.56
J9	5	0.00	32.00	0.00	34.03	0.00	24.83	0.00	34.03	0.00	34.03

Table 3.8b Test1: z_{J3} = 25 msl, z_{J5} = 105 msl, PDD_{min} = -80 mwc, pipe flows and friction losses

Pipe ID	From node	To node	D (mm)	Qdd (l/s)	hfdd (mwc)	Qpdd1 (l/s)	hfpdd1 (mwc)	Qpdd2 (l/s)	hfpdd2 (mwc)	Qpdd3 (l/s)	hfpdd3 (mwc)	Qpdd4 (l/s)	hfpdd4 (mwc)
P03a	R1	J6	250	45.00	6.26	41.88	5.44	36.96	4.26	41.88	5.44	41.88	5.44
P05a	J9	J4	150	3.15	0.51	4.52	1.01	-6.96	2.31	4.52	1.01	4.52	1.01
P01a	R1	J8	250	45.00	6.26	41.88	5.44	36.96	4.26	41.88	5.44	41.88	5.44
P04a	J7	J2	150	3.15	0.51	4.52	1.01	-6.96	2.31	4.52	1.01	4.52	1.01
P03b	J6	J4	200	26.85	7.25	25.48	6.54	36.96	13.59	25.48	6.54	25.48	6.54
P04b	J7	J3	200	15.00	2.32	11.88	1.48	6.96	0.53	11.88	1.48	11.88	1.48
P08a	J8	J5	200	18.15	3.37	16.40	2.77	0.00	0.00	16.40	2.77	16.40	2.77
P08b	J5	J9	200	18.15	3.37	16.40	2.77	0.00	0.00	16.40	2.77	16.40	2.77
P01b	J8	J2	200	26.85	7.25	25.48	6.54	36.96	13.59	25.48	6.54	25.48	6.54
P05b	J9	J3	200	15.00	2.32	11.88	1.48	6.96	0.53	11.88	1.48	11.88	1.48
P07a	J6	J5	200	18.15	3.37	16.40	2.77	0.00	0.00	16.40	2.77	16.40	2.77
P07b	J5	J7	200	18.15	3.37	16.40	2.77	0.00	0.00	16.40	2.77	16.40	2.77

Table 3.9a Test1: z_{J3} = 55 msl, z_{J5} = 105 msl, PDD_{min} = -80 mwc, node demands and pressures

Node ID	Z (msl)	Qdd (l/s)	Pdd (mwc)	Qpdd1 (l/s)	ppdd1 (mwc)	Qpdd2 (l/s)	ppdd2 (mwc)	Qpdd3 (l/s)	ppdd3 (mwc)	Qpdd4 (l/s)	ppdd4 (mwc)
J4	5	30.00	31.49	30.00	39.19	30.00	33.16	30.00	37.41	30.00	37.41
J3	55	30.00	-20.33	-16.25	-5.87	0.00	0.00	0.00	-9.32	0.00	-9.32
J2	5	30.00	31.49	30.00	39.19	30.00	33.16	30.00	37.41	30.00	37.41
J5	105	0.00	-64.63	0.00	-56.56	0.00	-62.34	0.00	-58.58	0.00	-58.58
J6	5	0.00	38.74	0.00	43.47	0.00	42.17	0.00	42.17	0.00	42.17
J7	5	0.00	32.00	0.00	43.42	0.00	33.16	0.00	40.68	0.00	40.68
J8	5	0.00	38.74	0.00	43.47	0.00	42.17	0.00	42.17	0.00	42.17
J9	5	0.00	32.00	0.00	43.42	0.00	33.16	0.00	40.68	0.00	40.68

Table 3.9b Test1: z_{J3} = 55 msl, z_{J5} = 105 msl, PDD_{min} = -80 mwc, pipe flows and friction losses

Pipe ID	From node	To node	D (mm)	Qdd (l/s)	hfdd (mwc)	Qpdd1 (l/s)	hfpdd1 (mwc)	Qpdd2 (l/s)	hfpdd2 (mwc)	Qpdd3 (l/s)	hfpdd3 (mwc)	Qpdd4 (l/s)	hfpdd4 (mwc)
P03a	R1	J6	250	45.00	6.26	21.87	1.53	30.00	2.83	30.00	2.83	30.00	2.83
P05a	J9	J4	150	3.15	0.51	9.49	4.22	0.00	0.00	8.32	3.27	8.32	3.27
P01a	R1	J8	250	45.00	6.26	21.87	1.53	30.00	2.83	30.00	2.83	30.00	2.83
P04a	J7	J2	150	3.15	0.51	9.49	4.22	0.00	0.00	8.32	3.27	8.32	3.27
P03b	J6	J4	200	26.85	7.25	20.51	4.28	30.00	9.01	21.68	4.77	21.68	4.77
P04b	J7	J3	200	15.00	2.32	-8.13	0.71	0.00	0.00	0.00	0.00	0.00	0.00
P08a	J8	J5	200	18.15	3.37	1.36	0.03	0.00	0.00	8.32	0.75	8.32	0.75
P08b	J5	J9	200	18.15	3.37	1.36	0.03	0.00	0.00	8.32	0.75	8.32	0.75
P01b	J8	J2	200	26.85	7.25	20.51	4.28	30.00	9.01	21.68	4.77	21.68	4.77
P05b	J9	J3	200	15.00	2.32	-8.13	0.71	0.00	0.00	0.00	0.00	0.00	0.00
P07a	J6	J5	200	18.15	3.37	1.36	0.03	0.00	0.00	8.32	0.75	8.32	0.75
P07b	J5	J7	200	18.15	3.37	1.36	0.03	0.00	0.00	8.32	0.75	8.32	0.75

Table 3.10a Test1: z_{J3} = 25 msl, z_{J2} = z_{J4} = z_{J5} = 105 msl, PDD_{min} = -80 mwc, node demands and pressures

Node ID	Z (msl)	Qdd (l/s)	Pdd (mwc)	Qpdd1 (l/s)	ppdd1 (mwc)	Qpdd2 (l/s)	ppdd2 (mwc)	Qpdd3 (l/s)	ppdd3 (mwc)	Qpdd4 (l/s)	ppdd4 (mwc)
J4	105	30	-68.51	-44.13	-43.27	0.00	0.00	0.00	-56.09	0.00	-56.09
J3	25	30	9.67	30.00	25.32	0.00	0.00	30.00	20.06	30.00	20.06
J2	105	30	-68.51	-44.13	-43.27	0.00	0.00	0.00	-56.09	0.00	-56.09
J5	105	0	-64.63	0.00	-52.34	0.00	-67.50	0.00	-56.68	0.00	-56.68
J6	5	0	38.74	0.00	47.67	0.00	45.00	0.00	44.26	0.00	44.26
J7	5	0	32.00	0.00	47.64	0.00	20.00	0.00	42.38	0.00	42.38
J8	5	0	38.74	0.00	47.67	0.00	45.00	0.00	44.26	0.00	44.26
J9	5	0	32.00	0.00	47.64	0.00	20.00	0.00	42.38	0.00	42.38

Table 3.10b Test1: z_{J3} = 25 msl, z_{J2} = z_{J4} = z_{J5} = 105 msl, PDD_{min} = -80 mwc, pipe flows and friction losses

Pipe ID	From node	To node	D (mm)	Qdd (l/s)	hfdd (mwc)	Qpdd1 (l/s)	hfpdd1 (mwc)	Qpdd2 (l/s)	hfpdd2 (mwc)	Qpdd3 (l/s)	hfpdd3 (mwc)	Qpdd4 (l/s)	hfpdd4 (mwc)
P03a	R1	J6	250	45.00	6.26	-29.13	2.67	0.00	0.00	15.00	0.74	15.00	0.74
P05a	J9	J4	150	3.15	0.51	-14.05	9.09	0.00	0.00	-5.61	1.53	-5.61	1.53
P01a	R1	J8	250	45.00	6.26	-29.13	2.67	0.00	0.00	15.00	0.74	15.00	0.74
P04a	J7	J2	150	3.15	0.51	-14.05	9.09	0.00	0.00	-5.61	1.53	-5.61	1.53
P03b	J6	J4	200	26.85	7.25	-30.08	9.06	0.00	0.00	5.61	0.35	5.61	0.35
P04b	J7	J3	200	15.00	2.32	15.00	2.32	0.00	0.00	15.00	2.32	15.00	2.32
P08a	J8	J5	200	18.15	3.37	0.95	0.01	0.00	0.00	9.39	0.94	9.39	0.94
P08b	J5	J9	200	18.15	3.37	0.95	0.01	0.00	0.00	9.39	0.94	9.39	0.94
P01b	J8	J2	200	26.85	7.25	-30.08	9.06	0.00	0.00	5.61	0.35	5.61	0.35
P05b	J9	J3	200	15.00	2.32	15.00	2.32	0.00	0.00	15.00	2.32	15.00	2.32
P07a	J6	J5	200	18.15	3.37	0.95	0.01	0.00	0.00	9.39	0.94	9.39	0.94
P07b	J5	J7	200	18.15	3.37	0.95	0.01	0.00	0.00	9.39	0.94	9.39	0.94

The following is the list of specific conclusions:

- There has been little difference in the results for normal range of elevations, i.e. as long the negative pressures are not occurring in the 'pdd1' variant. Consequently, the results in Table 3.1 showing the situation of flat terrain are almost identical. The results in Table 3.2 showing the situation of moderately elevated node *J3*, too.
- The negative pressure occurs in the case of more radical elevation of *J3*, which is shown in Table 3.3. The 'pdd2' reacts to it by disconnecting the pipes connected to that node. As a result, the EPANET calculation produces the result zero, both for the nodal demand and the pressure in *J3*. The 'pdd3' variant will have the same nodal demands as the 'pdd2' except that the pressure in *J3* is kept negative. Furthermore, the results of 'pdd4' are the same as those of 'pdd3' for the negative pressure is above the lower limit of -10 mwc. Finally, the results of 'pdd5' will comply with those of 'pdd3' and 'pdd4'. The implications on the demand reduction are that the pipe flow and friction loss distribution are the same in all cases.
- The difference in the results starts in the case of more radical change of the elevation of node *J5*, which is shown in Table 3.4. Due to very low negative pressure in *J5*, the results of 'pdd4' are here identical to those of 'pdd2', showing more radical loss of demand than in the case of 'pdd3' and 'pdd5'. The basic option of 'pdd1' also complies with 'pdd3' and 'pdd5' because the negative pressure has been registered in non-demand node. Accordingly, the pipe flow and friction loss distribution will be the same for 'pdd1', 'pdd3' and 'pdd5', while the results for 'pdd2' and 'pdd4' will be different (but mutually, the same). In the first group, the total demand drops from 90 to 83.76 l/s, while in the second group, the total demand shall be 73.92 l/s.
- The results in Table 3.5 also show the compliance between 'pdd2' and 'pdd4', and 'pdd3' and 'pdd5' but, unlike is the case in Table 3.4, the pipe flow and friction loss patterns will be the same in all four cases because the total flow has been uniformly reduced to 60 l/s, as a result of extreme elevation of node *J3*.
- In case of a 'high hill' along the diagonal *J4-J5-J2*, the flow pattern shown in Table 3.6 will be uninterrupted in 'pdd3' and 'pdd5'. The demands in *J2* and *J4* will be lost due to high elevation but the one in *J3* will be delivered 100% as the sufficient pressure (above 20 mwc) has been generated in that node resulting from the demand reduction in the other two nodes. This is not the picture as created by 'pdd2' and 'pdd4' where the demand in *J3* is also zero, because of the disconnected pipes. Also, the node pressures calculated in both cases do not look quite realistic.
- For the elevation of 55 msl in *J3*, all four variants in Table 3.7 show logical no-flow condition and the pressures based on the source head, except for 'pdd2'.
- The results in Tables 3.8 to 3.10 show that the lowering of the lower pressure limit actually makes the results of 'ppd4' equal to 'pdd3' instead of 'pdd2'.
- Finally, a general observation is that in all the calculations the 'pdd3' and 'pdd5' results complied suggesting the same numerical approach used in both algorithms.

3.6   CALCULATION OF AVAILABLE DEMAND

In the next round of testing and comparisons, the same network has been calculated for the same set of node elevations by failing each pipe in order to calculate the total available demand. These results are presented in Tables 3.11 to 3.17.

Table 3.11 Test1: Flat terrain (as in Figure 3.7), total available demand and pressure range

Mode	Qtot (l/s)	Pmin (mwc)	Node	Pmax (mwc)	Node	Mode	Qtot (l/s)	Pmin (mwc)	Node	Pmax (mwc)	Node	Mode	Qtot (l/s)	Pmin (mwc)	Node	Pmax (mwc)	Node
failure P01a	from		R1	to	J8	failure P03a	from		R1	to	J6	failure P04a	from		J2	to	J7
DD	90.00	-10.94	J2	20.39	J6	DD	90.00	-10.94	J4	20.39	J8	DD	90.00	29.45	J2	39.00	J6
PDD1	72.42	10.43	J2	28.99	J6	PDD1	72.42	10.43	J4	28.99	J8	PDD1	90.00	29.45	J2	39.00	J6
PDD2	72.42	10.43	J2	28.99	J6	PDD2	72.42	10.43	J4	28.99	J8	PDD2	90.00	29.45	J2	39.00	J6
PDD3	72.42	10.43	J2	28.99	J6	PDD3	72.42	10.43	J4	28.99	J8	PDD3	90.00	29.45	J2	39.00	J6
PDD4	72.42	10.43	J2	28.99	J6	PDD4	72.42	10.43	J4	28.99	J8	PDD4	90.00	29.45	J2	39.00	J6
PDD5	72.42	10.43	J2	28.99	J6	PDD5	72.42	10.43	J4	28.99	J8	PDD5	90.00	29.45	J2	39.00	J6
failure P01b	from		J8	to	J2	failure P03b	from		J6	to	J4	failure P04b	from		J7	to	J3
DD	90.00	-22.51	J2	41.38	J8	DD	90.00	-22.51	J4	41.38	J6	DD	90.00	18.75	J3	39.32	J8
PDD1	79.29	8.27	J2	42.22	J8	PDD1	79.29	8.27	J4	42.22	J6	PDD1	89.49	19.32	J3	39.38	J8
PDD2	79.29	8.27	J2	42.22	J8	PDD2	79.29	8.27	J4	42.22	J6	PDD2	89.49	19.32	J3	39.38	J8
PDD3	79.29	8.27	J2	42.22	J8	PDD3	79.29	8.27	J4	42.22	J6	PDD3	89.49	19.32	J3	39.38	J8
PDD4	79.29	8.27	J2	42.22	J8	PDD4	79.29	8.27	J4	42.22	J6	PDD4	89.49	19.32	J3	39.38	J8
PDD5	79.29	8.27	J2	42.22	J8	PDD5	79.29	8.27	J4	42.22	J6	PDD5	89.48	19.32	J3	39.38	J8
failure P05a	from		J4	to	J9	failure P07a	from		J6	to	J5	failure P08a	from		J8	to	J5
DD	90.00	29.45	J4	39.00	J8	DD	90.00	25.05	J3	41.18	J6	DD	90.00	25.05	J3	41.18	J8
PDD1	90.00	29.45	J4	39.00	J8	PDD1	90.00	25.05	J3	41.18	J6	PDD1	90.00	25.05	J3	41.18	J8
PDD2	90.00	29.45	J4	39.00	J8	PDD2	90.00	25.05	J3	41.18	J6	PDD2	90.00	25.05	J3	41.18	J8
PDD3	90.00	29.45	J4	39.00	J8	PDD3	90.00	25.05	J3	41.18	J6	PDD3	90.00	25.05	J3	41.18	J8
PDD4	90.00	29.45	J4	39.00	J8	PDD4	90.00	25.05	J3	41.18	J6	PDD4	90.00	25.05	J3	41.18	J8
PDD5	90.00	29.45	J4	39.00	J8	PDD5	90.00	25.05	J3	41.18	J6	PDD5	90.00	25.05	J3	41.18	J8
failure P05b	from		J9	to	J3	failure P07b	from		J5	to	J7	failure P08b	from		J5	to	J9
DD	90.00	18.75	J3	39.32	J6	DD	90.00	24.33	J3	39.12	J6	DD	90.00	24.33	J3	39.12	J8
PDD1	89.49	19.32	J3	39.38	J6	PDD1	90.00	24.33	J3	39.12	J6	PDD1	90.00	24.33	J3	39.12	J8
PDD2	89.49	19.32	J3	39.38	J6	PDD2	90.00	24.33	J3	39.12	J6	PDD2	90.00	24.33	J3	39.12	J8
PDD3	89.49	19.32	J3	39.38	J6	PDD3	90.00	24.33	J3	39.12	J6	PDD3	90.00	24.33	J3	39.12	J8
PDD4	89.49	19.32	J3	39.38	J6	PDD4	90.00	24.33	J3	39.12	J6	PDD4	90.00	24.33	J3	39.12	J8
PDD5	89.48	19.32	J3	39.38	J6	PDD5	90.00	24.33	J3	39.12	J6	PDD5	90.00	24.33	J3	39.12	J8

Table 3.12 Test1: z_{J3} = 25 msl (Figure 3.8 & 3.9 left), total available demand and pressure range

Mode	Qtot (l/s)	Pmin (mwc)	Node	Pmax (mwc)	Node	Mode	Qtot (l/s)	Pmin (mwc)	Node	Pmax (mwc)	Node	Mode	Qtot (l/s)	Pmin (mwc)	Node	Pmax (mwc)	Node
failure P01a	from		R1	to	J8	failure P03a	from		R1	to	J6	failure P04a	from		J2	to	J7
DD	90.00	-27.51	J3	20.39	J6	DD	90.00	-27.51	J3	20.39	J8	DD	90.00	10.59	J3	39.00	J6
PDD1	64.21	1.14	J3	32.38	J6	PDD1	64.21	1.14	J3	32.38	J8	PDD1	84.53	13.37	J3	39.79	J6
PDD2	64.21	1.14	J3	32.38	J6	PDD2	64.21	1.14	J3	32.38	J8	PDD2	84.53	13.37	J3	39.79	J6
PDD3	64.21	1.14	J3	32.38	J6	PDD3	64.21	1.14	J3	32.38	J8	PDD3	84.53	13.37	J3	39.79	J6
PDD4	64.21	1.14	J3	32.38	J6	PDD4	64.21	1.14	J3	32.38	J8	PDD4	84.53	13.37	J3	39.79	J6
PDD5	64.21	1.14	J3	32.38	J6	PDD5	64.21	1.14	J3	32.38	J8	PDD5	84.52	13.37	J3	39.79	J6
failure P01b	from		J8	to	J2	failure P03b	from		J6	to	J4	failure P04b	from		J7	to	J3
DD	90.00	-22.51	J2	41.38	J8	DD	90.00	-22.51	J4	41.38	J6	DD	90.00	-1.25	J3	39.32	J8
PDD1	71.19	9.20	J3	42.77	J8	PDD1	71.19	9.20	J3	42.77	J6	PDD1	79.78	8.69	J3	40.34	J8
PDD2	71.19	9.20	J3	42.77	J8	PDD2	71.19	9.20	J3	42.77	J6	PDD2	79.78	8.69	J3	40.34	J8
PDD3	71.19	9.20	J3	42.77	J8	PDD3	71.19	9.20	J3	42.77	J6	PDD3	79.78	8.69	J3	40.34	J8
PDD4	71.19	9.20	J3	42.77	J8	PDD4	71.19	9.20	J3	42.77	J6	PDD4	79.78	8.69	J3	40.34	J8
PDD5	71.19	9.20	J3	42.77	J8	PDD5	71.19	9.20	J3	42.77	J6	PDD5	79.78	8.69	J3	40.33	J8
failure P05a	from		J4	to	J9	failure P07a	from		J6	to	J5	failure P08a	from		J8	to	J5
DD	90.00	10.59	J3	39.00	J8	DD	90.00	5.05	J3	41.18	J6	DD	90.00	5.05	J3	41.18	J8
PDD1	84.53	13.37	J3	39.79	J8	PDD1	81.13	9.92	J3	41.80	J6	PDD1	81.13	9.92	J3	41.80	J8
PDD2	84.53	13.37	J3	39.79	J8	PDD2	81.13	9.92	J3	41.80	J6	PDD2	81.13	9.92	J3	41.80	J8
PDD3	84.53	13.37	J3	39.79	J8	PDD3	81.13	9.92	J3	41.80	J6	PDD3	81.13	9.92	J3	41.80	J8
PDD4	84.53	13.37	J3	39.79	J8	PDD4	81.13	9.92	J3	41.80	J6	PDD4	81.13	9.92	J3	41.80	J8
PDD5	84.52	13.37	J3	39.79	J8	PDD5	81.14	9.92	J3	41.80	J6	PDD5	81.14	9.92	J3	41.80	J8
failure P05b	from		J9	to	J3	failure P07b	from		J5	to	J7	failure P08b	from		J5	to	J9
DD	90.00	-1.25	J3	39.32	J6	DD	90.00	4.33	J3	39.12	J6	DD	90.00	4.33	J3	39.12	J8
PDD1	79.78	8.69	J3	40.34	J6	PDD1	80.81	9.62	J3	40.20	J6	PDD1	80.81	9.62	J3	40.20	J8
PDD2	79.78	8.69	J3	40.34	J6	PDD2	80.81	9.62	J3	40.20	J6	PDD2	80.81	9.62	J3	40.20	J8
PDD3	79.78	8.69	J3	40.34	J6	PDD3	80.81	9.62	J3	40.20	J6	PDD3	80.81	9.62	J3	40.20	J8
PDD4	79.78	8.69	J3	40.34	J6	PDD4	80.81	9.62	J3	40.20	J6	PDD4	80.81	9.62	J3	40.20	J8
PDD5	79.78	8.69	J3	40.33	J6	PDD5	80.81	9.62	J3	40.20	J6	PDD5	80.81	9.62	J3	40.20	J8

Table 3.13 Test1: z_{J3} = 55 msl (Figure 3.8 & 3.9 right), total available demand and pressure range

Mode	Qtot (l/s)	Pmin (mwc)	Node	Pmax (mwc)	Node	Mode	Qtot (l/s)	Pmin (mwc)	Node	Pmax (mwc)	Node	Mode	Qtot (l/s)	Pmin (mwc)	Node	Pmax (mwc)	Node
failure P01a	from		R1	to	J8	failure P03a	from		R1	to	J6	failure P04a	from		J2	to	J7
DD	90.00	-57.51	J3	20.39	J6	DD	90.00	-57.51	J3	20.39	J8	DD	90.00	-19.41	J3	39.00	J6
PDD1	60.00	-10.70	J3	40.49	J6	PDD1	60.00	-10.70	J3	40.49	J8	PDD1	60.00	-5.45	J3	44.09	J7
PDD2	59.08	0.00	J3	34.29	J6	PDD2	59.08	0.00	J3	34.29	J8	PDD2	60.00	0.00	J3	42.42	J6
PDD3	59.24	-24.84	J3	34.24	J6	PDD3	59.24	-24.84	J3	34.24	J8	PDD3	60.00	-8.38	J3	42.44	J6
PDD4	59.08	-25.07	J3	34.29	J6	PDD4	59.08	-25.07	J3	34.29	J8	PDD4	60.00	-8.38	J3	42.44	J6
PDD5	59.24	-24.84	J3	34.24	J6	PDD5	59.24	-24.84	J3	34.24	J8	PDD5	60.00	-8.38	J3	42.44	J6
failure P01b	from		J8	to	J2	failure P03b	from		J6	to	J4	failure P04b	from		J7	to	J3
DD	90.00	-32.17	J3	41.38	J8	DD	90.00	-32.17	J3	41.38	J6	DD	90.00	-31.25	J3	39.32	J8
PDD1	54.75	-7.00	J3	44.49	J8	PDD1	54.75	-7.00	J3	44.49	J6	PDD1	60.00	-4.44	J3	43.49	J9
PDD2	52.72	0.00	J3	43.82	J8	PDD2	52.72	0.00	J3	43.82	J6	PDD2	60.00	0.00	J3	42.17	J6
PDD3	53.46	-12.35	J3	43.81	J8	PDD3	53.46	-12.35	J3	43.81	J6	PDD3	60.00	-9.32	J3	42.17	J6
PDD4	52.72	-12.85	J3	43.82	J8	PDD4	52.72	-12.85	J3	43.82	J6	PDD4	60.00	-9.32	J3	42.17	J6
PDD5	53.46	-12.35	J3	43.81	J8	PDD5	53.46	-12.35	J3	43.81	J6	PDD5	60.00	-9.32	J3	42.17	J6
failure P05a	from		J4	to	J9	failure P07a	from		J6	to	J5	failure P08a	from		J8	to	J5
DD	90.00	-19.41	J3	39.00	J8	DD	90.00	-24.95	J3	41.18	J6	DD	90.00	-24.95	J3	41.18	J8
PDD1	60.00	-5.45	J3	44.09	J9	PDD1	60.00	-6.01	J3	43.62	J6	PDD1	60.00	-6.01	J3	43.62	J8
PDD2	60.00	0.00	J3	42.42	J8	PDD2	60.00	0.00	J3	43.13	J6	PDD2	60.00	0.00	J3	43.13	J8
PDD3	60.00	-8.38	J3	42.44	J8	PDD3	60.00	-11.29	J3	43.12	J6	PDD3	60.00	-11.29	J3	43.12	J8
PDD4	60.00	-8.38	J3	42.44	J8	PDD4	60.00	-11.29	J3	43.12	J6	PDD4	60.00	-11.29	J3	43.12	J8
PDD5	60.00	-8.38	J3	42.44	J8	PDD5	60.00	-11.29	J3	43.12	J6	PDD5	60.00	-11.29	J3	43.12	J8
failure P05b	from		J9	to	J3	failure P07b	from		J5	to	J7	failure P08b	from		J5	to	J9
DD	90.00	-31.25	J3	39.32	J6	DD	90.00	-25.67	J3	39.12	J6	DD	90.00	-25.67	J3	39.12	J8
PDD1	60.00	-4.44	J3	43.49	J7	PDD1	60.00	-6.02	J3	43.49	J6	PDD1	60.00	-6.02	J3	43.49	J8
PDD2	60.00	0.00	J3	42.17	J6	PDD2	60.00	0.00	J3	42.42	J6	PDD2	60.00	0.00	J3	42.42	J8
PDD3	60.00	-9.32	J3	42.17	J6	PDD3	60.00	-10.86	J3	42.20	J6	PDD3	60.00	-10.86	J3	42.20	J8
PDD4	60.00	-9.32	J3	42.17	J6	PDD4	60.00	-10.86	J3	42.20	J6	PDD4	60.00	-10.86	J3	42.20	J8
PDD5	60.00	-9.32	J3	42.17	J6	PDD5	60.00	-10.86	J3	42.20	J6	PDD5	60.00	-10.86	J3	42.20	J8

Table 3.14 Test1: z_{J3} = 25 msl, z_{J5} = 105 msl (Figure 3.10 left), total available demand and pressure range

Mode	Qtot (l/s)	Pmin (mwc)	Node	Pmax (mwc)	Node	Mode	Qtot (l/s)	Pmin (mwc)	Node	Pmax (mwc)	Node	Mode	Qtot (l/s)	Pmin (mwc)	Node	Pmax (mwc)	Node
failure P01a	from		R1	to	J8	failure P03a	from		R1	to	J6	failure P04a	from		J2	to	J7
DD	90.00	-102.60	J5	20.39	J6	DD	90.00	-102.60	J5	20.39	J8	DD	90.00	-64.17	J5	39.00	J6
PDD1	64.21	-77.88	J5	32.38	J6	PDD1	64.21	-77.88	J5	32.38	J8	PDD1	84.53	-62.79	J5	39.79	J6
PDD2	30.00	-81.17	J5	42.17	J6	PDD2	30.00	-81.17	J5	42.17	J8	PDD2	68.44	-68.37	J5	42.17	J8
PDD3	64.21	-77.88	J5	32.38	J6	PDD3	64.21	-77.88	J5	32.38	J8	PDD3	84.53	-62.79	J5	39.79	J6
PDD4	42.57	-83.04	J5	39.38	J6	PDD4	42.57	-83.04	J5	39.38	J8	PDD4	68.44	-68.37	J5	42.17	J8
PDD5	64.21	-77.88	J5	32.38	J6	PDD5	64.21	-77.88	J5	32.38	J8	PDD5	84.52	-62.79	J5	39.79	J6
failure P01b	from		J8	to	J2	failure P03b	from		J6	to	J4	failure P04b	from		J7	to	J3
DD	90.00	-70.17	J5	41.38	J8	DD	90.00	-70.17	J5	41.38	J6	DD	90.00	-64.30	J5	39.32	J8
PDD1	71.19	-64.33	J5	42.77	J8	PDD1	71.19	-64.33	J5	42.77	J6	PDD1	79.78	-62.20	J5	40.34	J8
PDD2	30.00	-69.92	J5	45.00	J8	PDD2	30.00	-69.92	J5	45.00	J6	PDD2	68.44	-65.48	J5	42.17	J8
PDD3	71.19	-64.33	J5	42.77	J8	PDD3	71.19	-64.33	J5	42.77	J6	PDD3	79.78	-62.20	J5	40.34	J8
PDD4	42.57	-72.67	J5	45.00	J8	PDD4	42.57	-72.67	J5	45.00	J6	PDD4	68.44	-65.48	J5	42.17	J8
PDD5	71.19	-64.33	J5	42.77	J8	PDD5	71.19	-64.33	J5	42.77	J6	PDD5	79.78	-62.20	J5	40.33	J8
failure P05a	from		J4	to	J9	failure P07a	from		J6	to	J5	failure P08a	from		J8	to	J5
DD	90.00	-64.17	J5	39.00	J8	DD	90.00	-70.89	J5	41.18	J6	DD	90.00	-70.89	J5	41.18	J8
PDD1	84.53	-62.79	J5	39.79	J8	PDD1	81.13	-67.55	J5	41.80	J6	PDD1	81.13	-67.55	J5	41.80	J8
PDD2	68.44	-68.37	J5	42.17	J6	PDD2	73.92	-67.21	J5	40.74	J6	PDD2	73.92	-67.21	J5	40.74	J6
PDD3	84.53	-62.79	J5	39.79	J8	PDD3	81.13	-67.55	J5	41.80	J6	PDD3	81.13	-67.55	J5	41.80	J8
PDD4	68.44	-68.37	J5	42.17	J6	PDD4	73.92	-67.21	J5	40.74	J6	PDD4	73.92	-67.21	J5	40.74	J6
PDD5	84.52	-62.79	J5	39.79	J8	PDD5	81.14	-67.55	J5	41.80	J6	PDD5	81.14	-67.55	J5	41.80	J8
failure P05b	from		J9	to	J3	failure P07b	from		J5	to	J7	failure P08b	from		J5	to	J9
DD	90.00	-64.30	J5	39.32	J6	DD	90.00	-62.98	J5	39.12	J6	DD	90.00	-62.98	J5	39.12	J8
PDD1	79.78	-62.20	J5	40.34	J6	PDD1	80.81	-61.42	J5	40.20	J6	PDD1	80.81	-61.42	J5	40.20	J8
PDD2	68.44	-65.48	J5	42.17	J6	PDD2	73.92	-67.21	J5	40.74	J6	PDD2	73.92	-67.21	J5	40.74	J6
PDD3	79.78	-62.20	J5	40.34	J6	PDD3	80.81	-61.42	J5	40.20	J6	PDD3	80.81	-61.42	J5	40.20	J8
PDD4	68.44	-65.48	J5	42.17	J6	PDD4	73.92	-67.21	J5	40.74	J6	PDD4	73.92	-67.21	J5	40.74	J6
PDD5	79.78	-62.20	J5	40.33	J6	PDD5	80.81	-61.42	J5	40.20	J6	PDD5	80.81	-61.42	J5	40.20	J8

Table 3.15 Test1: $z_{J3} = 55$ msl, $z_{J5} = 105$ msl (Figure 3.10 right), total available demand and pressure range

Mode	Qtot (l/s)	Pmin (mwc)	Node	Pmax (mwc)	Node	Mode	Qtot (l/s)	Pmin (mwc)	Node	Pmax (mwc)	Node	Mode	Qtot (l/s)	Pmin (mwc)	Node	Pmax (mwc)	Node
failure P01a		from	R1	to	J8	failure P03a		from	R1	to	J6	failure P04a		from	J2	to	J7
DD	90.00	-102.60	J5	20.39	J6	DD	90.00	-102.60	J5	20.39	J8	DD	90.00	-64.17	J5	39.00	J6
PDD1	60.00	-62.02	J5	40.49	J6	PDD1	60.00	-62.02	J5	40.49	J8	PDD1	60.00	-56.36	J5	44.09	J7
PDD2	30.00	-81.17	J5	42.17	J6	PDD2	30.00	-81.17	J5	42.17	J8	PDD2	60.00	-60.40	J5	42.17	J6
PDD3	59.24	-74.45	J5	34.24	J6	PDD3	59.24	-74.45	J5	34.24	J8	PDD3	60.00	-58.14	J5	42.44	J6
PDD4	30.00	-81.17	J5	42.17	J6	PDD4	30.00	-81.17	J5	42.17	J8	PDD4	60.00	-62.34	J5	42.17	J6
PDD5	59.24	-74.45	J5	34.24	J6	PDD5	59.24	-74.45	J5	34.24	J8	PDD5	60.00	-58.14	J5	42.44	J6
failure P01b		from	J8	to	J2	failure P03b		from	J6	to	J4	failure P04b		from	J7	to	J3
DD	90.00	-70.17	J5	41.38	J8	DD	90.00	-70.17	J5	41.38	J6	DD	90.00	-64.30	J5	39.32	J8
PDD1	54.75	-57.11	J5	44.49	J8	PDD1	54.75	-57.11	J5	44.49	J6	PDD1	60.00	-56.74	J5	43.49	J9
PDD2	30.00	-69.92	J5	45.00	J8	PDD2	30.00	-69.92	J5	45.00	J6	PDD2	60.00	-62.34	J5	42.17	J6
PDD3	53.46	-59.96	J5	43.81	J8	PDD3	53.46	-59.96	J5	43.81	J6	PDD3	60.00	-58.58	J5	42.17	J6
PDD4	30.00	-69.92	J5	45.00	J8	PDD4	30.00	-69.92	J5	45.00	J6	PDD4	60.00	-62.34	J5	42.17	J6
PDD5	53.46	-59.96	J5	43.81	J8	PDD5	53.46	-59.96	J5	43.81	J6	PDD5	60.00	-58.58	J5	42.17	J6
failure P05a		from	J4	to	J9	failure P07a		from	J6	to	J5	failure P08a		from	J8	to	J5
DD	90.00	-64.17	J5	39.00	J8	DD	90.00	-70.89	J5	41.18	J6	DD	90.00	-70.89	J5	41.18	J8
PDD1	60.00	-56.36	J5	44.09	J9	PDD1	60.00	-56.72	J5	43.62	J6	PDD1	60.00	-56.72	J5	43.62	J8
PDD2	60.00	-60.40	J5	42.17	J8	PDD2	60.00	-62.34	J5	42.17	J6	PDD2	60.00	-62.34	J5	42.17	J6
PDD3	60.00	-58.14	J5	42.44	J8	PDD3	60.00	-60.81	J5	43.12	J6	PDD3	60.00	-60.81	J5	43.12	J8
PDD4	60.00	-62.34	J5	42.17	J8	PDD4	60.00	-62.34	J5	42.17	J6	PDD4	60.00	-62.34	J5	42.17	J6
PDD5	60.00	-58.14	J5	42.44	J8	PDD5	60.00	-60.81	J5	43.12	J6	PDD5	60.00	-60.81	J5	43.12	J8
failure P05b		from	J9	to	J3	failure P07b		from	J5	to	J7	failure P08b		from	J5	to	J9
DD	90.00	-64.30	J5	39.32	J6	DD	90.00	-62.98	J5	39.12	J6	DD	90.00	-62.98	J5	39.12	J8
PDD1	60.00	-56.74	J5	43.49	J7	PDD1	60.00	-56.53	J5	43.49	J6	PDD1	60.00	-56.53	J5	43.49	J8
PDD2	60.00	-62.34	J5	42.17	J6	PDD2	60.00	-62.34	J5	42.17	J6	PDD2	60.00	-62.34	J5	42.17	J6
PDD3	60.00	-58.58	J5	42.17	J6	PDD3	60.00	-58.36	J5	42.20	J6	PDD3	60.00	-58.36	J5	42.20	J8
PDD4	60.00	-62.34	J5	42.17	J6	PDD4	60.00	-62.34	J5	42.17	J6	PDD4	60.00	-62.34	J5	42.17	J6
PDD5	60.00	-58.58	J5	42.17	J6	PDD5	60.00	-58.36	J5	42.20	J6	PDD5	60.00	-58.36	J5	42.20	J8

Table 3.16 Test1: $z_{J3} = 25$ msl, $z_{J5} = 105$ msl, $PDD_{min} = -80$ mwc, available demand and pressure range

Mode	Qtot (l/s)	Pmin (mwc)	Node	Pmax (mwc)	Node	Mode	Qtot (l/s)	Pmin (mwc)	Node	Pmax (mwc)	Node	Mode	Qtot (l/s)	Pmin (mwc)	Node	Pmax (mwc)	Node
failure P01a		from	R1	to	J8	failure P03a		from	R1	to	J6	failure P04a		from	J2	to	J7
DD	90.00	-102.60	J5	20.39	J6	DD	90.00	-102.60	J5	20.39	J8	DD	90.00	-64.17	J5	39.00	J6
PDD1	64.21	-77.88	J5	32.38	J6	PDD1	64.21	-77.88	J5	32.38	J8	PDD1	84.53	-62.79	J5	39.79	J6
PDD2	30.00	-81.17	J5	42.17	J6	PDD2	30.00	-81.17	J5	42.17	J8	PDD2	68.44	-68.37	J5	42.17	J8
PDD3	64.21	-77.88	J5	32.38	J6	PDD3	64.21	-77.88	J5	32.38	J8	PDD3	84.53	-62.79	J5	39.79	J6
PDD4	64.21	-77.88	J5	32.38	J6	PDD4	59.24	-74.45	J5	34.24	J8	PDD4	84.53	-62.79	J5	39.79	J6
PDD5	64.21	-77.88	J5	32.38	J6	PDD5	64.21	-77.88	J5	32.38	J8	PDD5	84.52	-62.79	J5	39.79	J6
failure P01b		from	J8	to	J2	failure P03b		from	J6	to	J4	failure P04b		from	J7	to	J3
DD	90.00	-70.17	J5	41.38	J8	DD	90.00	-70.17	J5	41.38	J6	DD	90.00	-64.30	J5	39.32	J8
PDD1	71.19	-64.33	J5	42.77	J8	PDD1	71.19	-64.33	J5	42.77	J6	PDD1	79.78	-62.20	J5	40.34	J8
PDD2	30.00	-69.92	J5	45.00	J8	PDD2	30.00	-69.92	J5	45.00	J6	PDD2	68.44	-65.48	J5	42.17	J8
PDD3	71.19	-64.33	J5	42.77	J8	PDD3	71.19	-64.33	J5	42.77	J6	PDD3	79.78	-62.20	J5	40.34	J8
PDD4	71.19	-64.33	J5	42.77	J8	PDD4	71.19	-64.33	J5	42.77	J6	PDD4	79.78	-62.20	J5	40.34	J8
PDD5	71.19	-64.33	J5	42.77	J8	PDD5	71.19	-64.33	J5	42.77	J6	PDD5	79.78	-62.20	J5	40.33	J8
failure P05a		from	J4	to	J9	failure P07a		from	J6	to	J5	failure P08a		from	J8	to	J5
DD	90.00	-64.17	J5	39.00	J8	DD	90.00	-70.89	J5	41.18	J6	DD	90.00	-70.89	J5	41.18	J8
PDD1	84.53	-62.79	J5	39.79	J8	PDD1	81.13	-67.55	J5	41.80	J6	PDD1	81.13	-67.55	J5	41.80	J8
PDD2	68.44	-68.37	J5	42.17	J6	PDD2	73.92	-67.21	J5	40.74	J6	PDD2	73.92	-67.21	J5	40.74	J6
PDD3	84.53	-62.79	J5	39.79	J8	PDD3	81.13	-67.55	J5	41.80	J6	PDD3	81.13	-67.55	J5	41.80	J8
PDD4	84.53	-62.79	J5	39.79	J8	PDD4	81.13	-67.55	J5	41.80	J6	PDD4	81.13	-67.55	J5	41.80	J8
PDD5	84.52	-62.79	J5	39.79	J8	PDD5	81.14	-67.55	J5	41.80	J6	PDD5	81.14	-67.55	J5	41.80	J8
failure P05b		from	J9	to	J3	failure P07b		from	J5	to	J7	failure P08b		from	J5	to	J9
DD	90.00	-64.30	J5	39.32	J6	DD	90.00	-62.98	J5	39.12	J6	DD	90.00	-62.98	J5	39.12	J8
PDD1	79.78	-62.20	J5	40.34	J6	PDD1	80.81	-61.42	J5	40.20	J6	PDD1	80.81	-61.42	J5	40.20	J8
PDD2	68.44	-65.48	J5	42.17	J6	PDD2	73.92	-67.21	J5	40.74	J6	PDD2	73.92	-67.21	J5	40.74	J6
PDD3	79.78	-62.20	J5	40.34	J6	PDD3	80.81	-61.42	J5	40.20	J6	PDD3	80.81	-61.42	J5	40.20	J8
PDD4	79.78	-62.20	J5	40.34	J6	PDD4	80.81	-61.42	J5	40.20	J6	PDD4	80.81	-61.42	J5	40.20	J8
PDD5	79.78	-62.20	J5	40.33	J6	PDD5	80.81	-61.42	J5	40.20	J6	PDD5	80.81	-61.42	J5	40.20	J8

Table 3.17 Test1: z_{J3} = 55 msl, z_{J5} = 105 msl, PDD_{min} = -80 mwc, available demand and pressure range

Mode	Qtot (l/s)	Pmin (mwc)	Node	Pmax (mwc)	Node	Mode	Qtot (l/s)	Pmin (mwc)	Node	Pmax (mwc)	Node	Mode	Qtot (l/s)	Pmin (mwc)	Node	Pmax (mwc)	Node
failure *P01a*	from		*R1*	to	*J8*	failure *P03a*	from		*R1*	to	*J6*	failure *P04a*	from		*J2*	to	*J7*
DD	90.00	-102.60	J5	20.39	J6	DD	90.00	-102.60	J5	20.39	J8	DD	90.00	-64.17	J5	39.00	J6
PDD1	60.00	-62.02	J5	40.49	J6	PDD1	60.00	-62.02	J5	40.49	J8	PDD1	60.00	-56.36	J5	44.09	J7
PDD2	30.00	-81.17	J5	42.17	J6	PDD2	30.00	-81.17	J5	42.17	J8	PDD2	60.00	-60.40	J5	42.17	J6
PDD3	59.24	-74.45	J5	34.24	J6	PDD3	59.24	-74.45	J5	34.24	J8	PDD3	60.00	-58.14	J5	42.44	J6
PDD4	59.24	-74.45	J5	34.24	J6	PDD4	59.24	-74.45	J5	34.24	J8	PDD4	60.00	-58.14	J5	42.44	J6
PDD5	59.24	-74.45	J5	34.24	J6	PDD5	59.24	-74.45	J5	34.24	J8	PDD5	60.00	-58.14	J5	42.44	J6
failure *P01b*	from		*J8*	to	*J2*	failure *P03b*	from		*J6*	to	*J4*	failure *P04b*	from		*J7*	to	*J3*
DD	90.00	-70.17	J5	41.38	J8	DD	90.00	-70.17	J5	41.38	J6	DD	90.00	-64.30	J5	39.32	J8
PDD1	54.75	-57.11	J5	44.49	J8	PDD1	54.75	-57.11	J5	44.49	J6	PDD1	60.00	-56.74	J5	43.49	J9
PDD2	30.00	-69.92	J5	45.00	J8	PDD2	30.00	-69.92	J5	45.00	J6	PDD2	60.00	-62.34	J5	42.17	J6
PDD3	53.46	-59.96	J5	43.81	J8	PDD3	53.46	-59.96	J5	43.81	J6	PDD3	60.00	-58.58	J5	42.17	J6
PDD4	53.46	-59.96	J5	43.81	J8	PDD4	53.46	-59.96	J5	43.81	J6	PDD4	60.00	-58.58	J5	42.17	J6
PDD5	53.46	-59.96	J5	43.81	J8	PDD5	53.46	-59.96	J5	43.81	J6	PDD5	60.00	-58.58	J5	42.17	J6
failure *P05a*	from		*J4*	to	*J9*	failure *P07a*	from		*J6*	to	*J5*	failure *P08a*	from		*J8*	to	*J5*
DD	90.00	-64.17	J5	39.00	J8	DD	90.00	-70.89	J5	41.18	J6	DD	90.00	-70.89	J5	41.18	J8
PDD1	60.00	-56.36	J5	44.09	J9	PDD1	60.00	-56.72	J5	43.62	J6	PDD1	60.00	-56.72	J5	43.62	J8
PDD2	60.00	-60.40	J5	42.17	J8	PDD2	60.00	-62.34	J5	42.17	J6	PDD2	60.00	-62.34	J5	42.17	J6
PDD3	60.00	-58.14	J5	42.44	J8	PDD3	60.00	-60.81	J5	43.12	J6	PDD3	60.00	-60.81	J5	43.12	J8
PDD4	60.00	-58.14	J5	42.44	J8	PDD4	60.00	-60.81	J5	43.12	J6	PDD4	60.00	-60.81	J5	43.12	J8
PDD5	60.00	-58.14	J5	42.44	J8	PDD5	60.00	-60.81	J5	43.12	J6	PDD5	60.00	-60.81	J5	43.12	J8
failure *P05b*	from		*J9*	to	*J3*	failure *P07b*	from		*J5*	to	*J7*	failure *P08b*	from		*J5*	to	*J9*
DD	90.00	-64.30	J5	39.32	J6	DD	90.00	-62.98	J5	39.12	J6	DD	90.00	-62.98	J5	39.12	J8
PDD1	60.00	-56.74	J5	43.49	J7	PDD1	60.00	-56.53	J5	43.49	J6	PDD1	60.00	-56.53	J5	43.49	J8
PDD2	60.00	-62.34	J5	42.17	J6	PDD2	60.00	-62.34	J5	42.17	J6	PDD2	60.00	-62.34	J5	42.17	J6
PDD3	60.00	-58.58	J5	42.17	J6	PDD3	60.00	-58.36	J5	42.20	J6	PDD3	60.00	-58.36	J5	42.20	J8
PDD4	60.00	-58.58	J5	42.17	J6	PDD4	60.00	-58.36	J5	42.20	J6	PDD4	60.00	-58.36	J5	42.20	J8
PDD5	60.00	-58.58	J5	42.17	J6	PDD5	60.00	-58.36	J5	42.20	J6	PDD5	60.00	-58.36	J5	42.20	J8

The tables show similar performance of the algorithms as in the previous set of simulations, namely:

- Insignificant differences in case of positive pressures.
- General compliance of the results between 'pdd3' and 'pdd5' variant.
- Compliance of the results between 'pdd3' and 'pdd4' variant in case the pressures are above the lower threshold value.
- Compliance of the results between 'pdd2' and 'pdd4' variant in case the pressures are below the lower threshold value.

Furthermore, it becomes obvious that the value of total available demand ('Qtot' in the tables) will depend on the selection of the lower threshold value, which was for the calculations presented in Tables 3.11 to 3.15 set at PDD_{min} = -10 mwc. The difference in the results can be observed comparing the 'pdd4' values in Table 3.14 and 3.16, and also those in Table 3.15 and 3.17.

3.7   NETWORKS OF COMBINED CONFIGURATION

The previous calculations have been conducted on a simple, yet a fully looped network. Such network cannot be decomposed during failure analyses in a way that would affect its integrity, as well as there is a lesser chance that this will happen in the PDD-calculations where the layout is reconfigured due to negative pressures. The problem of convergence and peculiar results can however arise in case of layouts where branched and looped configurations have been combined such as the one shown in Figure 3.15.

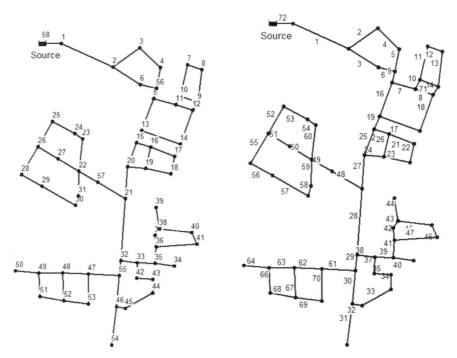

Figure 3.15 Combined gravity network layout: node IDs (left) and pipe IDs (right)

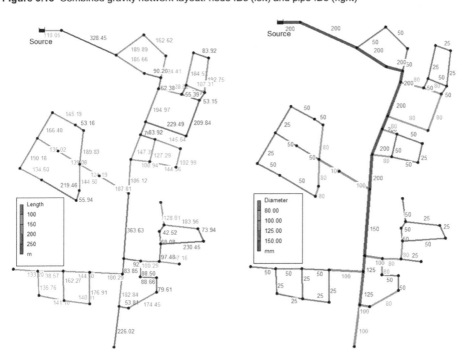

Figure 3.16 Combined gravity network layout: pipe lengths (m, left) and diameters (mm, right)

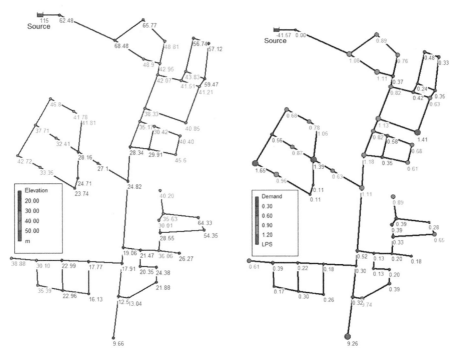

Figure 3.17 Combined gravity network layout: node elevations (msl, left) and demands (l/s, right)

The network main characteristics have been shown in Figures 3.16 and 3.17. The total demand is 41.57 l/s supplied by gravity from the reservoir with the head of 115 msl. The roughness factor has been set between 0.5 and 2 mm, based on the estimated pipe age.

Table 3.18 shows the results of the PDD hydraulic simulations for the PDD pressure range between -10 and 20, mwc. They are obviously all the same as the network is just a normal gravity network without high elevated nodes that could cause negative pressures.

More difference in the results will be obtained while failing the pipes. All 72 pipes have been failed, one by one, and the corresponding available demand calculated. The selection of results has been shown in Tables 3.19 to 3.21. Table 3.19 shows nine pipes: 72, 1, 3, 6, 19, 25, 27, 28, and 29, all belonging to the main route going through the middle, as can be seen in Figure 3.15. For these pipes, the convergence could have not been reached in the 'pdd2' and partially 'pdd4' variant (the highlighted cells), after 99 iterations. Furthermore, three peripheral pipes have been added: 41, 48, and 61, each one connecting a branch to the main route, where the convergence has been reached in all the variants. In all 12 cases, the closure of the pipe breaks the network integrity, the results for the part of the network behind the closed pipe then relying on the basic algorithm of EPANET. Without having a connection to alternative source, this part of the network would obviously have no supply in reality.

Table 3.18a Combined gravity network, PDD range from -10 to 20 mwc, node demands and pressures

Node ID	Z (msl)	Qdd (l/s)	Pdd (mwc)	Qpdd1 (l/s)	ppdd1 (mwc)	Qpdd2 (l/s)	ppdd2 (mwc)	Qpdd3 (l/s)	ppdd3 (mwc)	Qpdd4 (l/s)	ppdd4 (mwc)	Qpdd5 (l/s)	ppdd5 (mwc)
1	62.48	0.00	50.64	0.00	50.64	0.00	50.64	0.00	50.64	0.00	50.64	0.00	50.64
2	68.48	1.08	39.04	1.08	39.04	1.08	39.04	1.08	39.04	1.08	39.04	1.08	39.04
3	65.77	0.89	37.97	0.89	37.97	0.89	37.97	0.89	37.97	0.89	37.97	0.89	37.97
4	48.81	0.76	54.49	0.76	54.49	0.76	54.49	0.76	54.49	0.76	54.49	0.76	54.49
5	42.07	0.82	60.49	0.82	60.49	0.82	60.49	0.82	60.49	0.82	60.49	0.82	60.49
6	48.90	1.11	55.81	1.11	55.81	1.11	55.81	1.11	55.81	1.11	55.81	1.11	55.81
7	56.74	0.48	30.00	0.48	30.00	0.48	30.00	0.48	30.00	0.48	30.00	0.48	30.00
8	57.12	0.33	29.79	0.33	29.79	0.33	29.79	0.33	29.79	0.33	29.79	0.33	29.79
9	59.47	0.35	40.77	0.35	40.77	0.35	40.77	0.35	40.77	0.35	40.77	0.35	40.77
10	43.83	0.24	56.50	0.24	56.50	0.24	56.50	0.24	56.50	0.24	56.50	0.24	56.50
11	41.51	0.42	59.32	0.42	59.32	0.42	59.32	0.42	59.32	0.42	59.32	0.42	59.32
12	41.21	0.63	59.17	0.63	59.17	0.63	59.17	0.63	59.17	0.63	59.17	0.63	59.17
13	38.33	1.13	62.08	1.13	62.08	1.13	62.08	1.13	62.08	1.13	62.08	1.13	62.08
14	40.85	1.41	59.39	1.41	59.39	1.41	59.39	1.41	59.39	1.41	59.39	1.41	59.39
15	35.17	0.82	64.63	0.82	64.63	0.82	64.63	0.82	64.63	0.82	64.63	0.82	64.63
16	30.42	0.58	69.02	0.58	69.02	0.58	69.02	0.58	69.02	0.58	69.02	0.58	69.02
17	40.40	0.68	58.13	0.68	58.13	0.68	58.13	0.68	58.13	0.68	58.13	0.68	58.13
18	45.60	0.61	52.73	0.61	52.73	0.61	52.73	0.61	52.73	0.61	52.73	0.61	52.73
19	29.91	0.35	68.90	0.35	68.90	0.35	68.90	0.35	68.90	0.35	68.90	0.35	68.90
20	28.34	1.18	70.48	1.18	70.48	1.18	70.48	1.18	70.48	1.18	70.48	1.18	70.48
21	24.82	1.11	72.88	1.11	72.88	1.11	72.88	1.11	72.88	1.11	72.88	1.11	72.88
22	28.16	1.39	62.25	1.39	62.25	1.39	62.25	1.39	62.25	1.39	62.25	1.39	62.25
23	41.81	1.06	47.19	1.06	47.19	1.06	47.19	1.06	47.19	1.06	47.19	1.06	47.19
24	41.78	0.78	45.82	0.78	45.82	0.78	45.82	0.78	45.82	0.78	45.82	0.78	45.82
25	46.80	0.68	39.78	0.68	39.78	0.68	39.78	0.68	39.78	0.68	39.78	0.68	39.78
26	37.71	0.56	47.74	0.56	47.74	0.56	47.74	0.56	47.74	0.56	47.74	0.56	47.74
27	32.41	0.87	53.96	0.87	53.96	0.87	53.96	0.87	53.96	0.87	53.96	0.87	53.96
28	42.72	1.65	32.37	1.65	32.37	1.65	32.37	1.65	32.37	1.65	32.37	1.65	32.37
29	33.35	0.96	44.14	0.96	44.14	0.96	44.14	0.96	44.14	0.96	44.14	0.96	44.14
30	23.74	0.11	65.37	0.11	65.37	0.11	65.37	0.11	65.37	0.11	65.37	0.11	65.37
31	24.71	0.11	64.74	0.11	64.74	0.11	64.74	0.11	64.74	0.11	64.74	0.11	64.74
32	19.06	0.52	75.28	0.52	75.28	0.52	75.28	0.52	75.28	0.52	75.28	0.52	75.28
33	21.47	0.13	72.34	0.13	72.34	0.13	72.34	0.13	72.34	0.13	72.34	0.13	72.34
34	26.27	0.18	64.94	0.18	64.94	0.18	64.94	0.18	64.94	0.18	64.94	0.18	64.94
35	36.06	0.20	56.80	0.20	56.80	0.20	56.80	0.20	56.80	0.20	56.80	0.20	56.80
36	28.55	0.33	63.58	0.33	63.58	0.33	63.58	0.33	63.58	0.33	63.58	0.33	63.58
37	30.01	0.39	59.99	0.39	59.99	0.39	59.99	0.39	59.99	0.39	59.99	0.39	59.99
38	35.63	0.39	53.55	0.39	53.55	0.39	53.55	0.39	53.55	0.39	53.55	0.39	53.55
39	40.20	0.89	47.89	0.89	47.89	0.89	47.89	0.89	47.89	0.89	47.89	0.89	47.89
40	64.33	0.28	24.51	0.28	24.51	0.28	24.51	0.28	24.51	0.28	24.51	0.28	24.51
41	54.35	0.65	35.94	0.65	35.94	0.65	35.94	0.65	35.94	0.65	35.94	0.65	35.94
42	20.35	0.13	73.34	0.13	73.34	0.13	73.34	0.13	73.34	0.13	73.34	0.13	73.34
43	24.38	0.20	68.22	0.20	68.22	0.20	68.22	0.20	68.22	0.20	68.22	0.20	68.22
44	21.88	0.39	70.07	0.39	70.07	0.39	70.07	0.39	70.07	0.39	70.07	0.39	70.07
45	13.04	0.74	78.87	0.74	78.87	0.74	78.87	0.74	78.87	0.74	78.87	0.74	78.87
46	12.50	0.32	79.41	0.32	79.41	0.32	79.41	0.32	79.41	0.32	79.41	0.32	79.41
47	17.77	0.18	75.33	0.18	75.33	0.18	75.33	0.18	75.33	0.18	75.33	0.18	75.33
48	22.99	0.22	66.07	0.22	66.07	0.22	66.07	0.22	66.07	0.22	66.07	0.22	66.07
49	30.10	0.39	56.92	0.39	56.92	0.39	56.92	0.39	56.92	0.39	56.92	0.39	56.92
50	38.88	0.61	47.59	0.61	47.59	0.61	47.59	0.61	47.59	0.61	47.59	0.61	47.59
51	35.39	0.17	49.76	0.17	49.76	0.17	49.76	0.17	49.76	0.17	49.76	0.17	49.76
52	22.96	0.30	62.18	0.30	62.18	0.30	62.18	0.30	62.18	0.30	62.18	0.30	62.18
53	16.13	0.26	69.16	0.26	69.16	0.26	69.16	0.26	69.16	0.26	69.16	0.26	69.16
54	9.66	9.26	77.22	9.26	77.22	9.26	77.22	9.26	77.22	9.26	77.22	9.26	77.22
55	17.91	0.30	75.42	0.30	75.42	0.30	75.42	0.30	75.42	0.30	75.42	0.30	75.42
56	42.95	0.37	60.47	0.37	60.47	0.37	60.47	0.37	60.47	0.37	60.47	0.37	60.47
57	27.10	0.63	65.96	0.63	65.96	0.63	65.96	0.63	65.96	0.63	65.96	0.63	65.96

Table 3.18b Combined gravity network, PDD range from -10 to 20 mwc, pipe flows and friction losses

Pipe ID	From node	To node	D (mm)	Qdd (l/s)	hfdd (mwc)	Qpdd1 (l/s)	hfpdd1 (mwc)	Qpdd2 (l/s)	hfpdd2 (mwc)	Qpdd3 (l/s)	hfpdd3 (mwc)	Qpdd4 (l/s)	hfpdd4 (mwc)	Qpdd5 (l/s)	hfpdd5 (mwc)
1	1	2	200	41.57	5.60	41.57	5.60	41.57	5.60	41.57	5.60	41.57	5.60	41.57	5.60
2	2	3	50	1.37	3.78	1.37	3.78	1.37	3.78	1.37	3.78	1.37	3.78	1.37	3.78
3	2	6	200	39.12	2.81	39.12	2.81	39.12	2.81	39.12	2.81	39.12	2.81	39.12	2.81
4	3	4	50	0.48	0.43	0.48	0.43	0.48	0.43	0.48	0.43	0.48	0.43	0.48	0.43
5	4	56	50	-0.28	0.12	-0.28	0.12	-0.28	0.12	-0.28	0.12	-0.28	0.12	-0.28	0.12
6	6	56	200	38.01	1.29	38.01	1.29	38.01	1.29	38.01	1.29	38.01	1.29	38.01	1.29
7	5	11	80	3.15	1.73	3.15	1.73	3.15	1.73	3.15	1.73	3.15	1.73	3.15	1.73
8	11	12	80	1.81	0.45	1.81	0.45	1.81	0.45	1.81	0.45	1.81	0.45	1.81	0.45
9	5	56	200	-37.36	0.86	-37.36	0.86	-37.36	0.86	-37.36	0.86	-37.36	0.86	-37.36	0.86
10	11	10	50	0.92	0.50	0.92	0.50	0.92	0.50	0.92	0.50	0.92	0.50	0.92	0.50
11	10	7	25	0.41	13.59	0.41	13.59	0.41	13.59	0.41	13.59	0.41	13.59	0.41	13.59
12	7	8	25	-0.07	0.16	-0.07	0.16	-0.07	0.16	-0.07	0.16	-0.07	0.16	-0.07	0.16
13	8	9	25	-0.40	13.33	-0.40	13.33	-0.40	13.33	-0.40	13.33	-0.40	13.33	-0.40	13.33
14	9	12	50	-0.49	0.14	-0.49	0.14	-0.49	0.14	-0.49	0.14	-0.49	0.14	-0.49	0.14
16	5	13	200	33.40	2.15	33.40	2.15	33.40	2.15	33.40	2.15	33.40	2.15	33.40	2.15
17	13	14	80	0.72	0.17	0.72	0.17	0.72	0.17	0.72	0.17	0.72	0.17	0.72	0.17
18	14	12	80	-0.69	0.14	-0.69	0.14	-0.69	0.14	-0.69	0.14	-0.69	0.14	-0.69	0.14
19	13	15	200	31.55	0.61	31.55	0.61	31.55	0.61	31.55	0.61	31.55	0.61	31.55	0.61
20	15	16	80	1.99	0.36	1.99	0.36	1.99	0.36	1.99	0.36	1.99	0.36	1.99	0.36
21	16	17	50	0.75	0.90	0.75	0.90	0.75	0.90	0.75	0.90	0.75	0.90	0.75	0.90
22	17	18	25	0.07	0.20	0.07	0.20	0.07	0.20	0.07	0.20	0.07	0.20	0.07	0.20
23	18	19	50	-0.54	0.48	-0.54	0.48	-0.54	0.48	-0.54	0.48	-0.54	0.48	-0.54	0.48
24	19	20	80	-0.23	0.01	-0.23	0.01	-0.23	0.01	-0.23	0.01	-0.23	0.01	-0.23	0.01
25	20	15	200	-28.74	0.98	-28.74	0.98	-28.74	0.98	-28.74	0.98	-28.74	0.98	-28.74	0.98
26	16	19	50	0.66	0.62	0.66	0.62	0.66	0.62	0.66	0.62	0.66	0.62	0.66	0.62
27	20	21	200	27.33	1.12	27.33	1.12	27.33	1.12	27.33	1.12	27.33	1.12	27.33	1.12
28	21	32	150	17.42	3.36	17.42	3.36	17.42	3.36	17.42	3.36	17.42	3.36	17.42	3.36
29	32	55	125	12.26	1.01	12.26	1.01	12.26	1.01	12.26	1.01	12.26	1.01	12.26	1.01
30	55	46	125	9.83	1.43	9.83	1.43	9.83	1.43	9.83	1.43	9.83	1.43	9.84	1.43
31	46	54	100	9.26	5.03	9.26	5.03	9.26	5.03	9.26	5.03	9.26	5.03	9.26	5.03
32	46	45	100	0.26	0.00	0.26	0.00	0.26	0.00	0.26	0.00	0.26	0.00	0.26	0.00
33	45	44	80	-0.48	0.04	-0.48	0.04	-0.48	0.04	-0.48	0.04	-0.48	0.04	-0.48	0.04
34	44	43	50	-0.87	0.65	-0.87	0.65	-0.87	0.65	-0.87	0.65	-0.87	0.65	-0.87	0.65
35	43	42	50	-1.07	1.09	-1.07	1.09	-1.07	1.09	-1.07	1.09	-1.07	1.09	-1.07	1.09
37	42	33	80	-1.20	0.12	-1.20	0.12	-1.20	0.12	-1.20	0.12	-1.20	0.12	-1.20	0.12
38	33	32	100	-4.64	0.53	-4.64	0.53	-4.64	0.53	-4.64	0.53	-4.64	0.53	-4.64	0.53
39	33	35	80	3.31	0.95	3.31	0.95	3.31	0.95	3.31	0.95	3.31	0.95	3.31	0.95
40	35	34	25	0.18	1.65	0.18	1.65	0.18	1.65	0.18	1.65	0.18	1.65	0.18	1.65
41	35	36	80	2.92	0.73	2.92	0.73	2.92	0.73	2.92	0.73	2.92	0.73	2.92	0.73
42	36	37	50	1.73	2.14	1.73	2.14	1.73	2.14	1.73	2.14	1.73	2.14	1.73	2.14
43	37	38	50	1.34	0.81	1.34	0.81	1.34	0.81	1.34	0.81	1.34	0.81	1.34	0.81
44	38	39	50	0.89	1.10	0.89	1.10	0.89	1.10	0.89	1.10	0.89	1.10	0.89	1.10
45	38	40	25	0.07	0.35	0.07	0.35	0.07	0.35	0.07	0.35	0.07	0.35	0.07	0.35
46	40	41	25	-0.21	1.46	-0.21	1.46	-0.21	1.46	-0.21	1.46	-0.21	1.46	-0.21	1.46
47	36	41	50	0.86	1.84	0.86	1.84	0.86	1.84	0.86	1.84	0.86	1.84	0.86	1.84
48	21	57	100	8.80	4.64	8.80	4.64	8.80	4.64	8.80	4.64	8.80	4.64	8.80	4.64
49	57	22	100	8.17	2.65	8.17	2.65	8.17	2.65	8.17	2.65	8.17	2.65	8.17	2.65
50	22	27	50	1.66	4.04	1.66	4.04	1.66	4.04	1.66	4.04	1.66	4.04	1.66	4.04
51	27	26	50	0.79	0.92	0.79	0.92	0.79	0.92	0.79	0.92	0.79	0.92	0.79	0.92
52	26	25	25	-0.12	1.12	-0.12	1.12	-0.12	1.12	-0.12	1.12	-0.12	1.12	-0.12	1.12
53	25	24	50	-0.80	1.02	-0.80	1.02	-0.80	1.02	-0.80	1.02	-0.80	1.02	-0.80	1.02
54	24	23	50	-1.58	1.40	-1.58	1.40	-1.58	1.40	-1.58	1.40	-1.58	1.40	-1.58	1.40
55	28	26	25	-0.35	10.36	-0.35	10.36	-0.35	10.36	-0.35	10.36	-0.35	10.36	-0.35	10.36
56	28	29	50	-1.29	2.40	-1.29	2.40	-1.29	2.40	-1.29	2.40	-1.29	2.40	-1.29	2.40
57	29	30	50	-2.26	11.63	-2.26	11.63	-2.26	11.63	-2.26	11.63	-2.26	11.63	-2.26	11.63
58	30	31	80	-2.37	0.34	-2.37	0.34	-2.37	0.34	-2.37	0.34	-2.37	0.34	-2.37	0.34
59	31	22	80	-2.48	0.96	-2.48	0.96	-2.48	0.96	-2.48	0.96	-2.48	0.96	-2.48	0.96
60	22	23	80	2.64	1.42	2.64	1.42	2.64	1.42	2.64	1.42	2.64	1.42	2.64	1.42
61	55	47	100	2.13	0.23	2.13	0.23	2.13	0.23	2.13	0.23	2.13	0.23	2.13	0.23
62	47	48	50	1.63	4.04	1.63	4.04	1.63	4.04	1.63	4.04	1.63	4.04	1.63	4.04
63	48	49	50	1.17	2.04	1.17	2.04	1.17	2.04	1.17	2.04	1.17	2.04	1.17	2.04
64	49	50	50	0.61	0.56	0.61	0.56	0.61	0.56	0.61	0.56	0.61	0.56	0.61	0.56
66	51	49	25	-0.17	1.87	-0.17	1.87	-0.17	1.87	-0.17	1.87	-0.17	1.87	-0.17	1.87
67	48	52	25	0.23	3.92	0.23	3.92	0.23	3.92	0.23	3.92	0.23	3.92	0.23	3.92
68	52	51	25	-0.01	0.01	-0.01	0.01	-0.01	0.01	-0.01	0.01	-0.01	0.01	-0.01	0.01
69	52	53	25	-0.06	0.15	-0.06	0.15	-0.06	0.15	-0.06	0.15	-0.06	0.15	-0.06	0.15
70	53	47	25	-0.32	7.81	-0.32	7.81	-0.32	7.81	-0.32	7.81	-0.32	7.81	-0.32	7.81
71	10	9	50	0.26	0.09	0.26	0.09	0.26	0.09	0.26	0.09	0.26	0.09	0.26	0.09
72	58	1	200	41.57	1.88	41.57	1.88	41.57	1.88	41.57	1.88	41.57	1.88	41.57	1.88

Table 3.19 Combined gravity network, main pipes

Mode	Qtot (l/s)	Pmin (mwc)	Node	Pmax (mwc)	Node	Mode	Qtot (l/s)	Pmin (mwc)	Node	Pmax (mwc)	Node	Mode	Qtot (l/s)	Pmin (mwc)	Node	Pmax (mwc)	Node
failure	72	from	58	to	1	failure	1	from	1	to	2	failure	3	from	2	to	6
DD	41.57	-4E+07	40	-1E+06		DD	41.57	-4E+07	40	50.64	1	DD	41.57	-7545	40	50.64	1
PDD1	14.73	-35.48	2	17.43	46	PDD1	14.73	-35.48	2	52.52	1	PDD1	16.90	-29.29	40	52.48	1
PDD2	1.57	-6.83	29	29.26	1	PDD2	1.57	-6.83	29	52.52	1	PDD2	6.99	-4.79	38	52.49	1
PDD3	0.00	-48.59	40	25.50	1	PDD3	0.00	-48.59	40	52.52	1	PDD3	5.86	-49.51	40	52.48	1
PDD4	0.00	-58.82	2	0.00	54	PDD4	26.85	40.69	40	91.01	46	PDD4	2.42	-54.67	40	55.80	6
PDD5	0.00	-58.82	2	0.00	54	PDD5	0.00	-58.52	2	52.52	1	PDD5	5.86	-49.51	40	52.48	1
failure	6	from	6	to	56	failure	19	from	13	to	15	failure	25	from	15	to	20
DD	41.57	-7128	40	58.61	6	DD	41.57	-3E+07	40	62.08	13	DD	41.57	-958	40	64.63	15
PDD1	17.44	-29.80	40	65.90	6	PDD1	19.81	-33.88	40	76.01	13	PDD1	25.98	-30.23	40	76.58	15
PDD2	9.29	-13.95	49	66.01	6	PDD2	10.81	-3.32	45	76.01	13	PDD2	22.36	-9.93	35	76.83	15
PDD3	4.21	-54.90	40	66.02	6	PDD3	10.02	-35.78	40	76.01	13	PDD3	19.75	-42.02	40	76.78	15
PDD4	27.74	40.41	40	90.73	46	PDD4	31.61	35.69	8	88.83	46	PDD4	31.61	35.69	8	88.83	46
PDD5	6.97	-49.51	40	65.88	6	PDD5	10.02	-54.67	40	76.01	13	PDD5	20.69	-40.34	40	76.44	15
failure	27	from	20	to	21	failure	28	from	21	to	32	failure	29	from	32	to	55
DD	41.57	-3E+07	40	70.48	20	DD	41.57	-2E+07	40	72.88	21	DD	41.57	-295	50	75.28	32
PDD1	22.84	-35.35	40	85.19	20	PDD1	29.38	-41.15	40	85.10	21	PDD1	35.74	-15.54	50	83.08	32
PDD2	14.93	-3.92	48	85.19	20	PDD2	24.42	-5.61	49	85.10	21	PDD2	34.80	0.00	51	83.00	32
PDD3	14.26	-35.78	40	85.19	20	PDD3	24.27	-31.06	40	85.10	21	PDD3	34.80	-8.86	51	83.00	32
PDD4	32.90	35.02	8	88.04	46	PDD4	37.88	32.17	8	84.23	46	PDD4	40.40	26.01	40	81.14	46
PDD5	14.24	-54.67	40	85.19	20	PDD5	24.15	-54.67	40	85.10	21	PDD5	34.80	-20.44	50	83.00	32
failure	41	from	35	to	36	failure	48	from	21	to	57	failure	55	from	55	to	47
DD	41.57	-3E+06	40	79.41	46	DD	41.57	-9E+06	28	79.41	46	DD	41.57	-2E+06	50	79.41	46
PDD1	39.46	-22.06	40	83.04	46	PDD1	35.08	-9.32	25	86.51	46	PDD1	40.04	-9.46	50	82.35	46
PDD2	38.84	-4.99	38	83.04	46	PDD2	33.22	-4.33	29	86.51	46	PDD2	39.71	-7.19	49	82.35	46
PDD3	39.08	-29.01	40	83.04	46	PDD3	32.86	-20.02	25	86.51	46	PDD3	39.50	-22.18	50	82.35	46
PDD4	38.84	-33.72	40	83.04	46	PDD4	32.86	-20.02	25	86.51	46	PDD4	39.71	-16.02	50	82.35	46
PDD5	38.65	-35.78	40	83.04	46	PDD5	32.77	-23.06	25	86.51	46	PDD5	39.44	-22.75	50	82.35	46

Table 3.20 Combined gravity network, main pipes

Mode	Qtot (l/s)	Pmin (mwc)	Node	Pmax (mwc)	Node	Mode	Qtot (l/s)	Pmin (mwc)	Node	Pmax (mwc)	Node	Mode	Qtot (l/s)	Pmin (mwc)	Node	Pmax (mwc)	Node
failure	2	from	2	to	3	failure	4	from	3	to	4	failure	12	from	7	to	8
DD	41.57	24.21	40	79.11	46	DD	41.57	24.40	40	79.31	46	DD	41.57	24.51	40	79.41	46
PDD1	41.57	24.21	40	79.11	46	PDD1	41.57	24.40	40	79.31	46	PDD1	41.57	24.51	40	79.41	46
PDD2	41.57	24.21	40	79.11	46	PDD2	41.57	24.40	40	79.31	46	PDD2	41.57	24.51	40	79.41	46
PDD3	41.57	24.21	40	79.11	46	PDD3	41.57	24.40	40	79.31	46	PDD3	41.57	24.51	40	79.41	46
PDD4	41.57	24.21	40	79.11	46	PDD4	41.57	24.40	40	79.31	46	PDD4	41.57	24.51	40	79.41	46
PDD5	41.57	24.21	40	79.11	46	PDD5	41.57	24.51	40	79.41	46	PDD5	41.57	24.51	40	79.41	46
failure	13	from	8	to	9	failure	18	from	14	to	12	failure	22	from	17	to	18
DD	41.57	-12.37	8	79.41	46	DD	41.57	24.42	40	79.32	46	DD	41.57	24.50	40	79.40	46
PDD1	41.38	10.44	8	79.53	46	PDD1	41.57	24.42	40	79.32	46	PDD1	41.57	24.50	40	79.40	46
PDD2	41.38	10.44	8	79.53	46	PDD2	41.57	24.42	40	79.32	46	PDD2	41.57	24.50	40	79.40	46
PDD3	41.38	10.44	8	79.53	46	PDD3	41.57	24.42	40	79.32	46	PDD3	41.57	24.50	40	79.40	46
PDD4	41.43	6.63	8	79.50	46	PDD4	41.57	24.42	40	79.32	46	PDD4	41.57	24.50	40	79.40	46
PDD5	41.38	10.43	8	79.53	46	PDD5	41.57	24.42	40	79.32	46	PDD5	41.57	24.50	40	79.40	46
failure	46	from	40	to	41	failure	34	from	44	to	43	failure	68	from	52	to	51
DD	41.57	17.86	40	79.41	46	DD	41.57	24.69	40	79.00	46	DD	41.57	24.51	40	79.41	46
PDD1	41.56	18.42	40	79.42	46	PDD1	41.57	24.69	40	79.00	46	PDD1	41.57	24.51	40	79.41	46
PDD2	41.56	18.42	40	79.42	46	PDD2	41.57	24.69	40	79.00	46	PDD2	41.57	24.51	40	79.41	46
PDD3	41.56	18.42	40	79.42	46	PDD3	41.57	24.69	40	79.00	46	PDD3	41.57	24.51	40	79.41	46
PDD4	41.56	18.42	40	79.42	46	PDD4	41.57	24.69	40	79.00	46	PDD4	41.57	24.51	40	79.41	46
PDD5	41.56	18.41	40	79.42	46	PDD5	41.57	24.68	40	79.00	46	PDD5	41.57	24.51	40	79.41	46
failure	66	from	51	to	49	failure	52	from	26	to	25	failure	55	from	28	to	26
DD	41.57	24.51	40	79.41	46	DD	41.57	24.51	40	79.41	46	DD	41.57	24.51	40	79.41	46
PDD1	41.57	24.51	40	79.41	46	PDD1	41.57	24.51	40	79.41	46	PDD1	41.57	24.51	40	79.41	46
PDD2	41.57	24.51	40	79.41	46	PDD2	41.57	24.51	40	79.41	46	PDD2	41.57	24.51	40	79.41	46
PDD3	41.57	24.51	40	79.41	46	PDD3	41.57	24.51	40	79.41	46	PDD3	41.57	24.51	40	79.41	46
PDD4	41.57	24.51	40	79.41	46	PDD4	41.57	24.51	40	79.41	46	PDD4	41.57	24.51	40	79.41	46
PDD5	41.57	24.51	40	79.41	46	PDD5	41.57	24.51	40	79.41	46	PDD5	41.57	24.51	40	79.41	46

With a few exceptions, the results of 'pdd3' and 'pdd5' show reasonable compliance. In cases where the minimum pressures differ for the same available demand, the referring nodes will be positioned downstream of the closed pipe, making this difference irrelevant, as e.g. in the case of failure of pipes 19, or 27 to 29. Yet, it shows that the concept of disconnection i.e. reconfiguration of the network as a result of negative pressures maintained in 'pdd2' and 'pdd4' can create convergence problems. One possible reason is that the pressure build-up above the upper threshold, resulting from the demand reduction, actually calls for the restoration of original configuration and demand in the next iteration, resulting eventually in never-ending loop. This has been observed in couple of cases and would ask for improvement of the algorithm. Moreover, even in some of the cases where the convergence of 'pdd4' has been reached, the results appear to be significantly different than in case of 'pdd3' and 'pdd5'. This is visible in all the cases where 'pdd2' have not converged. Moreover, being in fact a part of the 'pdd4' concept, makes those results also questionable.

The second sample of the results, shown in Table 3.20, gives the situation of the failure of 12 peripheral pipes: 2, 4, 12, 13, 18, 22, 46, 34, 68, 66, 52, and 55. All of these have been positioned at higher elevations (the network is located in a valley). Nevertheless, the results show full compliance, the reason being the low flows carried by those pipes; no failure of any of these would cause a drop of the demand.

Table 3.21 Combined gravity network, pipes carrying larger flows

Mode	Qtot (l/s)	Pmin (mwc)	Node	Pmax (mwc)	Node	Mode	Qtot (l/s)	Pmin (mwc)	Node	Pmax (mwc)	Node	Mode	Qtot (l/s)	Pmin (mwc)	Node	Pmax (mwc)	Node
failure	16	from	5	to	13	failure	30	from	55	to	46	failure	7	from	5	to	11
DD	41.57	-965.34	40	60.49	5	DD	41.57	-150.00	54	76.39	55	DD	41.57	22.72	8	78.99	46
PDD1	23.57	-29.24	40	71.02	5	PDD1	37.12	2.74	44	81.69	55	PDD1	41.57	22.72	8	78.99	46
PDD2	17.31	-15.08	35	72.26	5	PDD2	31.64	0.00	54	86.98	55	PDD2	41.57	22.72	8	78.99	46
PDD3	17.14	-39.88	40	70.93	5	PDD3	37.12	2.74	44	81.69	55	PDD3	41.57	22.72	8	78.99	46
PDD4	15.72	-29.21	40	71.26	5	PDD4	37.10	3.29	44	81.71	55	PDD4	41.57	22.72	8	78.99	46
PDD5	17.14	-39.87	40	70.93	5	PDD5	37.12	2.75	44	81.69	55	PDD5	41.57	22.72	8	78.99	46
failure	20	from	15	to	16	failure	59	from	31	to	22	failure	60	from	22	to	23
DD	41.57	24.37	40	79.27	46	DD	41.57	-617.51	28	79.41	46	DD	41.57	-425.4	23	79.41	46
PDD1	41.57	24.37	40	79.27	46	PDD1	39.56	-0.05	28	81.26	46	PDD1	39.84	-0.21	25	81.01	46
PDD2	41.57	24.37	40	79.27	46	PDD2	39.48	0.00	31	81.26	46	PDD2	39.84	-0.21	25	81.01	46
PDD3	41.57	24.37	40	79.27	46	PDD3	39.49	-1.70	28	81.25	46	PDD3	39.78	-0.96	25	81.01	46
PDD4	41.57	24.37	40	79.27	46	PDD4	39.47	1.89	28	81.28	46	PDD4	39.78	-0.96	25	81.01	46
PDD5	41.57	24.51	40	79.41	46	PDD5	39.49	-1.71	28	81.13	46	PDD5	39.78	-0.97	25	80.88	46
failure	38	from	33	to	32	failure	42	from	36	to	37	failure	62	from	47	to	48
DD	41.57	-6.89	40	76.90	46	DD	41.57	-293.66	39	79.41	46	DD	41.57	-431.1	50	79.41	46
PDD1	41.19	0.86	40	77.62	46	PDD1	40.41	0.50	39	80.90	46	PDD1	40.58	-2.32	50	81.09	46
PDD2	41.19	0.86	40	77.62	46	PDD2	40.41	0.50	39	80.90	46	PDD2	40.38	0.00	51	81.08	46
PDD3	41.19	0.86	40	77.62	46	PDD3	40.41	0.50	39	80.90	46	PDD3	40.41	-6.01	50	81.06	46
PDD4	41.19	0.86	40	77.62	46	PDD4	40.39	4.38	39	80.93	46	PDD4	40.39	-6.41	50	81.06	46
PDD5	41.19	0.80	40	77.49	46	PDD5	40.41	0.50	39	80.77	46	PDD5	40.39	-6.42	50	80.93	46
failure	31	from	46	to	54	failure	39	from	33	to	35	failure	50	from	22	to	27
DD	41.57	-1E+07	54	79.41	46	DD	41.57	-4E+06	40	79.41	46	DD	41.57	-6.87	26	79.41	46
PDD1	32.31	0.00	54	91.70	46	PDD1	39.20	-22.58	40	83.49	46	PDD1	41.26	9.99	26	79.69	46
PDD2	32.31	0.00	54	91.70	46	PDD2	38.56	-5.01	35	83.49	46	PDD2	41.26	9.99	26	79.69	46
PDD3	32.31	0.00	54	91.70	46	PDD3	38.37	-35.68	40	83.49	46	PDD3	41.26	9.99	26	79.69	46
PDD4	32.31	0.00	54	91.70	46	PDD4	38.56	-33.27	40	83.49	46	PDD4	41.26	9.99	26	79.69	46
PDD5	32.31	0.00	54	91.62	46	PDD5	38.26	-38.26	40	83.37	46	PDD5	41.26	9.99	26	79.55	46

Finally, the last batch of the results in Table 3.21 shows the consequences of the failure of pipes that carry comparatively larger flows. Those are: 16, 30, 7, 20, 59, 60, 38, 42, 62, 31, 39 and 50. With few exceptions, the calculations of 'pdd2' to 'pdd5' point similar results in terms of the available demand, critical nodes and the maximum pressure. Again the differences in the minimum (negative) pressure would result from the position of the node compared to the failed pipe, which would not quite affect the value of available demand. The better equality of the maximum pressures is like in all previous cases a consequence of upstream location of these nodes, towards the failed pipe.

3.8 CONCLUSIONS

The discussion presented in this chapter assesses four ways of using EPANET emitters in running pressure-demand driven simulations, namely:
- The basic mode where emitters are used for all the pressure below the threshold, allowing negative nodal pressures and demands (coded as 'pdd1').
- The mode where all pipes connecting negative pressure nodes are disconnected ('pdd2').
- The mode where all negative demands will be set to zero, allowing negative pressures as the result of extreme topographic conditions ('pdd3').
- The mode where lower pressure threshold has been introduced for pipe disconnection or alternative setting of negative nodal demands to zero ('pdd4').

The results of these four modalities have been compared with the PDD version of EPANET developed by Pathirana (2010), as option 'pdd5'. The analyses have primarily focused to the boundary conditions that could yield peculiar results and have briefly explored a few simple options of bringing them closer to the reality without massive programming interventions. For this purpose, an algorithm has been developed using the EPANET toolkit functions in C++ programming environment. Following is the list of conclusions based on the results of analysed network layouts:
- Negative nodal pressures in the PDD mode of hydraulic calculation will occur only in the cases of really extreme topographic conditions and in normal situations even the basic PDD mode ('pdd1') should be yielding satisfactory results.
- In cases of negative nodal pressures, setting the corresponding nodal demands to zero ('pdd3') yields more stable computation, but avoids consideration of negative pressures both in demand- and non-demand nodes potentially leading to the reduction of conveying capacity i.e. further demand reduction.
- Network reconfiguration as the result of pipe disconnection, be it a result of negative pressures ('pdd2') or the pressures below the lower threshold ('pdd4'), offers more conservative and likely more accurate results but the simplified algorithm can create convergence problems if integrity of the network layout has been severely affected. To arrive at more robust algorithm, additional methodology is necessary that can identify the network 'islands' without connection to any of the sources.

The final conclusion of this study is that in case of calculations of large networks, the 'pdd1' and 'pdd3' approaches are likely to create fewer troubles than the 'pdd2' and/or 'pdd4' approach, although in extreme topographic conditions those may also yield less accurate results. Nevertheless, the concept of upper and lower pressure threshold looks valid and should be further tested on more robust PDD algorithm.

The remaining but also important general issue is the one of calculation of emitter coefficients based on the upper pressure threshold, which in itself is an approximation. The model calibration in low-pressure conditions is commonly a problem. Nevertheless, reasonably estimated PDD-thresholds make the presented approach of emitter calculations the best possible simulation of reality.

REFERENCES

1. Ent, W.I. van der, (1993). *Aleid 7.0 - User's Guide*. KIWA, Nieuwegein
2. Gupta, R. and Bhave, R. (1996). *Reliability-based Design of Water Distribution Systems*, Journal of Environmental Engineering Division, ASCE, 122(1), 51-54.
3. Pathirana, A. (2010). *EPANET2 Desktop Application for Pressure Driven Demand Modeling.* In conference proceedings of Water Distribution System Analysis 2010 - WDSA2010, Sept. 12-15, Tucson, AZ, USA.
4. Piller, O., Van Zyl, J.E.(2009). *Pressure-driven analysis of network sections supplied via high-lying nodes.* In Proceedings of the Tenth International Conference on Computing and Control for the Water Industry, CCWI 2009 - Integrating Water Systems, 257-262, 1-3 September, Sheffield, UK.
5. Rossman, A. L. (2000). *EPANET2 Users Manual.* Water Supply and Water Resources Division, National Risk Management Research Laboratory of Computer and System Sciences, Cincinnati OH 45268.
6. Tanyimboh, T. T., Tabesh, M., and Burrows, R. (2001). *Appraisal of source head methods for calculating reliability of water distribution networks*, J. Water Resources Planning and Management, ASCE, 127(4), 206-213.
7. Trifunović, N. (2006). *Introduction to Urban Water Distribution*, Taylor & Francis Group, London, UK, 509 p.

CHAPTER 4

Spatial Network Generation Tool for Performance Analysis of Water Distribution Networks[1]

The research presented in this chapter aims at support tool for generation of multiple networks with preset or randomised properties. The purpose is twofold: the use for research, and more practically, for design of water distribution networks and their extensions. To explore particular phenomena, water distribution research may require a coherent set of cases. Readily available in the literature are simple networks used for benchmarking, either real-life cases that are too diverse in size and configuration. The *network generation tool* (NGT) developed on the principles of graph theory connects any seed of nodes prepared in EPANET, by applying the rule of maximum number of (closest) connections, and avoiding pipe crossings or unnecessary duplications. The pipe properties are assigned in different ways: by specifying a range of arbitrary lengths and diameters, by using coordinates to calculate the lengths, or by GA optimisation of initial diameters. Equally, the nodal elevations and demands can be arbitrarily assigned when not predefined in EPANET. Several sets of networks have been generated, up to 200 junctions, some of them used in the analyses described in Chapters 6 and 7. The upgrade and the use of NGT for design purposes have been further discussed in Chapter 9.

---

[1] Paper submitted by Trifunović, N., Maharjan, B., Bang, S.J. and Vairavamoorthy, K., (2012) under the title *Spatial Network Generation Tool for Water Distribution Network Design and Performance Analysis*, to the Journal of Water Science and Technology; under review.

4.1   INTRODUCTION

Enormous increase of computational speed in the past decade has opened space for new methods in analysis of water distribution network hydraulic performance by using models. Insufficient number of case studies to test these methods has consequently created a need for tools that can generate synthetic networks with properties that resemble real-life situations and enable analysis of correlations between the network performance parameters. Recent developments of algorithms for network generation have opened possibilities to create large samples of networks whose properties can be altered for particular research needs. The impact on the network hydraulic performance can be further assessed and conclusions drawn based on the results of stochastic analyses. Various approaches from the literature point graph theory as the most common concept for looped network generation and analyses of node/pipe connectivity.

Möderl et al. (2007) have developed a MATLAB based tool named *Modular Design System* (MDS) using graph matrices. The nodal connectivity is defined by binary-coded horizontal and vertical links as shown in Figure 4.1. Taking such coding into consideration, the set of typical modules representing different network sections is developed and various networks are generated by combining them in a 'LEGO building bricks fashion'. Eventually, any grid constructed in this way has a corresponding matrix with unique elements, as shown in Figure 4.2. The result of application is the set of 2280 EPANET networks between 27 and 3977 nodes and 26 and 5704 pipes, supplied from one to six sources. The layouts have been generated to resemble different level of complexity typical for urban or rural areas, combining looped- and tree sections with various population densities i.e. demand distribution. The MDS approach allows for relatively fast generation of large sample of networks. The generated networks will be visually squared, which is more of an aesthetic drawback; three typical layouts are shown in Figure 4.3. More importantly, the definition of network properties is rather limited and modification of model input parameters time consuming, especially in larger networks. Thus, the tool is still in the stage that makes it difficult to apply for design purposes.

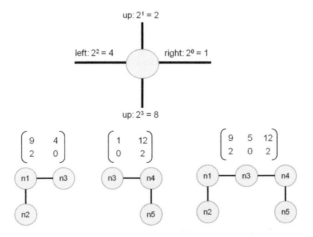

Figure 4.1 Binary code of nodal connections (according to Möderl et al.,2007)

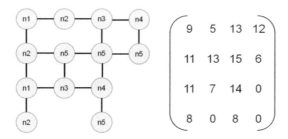

Figure 4.2 Network matrix represented by binary code (according to Möderl et al.,2007)

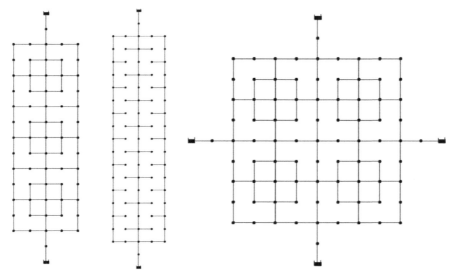

Figure 4.3 Typical layouts of MDS network sample (Möderl et al.,2007)

The problem of network generation is also known in the field of river and drainage/sewer networks as well as in the wider context in mathematics and computer science. In their further research, Möderl et al. (2009) propose a MATLAB tool called *Case Study Generator* (CSG) that creates virtual urban drainage networks. Having the features of random generation of nodal coordinates and elevations, and preventing the link intersection, CSG brings the network configuration closer to the reality. A sample of 10,000 synthetic networks has been generated and analysed for hydraulic performance under surface flooding.

A few other applications related to drainage networks can be found for instance in the work of Ghosh et al. (2006) who developed a program named *Artificial Network Generator* (ANGel). This is an ArcGIS tool that can develop fractal trees for reference network or line and is able to assess travelling time distribution and does various hydrological statistics for generated layouts. Furthermore, Sitzenfrei et al. (2010) present *Virtual Infrastructure Benchmarking* (VIBe) as methodology for algorithmic generation of *Virtual Case Studies* (VCSs) especially for urban water systems. VIBe provides input files for EPANET and SWMM software using parameters based on real world case studies and literature. The generated sample of 1000 VCSs was simulated and evaluated in this study.

Similarly, Urich et al. (2010) investigated the generation of virtual sewer systems using the agent based modelling approach. In this approach virtual combined sewer system are generated on the basis of virtual environment (digital elevation model, land use, population density etc) provided by VIBe. The infrastructure generation module exports the urban drainage system in the format used by SWMM for further simulations. Stochastic investigations and benchmarking of combined sewer systems can further be conducted based on the results of computer simulations and the layout properties of virtual sewer systems.

The synthetic networks generated in the field of river and drainage networks are mostly configured as fractal tree or system of branches. Water distribution networks are more frequently looped rather than branched, actually quite often a combination of lopes and branches. This makes the application of concepts used for drainage networks rather limited. The concepts of network generation based on graph theory initiated in mathematics and computer science offer more similarities to the network configurations typical for water distribution.

For instance, Rodionov and Choo (2004) present in their paper of computational mathematics the set of algorithms for generating connected random graphs (RG) based on specified limitations, such as given node connectivity or different probabilities of links existence. The process of sequential growth starts by generating spanning trees and then allocating remaining links. If node connectivity has achieved its maximum number of connections, free links are not considered to connect to it any more. All nodes are grouped around one or several centres with restriction of density with distance. The intersection of links may occur and there is a limitation to work with gradual increase in complexity for the number of virtual cases. This graph generation algorithms are mostly useful to create the Internet structures but some of the aspects are also applicable in water distribution; for instance, the planarity, non uniformity of distribution of node's coordinates, or limitation of maximum number of links connected to a node.

4.2   GRAPH THEORY TERMINOLOGY AND APPLICATION

Water distribution networks consisting of nodes and links are in graph theory presented as graphs composed of *vertices* and *edges*, respectively. A network is usually mapped as a *directed graph* (or *digraph*) with edges that have capacities (e.g. flow rates), and vertices that have demands, pressures etc. In water networks, the flow is delivered from sources to sinks. When the flow direction is not considered, the network is an *undirected graph*.

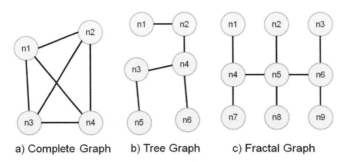

a) Complete Graph   b) Tree Graph   c) Fractal Graph

Figure 4.4 Various types of graphs

A graph can be complete, tree or fractal, as shown in Figure 4.4. A *complete graph* contains every possible pair of vertices connected by an edge. It is therefore impossible to add an edge to a complete graph without creating duplications or self-connected vertices. Graphs where this is not done are called *simple graphs*. Where multiple connections between the same pair of vertices or self-connected vertices are possible, the graphs are called *pseudo graphs* or *degenerated graphs*. The complete graph of *n* vertices will have total $n(n-1)/2$ edges. A *tree graph* is a graph without loops where any two vertices are connected by unique single path. Finally, a *fractal graph* specifies a geometric shape that can be split into parts, each of which is a reduced size copy of it (Möderl et al., 2009). Hence, the fractal is a geometric pattern that is composed of equal smaller shapes.

Furthermore, graph can be planar or non-planar. A *planar graph* has edges that only touch each other at vertices. If the edges cross, as is the case in Figure 4.4a, the graph will have *non-planar representation*. Non-planar graphs are rare, although not impossible in water distribution practice. It is therefore common to place a node at the intersection of pipes, which implies that network generation tool should avoid non-planar graphs.

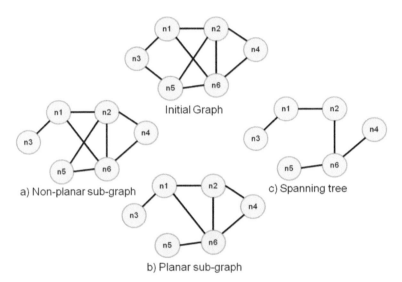

Figure 4.5 Spanning sub-graphs

Graph normally consists of spanning sub-graphs, as shown in Figure 4.5. A graph is a *spanning sub-graph* or a *factor* of another graph if both share the same vertex set. Moreover, the sub-graph will be *induced* in case it only contains the edges also available in the original graph.

Both simple graphs and pseudo graphs can be represented by an *adjacency matrix*. For graphs with *n* nodes, this matrix has $n \times n$ elements where A_{ij} is equal to the number of links connecting nodes *i* and *j* (for $i \neq j$). Diagonal elements A_{ii} will count twice the number of self-connecting links attached to node i, because each endpoint of the link is to be counted once.

Applying the terminology of graph theory for the simple network shown in Figure 4.6, an undirected graph G (V, E) is the object consisting of set V (vertices) and set E (edges) such that V(G) = (R, n1, n2, n3, n4, n5, n6) and E(G) = (1, 2, 3, 4, 5, 6, 7, 8) . The magnitude of graph G is characterized by the number of vertices | V |, which is seven (called the *order* of G) and number of edges | E |, which is eight (the *size* of G). The content of the adjacency matrix is shown in Table 4.1.

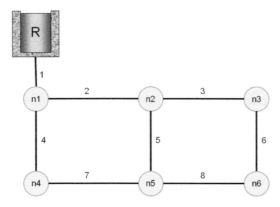

Figure 4.6 Water distribution network as a graph

Table 4.1 Adjacency matrix for the graph shown in Figure 4.6

$$
A = \begin{Bmatrix}
0 & 1 & 0 & 0 & 0 & 0 & 0 \\
1 & 0 & 1 & 0 & 1 & 0 & 0 \\
0 & 1 & 0 & 1 & 0 & 1 & 0 \\
0 & 0 & 1 & 0 & 0 & 0 & 1 \\
0 & 1 & 0 & 0 & 0 & 1 & 0 \\
0 & 0 & 1 & 0 & 1 & 0 & 1 \\
0 & 0 & 0 & 1 & 0 & 1 & 0
\end{Bmatrix}
$$

The matrix will show only the vertex connectivity. For simple graphs, all diagonal elements are zero, while all other elements can be zero or one. The sum of elements in the ith row (or in the ith column) gives the *degree* (or the *order*) of vertex i i.e. the number of edges connected to the vertex.

Table 4.2 Node-Link matrix for the graph shown in Figure 4.6

From node	To node						
	R	n1	n2	n3	n4	n5	n6
R	0	1	0	0	0	0	0
n1	1	0	2	0	4	0	0
n2	0	2	0	3	0	5	0
n3	0	0	3	0	0	0	6
n4	0	4	0	0	0	7	0
n5	0	0	5	0	7	0	8
n6	0	0	0	6	0	8	0

The matrix created in the same way but with the IDs of available edges as elements is called *node-link matrix*. For the graph in Figure 4.6, it will look as in Table 4.2.

Graph theory concepts are applicable in different fields such as communication, circuits, mechanical systems, hydraulics, finance, transportation, etc. with different vertices, edges and flow parameters, as illustrated in Table 4.3 (Ráez, 2003).

Table 4.3 Graph theory applications in various fields (Ráez, 2003)

Graph (Network)	Vertexes (Nodes)	Edges (Arcs)	Flow
Communications	Telephones exchanges, computers, satellites	Cables, fiber optics, microwave relays	Voice, video, packets
Circuits	Gates, registers, processors	Wires	Current
Mechanical	Joints	Rods, beams, springs	Heat, energy
Hydraulic	Reservoirs, pumping stations, lakes	Pipelines	Fluid, oil
Financial	Stocks, currency	Transactions	Money
Transportation	Airports, rail yards, street intersections	Highways, rail beds, airway routes	Freight, vehicles, passengers

Summarised, the graph theory terminology applied in the field of water distribution shall look as given in Table 4.4.

Table 4.4 Graph theory terminology in water distribution

Graph theory	Water distribution
Graph	Network
Tree graph	Branched network
Simple graph	Network without parallel (and/or self-connected) pipes
Pseudo/degenerated graph	Network with parallel (and/or self-connected) pipes
Planar graph	Network without unconnected pipe crossings
Non-planar graph	Network with unconnected pipe crossings
(Spanning) sub-graph	Alternative network with the same set of nodes and arbitrary links
Induced (spanning) sub-graph	Alternative network with the same set of nodes and common links
Vertex	Reservoir, tank, (demand) node
Edge	Links: pipe, pump, valve[1]
Degree/order of vertex	Number of links connected to a reservoir/tank/(demand) node
Order of graph	Total number of reservoirs/tanks/(demand) nodes
Size of graph	Total number of links

1) Pumps and valves will be commonly presented as edges because of the change of head they inflict between the two vertices. In newer computer software in water distribution, they can also be defined as vertices.

4.3 GRAPH THEORY CONCEPTS USED IN NETWORK GENERATION ALGORITHM

The *network generation tool* (NGT) presented in this chapter aims to generate sufficiently large samples of synthetic networks that resemble water distribution practice. For this purpose, an algorithm has been developed where links connect a set of reservoirs, tanks, and/or nodes prepared initially in EPANET, with spatial and topographic characteristics (x, y and z coordinates), as well as the initial nodal demands, of users' choice. To enhance the process and avoid configurations that are not typical in reality, the following four principles have been built into the algorithm: (1) to respect the maximum defined number of nodal connections, (2) to give priority to the closest available node for connection, (3) to avoid crossings between pipes (without connecting them), and (4) to avoid pipe duplication that could occur by connecting the same nodes in reversed order. These principles have implications on the way the graph matrices are manipulated during the programme execution. For the sake of consistency, the generated networks are exclusively simple graphs i.e. do not have parallel pipes; those should be added manually afterwards, if necessary.

An example of graph/network and its corresponding matrix is shown in Figure 4.7 and Table 4.5, respectively.

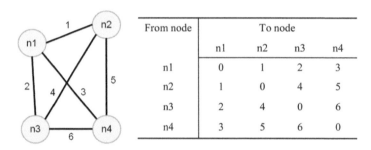

From node	To node			
	n1	n2	n3	n4
n1	0	1	2	3
n2	1	0	4	5
n3	2	4	0	6
n4	3	5	6	0

Figure 4.7/Table 4.5 Matrix representation of a graph

Not containing self-connected nodes, the diagonal of the matrix will be filled with zero values, splitting it into two symmetrical parts. The upper triangular matrix and lower triangular matrix are created, as shown in Table 4.6, each of them sufficiently describing the network connectivity and reducing the original matrix to a simpler form.

Table 4.6 Upper triangular matrix (left) and lower triangular matrix (right)

From node	To node			
	n1	n2	n3	n4
n1	0	1	2	3
n2		0	4	5
n3			0	6
n4				0

From node	To node			
	n1	n2	n3	n4
n1	0			
n2	1	0		
n3	2	4	0	
n4	3	5	6	0

Like in Figure 4.4a, the network in Figure 4.7 is a complete- and simple graph with total 12 connections out of six links, and the node degree ranging between one and three.

Nodes in water distribution networks are rarely connected with more than four pipes. Having a restriction on the maximum node degree can therefore drastically reduce the number of non-zero matrix elements, which limits the number of combinations and makes the network generation more efficient. Having this in mind, the algorithm has been developed with the limitation that each node can be connected to a maximum of three additional nodes, in the process of network generation. The choice for one, two or three closest connections will obviously influence the complexity of generated networks. Eventually, the actual nodal degree will occasionally exceed three, which can be further maximised in the settings of the programme. In how many cases and where in the network this will happen, depends in general on the ratio between the selected number of nodes and links, and in particular on the order of nodes in the process of generation and their coordinates, which is for larger sets not easy to predict in advance. Next to easier manipulation of matrices, this somewhat erratic generation process is believed to offer more variety of network layouts. Applying no limitation, the absolute maximum of nodal connections will rarely exceed six.

The planarization is applied in the next step of network generation, where all created pipe crossings are eliminated. The separation of edges that cross each other will be done by constructing two sub-graphs each with one of the edges, as shown in Figure 4.8. The result in the corresponding matrices is the change of one single element, as done in Table 4.7.

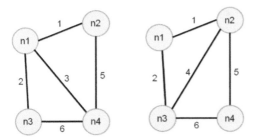

Figure 4.8 Spanning sub-graphs of network in Figure 4.7

Table 4.7 Upper triangular matrix of spanning sub-graphs in Figure 4.7

From node	To node				From node	To node			
	n1	n2	n3	n4		n1	n2	n3	n4
n1	0	1	2	3	n1	0	1	2	0
n2		0	0	5	n2		0	4	5
n3			0	6	n3			0	6
n4				0	n4				0

There are several methods of creating planar graphs known in the literature such as, *Path Addition Method* of Hopcroft and Tarjan (1974), *Tree Data Structure Method* developed by Booth and Lueker (1976), and more recently developed *Edge Addition Method* of de

Fraysseix, et al. (2006). The assessment of planarity applied in this study uses spatial coordinates to find possible intersection point. The intersection point (x, y) of two edges that are connecting the vertices (x_1, y_1) and (x_2, y_2), and vertices (x_3, y_3) and (x_4, y_4), respectively, is given by the following equations:

$$x = \frac{1}{m_1 - m_2}(y_3 - y_1 - m_2 x_3 + m_1 x_1) \quad ; \quad y = m_1 x - m_1 x_1 + y_1 \qquad 4.1$$

where m_1 and m_2 are slopes of the edges that are calculated as:

$$m_1 = \frac{y_2 - y_1}{x_2 - x_1} \quad ; \quad m_2 = \frac{y_4 - y_3}{x_4 - x_3} \qquad 4.2$$

If the two edges intersect, they are divided by the intersection point (x, y) in the proportions k_1 and k_2 in the following way:

$$k_1 = \frac{x_1 - x}{x - x_2} = \frac{y_1 - y}{y - y_2} \quad ; \quad k_2 = \frac{x_3 - x}{x - x_4} = \frac{y_3 - y}{y - y_4} \qquad 4.3$$

Positive values of k indicate that the intersection point cuts the edge internally while the negative values indicate external cuts, meaning that the intersection point is outside the edge.

Both situations are illustrated in Figure 4.9, left and right, respectively. Thus, the condition for two edges not to be intersected is that the k-value should be negative for both edges.

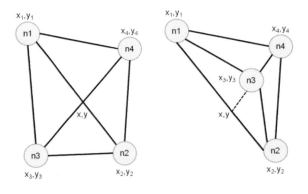

a) Non-planar Graph	a) Planar Graph

Figure 4.9 Intersection point in non-planar graph

4.4   GENERATION PROCESS

A set of generated networks is in fact a set of planar sub-graphs connecting all available vertices. Eliminating any discovered crossing of edges will actually lead to two alternative planar layouts, as has been shown in Figure 4.8 and Table 4.7.

In general, the process is conducted by manipulating the elements of the matrix constructed to satisfy selected boundary conditions. All possibilities will be taken into consideration initially meaning one sub-graph created for each degree of vertex i.e. each edge that connects it. For instance, in Table 4.7 on the left, the degree of vertex $n1$ is three when it is connected with all the edges in that row. On top of it, sub-graphs can also be produced with vertex degree one and two. Consequently, the total set of pipe combinations for $n1$ is: {1, 2, 3,(1,2), (1,3),(2,3),(1,2,3)}, total seven. This total number can be calculated by Equation 4.4 where n stands for the maximum number of connected edges to the vertices i.e. nodes.

$$nCr = \sum_{r=1}^{n} \frac{n!}{r!(n-r)!} = \frac{3!}{1!(3-1)!} + \frac{3!}{2!(3-2)!} + \frac{3!}{3!(3-3)!} = 7 \qquad 4.4$$

As the matrix table further shows, nodes $n2$ and $n3$ have one edge each, with combinations {5} and {6}, respectively. The total number of combinations for four nodes is therefore $7 \times 1 \times 1 = 7$. Finally, the total number of sub-graphs created by subsequently withdrawing matrix elements will be 49.

4.4.1 Non-Random Generation

Non-random generation will consider all possible combinations of matrix elements. This can be done by the recursive combination process starting from the last row of the matrix. For the sub-graph in Figure 4.8/Table 4.7 on the left, the list of all the possibilities will look as in Table 4.8. The process starts by making a combination of the first column {1,5,6}, then continues by searching the next element in the last row to replace the last element in this first combination. In the particular example, that one is not available as well is the case with the second (last) row. Finally, the successively made combinations are {2,5,6}, {3,5,6}, {1,2,5,6}, {1,3,5,6}, {2,3,5,6} and {1,2,3,5,6}.

Table 4.8 List of combinations for the sub-graph in Figure 4.8 - left

	Column 1	Column 2	Column 3	Column 4	Column 5	Column 6	Column 7
Row 1	1	2	3	1,2	1,3	2,3	1,2,3
Row 2	5	-	-	-	-	-	-
Row 3	6	-	-	-	-	-	-

The above seven combinations are generated while taking the elements from all three rows. The same process is to be repeated for the subsets of two rows and one row. After eliminating combinations with number of edges smaller than n-1, those that would qualify from the subset of two rows are {1,2,5}, {1,3,5}, {2,3,5}, {1,2,3,5}, {1,2,6}, {1,3,6}, {2,3,6}, {1,2,3,6}, and from one row is only {1,2,3}.

The drawback of the process is that it deals with huge number of possible network layouts, which grows exponentially with the number of nodes, as illustrated in Figure 4.10. Although many of these networks will be eliminated during the screening step, the total number of 'acceptable' layouts will still be big, probably too big for any sensible water distribution analysis, inflicting unnecessarily long computational times.

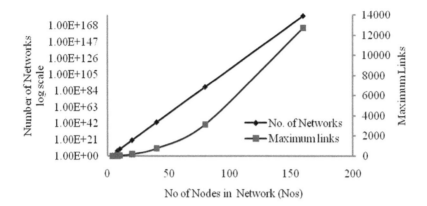

Figure 4.10 Theoretical number of network layouts/maximum number of links for given number of nodes

One way to reduce the number of generated networks and shorten the process is to skip some of the columns from Table 4.7 in the generation loop, which will eventually impact the complexity of generated networks. For instance, selecting the columns 1 to 3 and skipping the rest will result in pipe combinations {1,5,6}, {2,5,6}, and {3,5,6}, which depict branched configurations, as shown in Figure 4.11. On the other hand, picking the columns 4, 5 and 7 gives pipe combinations {1,2,5,6}, {1,3,5,6} and {1,2,3,5,6},which create loops as shown in Figure 4.12. Branched and looped networks are however not clearly distinguished by the choice of columns in the process of generation; column 7 can also produce combination {1,2,3} combining three rows where two are empty, which will again generate branched layout. Nevertheless, the degree of complexity is generally ascending by selecting more columns with two or three pipes.

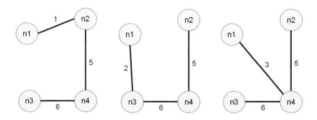

Figure 4.11 Sub-graphs of branched configurations

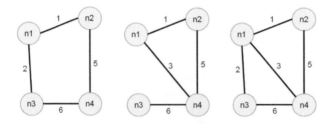

Figure 4.12 Sub-graphs of looped configurations

4.4.2 Random Generation

In case of larger number of nodes, a choice can be made to connect them entirely randomly. Pseudo-random number generator (PRNG) can be linked with serial number of combinations in each row, as it is shown in Table 4.9; the random number i picks the $i+1^{th}$ combination of edges Skipping of the rows is also possible in this case; an example of row 2 is shown in Table 4.10.

Randomly generated networks will be erratic in terms of complexity and connectivity. Also, the number of useless layouts within the number of selected generations can be high i.e. the real number of generated networks that can be of proper use is unpredictable. This becomes clear only after the screening process has been completed.

Table 4.9 Random generation of a sub-graph

	Combination of edges	Random number	Selected edge(s)	Selected connectivity	Unselected edge combinations
Row 1	1, 2, 3, (1,2), (1,3), (2,3), (1,2,3)	2	3	{3}	1, 2, (1,2), (1,3), (2,3), (1, 2, 3)
Row 2	5	0	5	{3,5}	-
Row 3	6	0	6	{3,5,6}	-

Table 4.10 Random generation of a sub-graph

	Combination of edges	Random number	Selected edge(s)	Selected connectivity	Unselected edge combinations
Row 1	1, 2, 3, (1,2), (1,3), (2,3), (1,2,3)	4	(1,3)	{1,3}	1,2,3, (1,2), (2,3), (1,2,3)
Row 2 (skipped)	-	-	-	-	-
Row 3	6	0	6	{1,3,6}	-

4.5 ALGORITHM OF NETWORK GENERATION TOOL

The NGT works with EPANET file in INP-format that contains preferred set of junctions with default or user-defined values; that file is prepared in advance. Two additional text files with unique names, *Rep1.txt* and *Rep2.txt*, are also to be prepared in advance based on the total number of nodes in the **.inp* file. *Rep1.txt* contains codes *MX* (stored one per row) used to represent dual edges from the combination list in Table 4.8, while *Rep2.txt* contains codes *NX* (also stored one per row) representing triple edge combinations. Both files can be used for multiple **.inp* files provided that *MX* and *NX* are not less than $6(n-1)$ and $3(n-1)$, respectively. For example, for networks up to 200 nodes *Rep1.txt* should contain the column of codes between *M1* and *M1200*, and *Rep2.txt* the codes from *N1* to *N600*, as shown in Table 4.11.

Table 4.11 Contents of *Rep1.txt* and *Rep2.txt* support files for networks up to 200 nodes

M1	N1
M2	N2
M3	N3
M4	N4
M5	N5
.....	.....
M1199	N599
M1200	N600

The output of NGT is the set of generated *.inp* files with different nodal connectivity, which are ready for hydraulic simulation in EPANET. The pipe properties in those files are defined either randomly within specified range, or by using coordinates to calculate the lengths. Redefinition of nodal elevations and demands is also possible. NGT can also prepare additional files that are necessary if diameters of generated networks are to be GA-optimised, which is done in separate programme discussed in Chapter 9.

The flow chart of NGT is given in Figure 4.13. The algorithm has ten distinct steps:

Step 1: Reading of initially prepared EPANET *.inp* file containing information about reservoirs/tanks/(demand) nodes; reading of supporting files *Rep1.txt* and *Rep2.txt*.

Step 2: The distances between the nodes are calculated using their coordinates and the values sorted in ascending order.

Step 3: Three nearest nodes to the source reservoir are used to start up the list of combinations by occupying the first three columns of the first row; these are having the vertex degree 1. The combinations (1,2), (1,3) and (2,3) will have vertex degree 2 and the combination (1,2,3) in the set has vertex degree 3.

Step 4: The following nearest node to the reservoir fills the next row of combinations made from its own three closest nodes. Each of those nodes is assigned a unique serial number and its edge is tested for planarity with already created edges in the adjacency lists. If an intersection exists, the numbered edge is put into a library of intersections maintained by two supporting matrices.

The steps 2 to 4 are repeated until the list of nodes has been fully exhausted.

Step 5: One set of edges creates a planar sub-graph, which is a combination from elements taken from each row. In non-random generation process, the algorithm generates recursively all possible combinations for rows and their subsets. In random generation process, each element is randomly chosen to make a set of edges.

Step 6: The edges of each created sub-graph are compared with those stored in the library of intersections. In cases where the IDs match, the other edge of the intersection shall replace the current one to create alternative sub-graph.

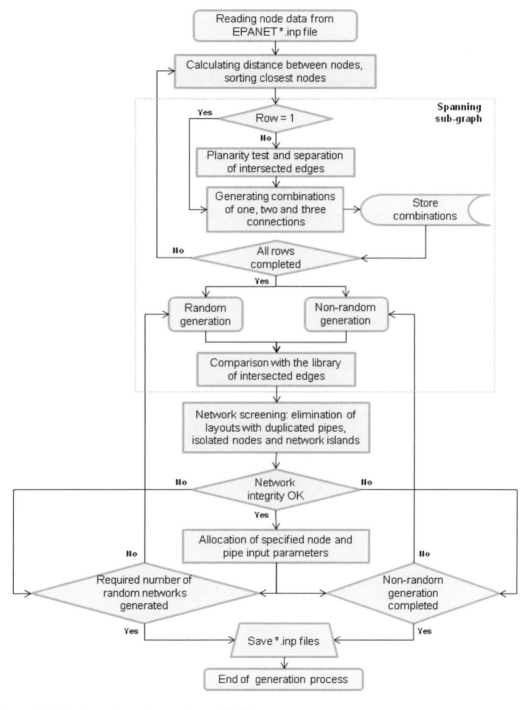

Figure 4.13 Flow chart of network generation tool (NGT)

Step 7: The IDs for each edge in the generated sets are detected representing two connected nodes, a start node ('NodeFrom') and an end node ('NodeTo').

Step 8: The screening of sub-graphs takes place in this step; the layouts of duplicated pipes, isolated nodes or islands of pipes will be eliminated.

Step 9: The attributes of finally selected sub-graphs are saved in a text file, after assigning the values for length, diameter and roughness from selected range of values (or pipe coordinates for the lengths).

Steps 5 to 9 are repeated until the process of generation has been completed. In non-random generation, the process ends when all elements in the rows and columns have been visited, or the (lesser) specified number of networks to be generated has been reaches. In random generation, that is the moment when the selected number of networks has been generated.

Step 10: The *.inp files are created from the generated text file for each network according to the required EPANET format.

4.5.1 Screening of Sub-Graphs

The need to eliminate the generated sub-graphs exists in the following three cases:
1. If the number of edges is less than *n*-1, some of the *n* vertices will not be connected, e.g. in case of combination {1,2} on Figure 4.15, left.
2. The number of edges can be equal or greater than *n*-1, and still some of the *n* vertices will not be connected, e.g. in case of combination {1,3,5} on Figure 4.15, middle.
3. The sub-graph may contain the number of isolated edges creating islands, e.g. in case of combination {2,5,7,8} on Figure 4.14, right.

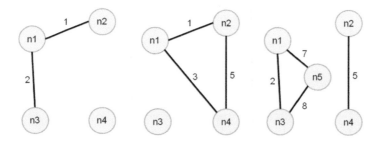

Figure 4.14 Faulty sub-graphs

The flow chart for the screening of sub-graphs mentioned above as Step 8 of the NGT algorithm described above is given in Figure 4.15. The screening process is similar to the Breadth First Search algorithm (BFS) discussed by Sempewo et al. (2008). The algorithm applied in NGT can be described in six steps:

Step 1: After a (non)-random generation process has been completed, a set of links that need to be screened on the above features will be stored in an array, which is a new input along with the array of nodes.

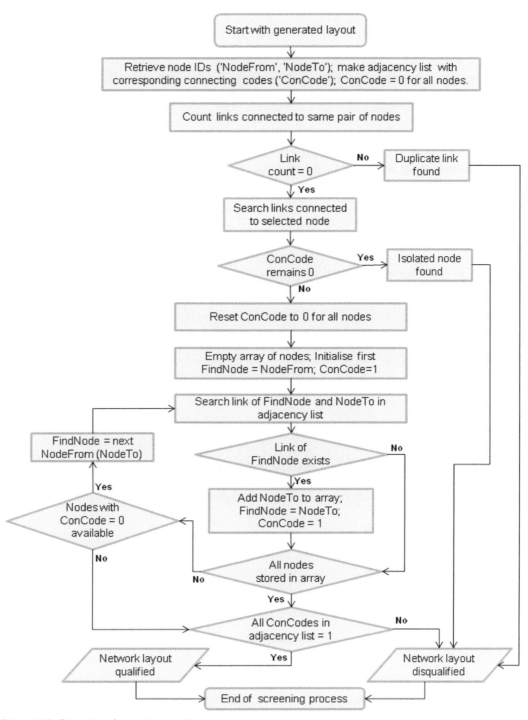

Figure 4.15 Flow chart of network screening process

Step 2: The nodes for every link are retrieved by respecting their order ('NodeFrom' and 'NodeTo'). Furthermore, an adjacency list is prepared with connectivity code assigned for each node ('ConCode', set as default to 0 and reset to 1 if the node link has been detected).

Step 3: The array of nodes is compared with the array of links. The node found to exist in any of the links from the array of links will change its connectivity code to 1. This kind of search is sufficient to detect and eliminate the layouts, such as those shown in Figure 4.14, left and middle.

Step 4: With no disjoint nodes found in the previous step, the screening process continues with the second stage, where the layouts with possible islands, as the one shown in Figure 4.14, right, are to be detected and eliminated. The connectivity code of all the nodes is again set to 0 and new array created where all the nodes found to have connections in the repeated search process will be stored.

Step 5: The node that has been detected in the adjacency list will have its connectivity code reset to 1 as well as all the other nodes found to be connected to it.

Step 6: All the nodes detected in the previous step will be assigned to the array of found nodes, if not existing there. The first of the adjacent nodes stored in this array is than taken for the repetitive search in Step 5.

Steps 5 and 6 are repeated until all the nodes with 'ConCode' of 0 have been checked for adjacency nodes. The program can also end its iteration when the array of found nodes has been filled with all the existing nodes. In this case all the nodes have changed their connectivity code to 1. Having any of the nodes with the connectivity code still at 0, after the check has been done, will indicate island(s) in the network.

4.5.2 Assigning of Network Parameters

After the generation process has been completed, the pipe parameters have to be specified for each layout in order to provide minimum input needed for EPANET to run hydraulic simulation with sensible results. Three options available in NGT are shown in Table 4.12.

Table 4.12 Assigning of network parameters

	Parameters	Option 1	Option 2: Uniform (user defined or default value)	Option 3: Random values from preselected range
Pipe properties	Length	Node coordinates from *.inp file	✓	✓
	Diameter	GA optimised diameter (constraint: minimum pressure, objective function: minimum cost)	✓	✓
	Roughness	-	✓	✓
Node properties	Elevation	Same as in *.inp file	-	✓
	Demand	Same as in *.inp file	-	✓

It is important to remember that the units of the parameters assigned by NGT will comply with default units of the initial EPANET file prepared for network generation process, which without any change will be the Imperial units and Hazen-Wiliams roughness constant. Possible conversion of input data will have to be done manually as EPANET does not recalculate these figures automatically.

4.6   NETWORK GENERATION TOOL IN USE

The network generation tool has been developed in C++ programming language using the EPANET toolkit function library. The tool runs within Microsoft Visual C++ 2010 Express platform and manipulates EPANET input file prepared in INP-format, which contains initial information about reservoirs/tanks/(demand) nodes. One example of such set of junctions consisting of single source and 19 nodes is shown in Figure 4.16. The shortened layout of the corresponding *.inp file is given in Table 4.13.

Figure 4.16 Layout of EPANET input prepared for network generation

Table 4.13 Network parameters in EPANET INP-format

```
[TITLE]

[JUNCTIONS]
;ID                Elev           Demand          Pattern
1                  0              0                               ;
4                  0              0                               ;
5                  0              0                               ;
.....etc.

[RESERVOIRS]
;ID                Head           Pattern
2                  0                                        ;
.....etc.

[TANKS]
;ID  Elevation   InitLevel   MinLevel MaxLevel Diameter MinVol VolCurve
```

```
[PIPES]
;ID    Node1    Node2  Length    Diameter    Roughness  MinorLoss    Status

[PUMPS]
;ID                    Node1               Node2            Parameters

[VALVES]
;ID      Node1       Node2     Diameter    Type   Setting    MinorLoss

[OPTIONS]
Units                 LPS
Headloss              D-W
.....etc.

[COORDINATES]
;Node                 X-Coord             Y-Coord
1                     4899.54             6877.90
4                     3508.50             4544.05
5                     2612.06             4296.75
.....etc.

[END]
```

A number of options are to be selected before the network generation process starts, which will depend from the selected mode of generation: random or non-random. Those include the number and the range of networks to be generated, the maximum number of nodal connections (can also be unlimited) and the handling of the matrix columns (in the non-random generation) influencing the complexity and variety of generated networks. The second group of the input settings is related to those shown in Table 4.12.

4.6.1 Random Generation

The input settings available in random generation process are listed in Figure 4.17. This is more straight forward process to launch but also more erratic to end. Therefore, a couple of options in the menu are made to control it more efficiently. These are described in the following bullets:

- *Required number of networks.* The preferred number of networks to be generated is specified first. Whether this number will be eventually reached depends on the total number of junctions and links, selected maximum junction degree and finally, the duplications of layouts that can be caused by random generation.
- *Range of links.* The generation can be conducted without any limitation in number of links for given number of junctions ('full range of links'). Alternatively, only the networks with number of links specified within certain range ('links from' - 'links to') shall be taken into consideration.
- *Limit of duplication.* Random generation process can cause duplication of layouts. All the duplicates will of course be eliminated in the screening process, but a boundary of duplication is needed to terminate the generation process in case other run parameters have been selected so that the selected number of generated networks is infeasible. Otherwise, the generation process could run endlessly.

- *Maximum links at node.* The node degree can also be maximised, which is influencing the complexity of generated layouts.
- *PRNG mode.* The random generation can be set in two modes: (1) repetitive simulations yield the same set of generated networks, and (2) each new simulation will results in different set of generated layouts.

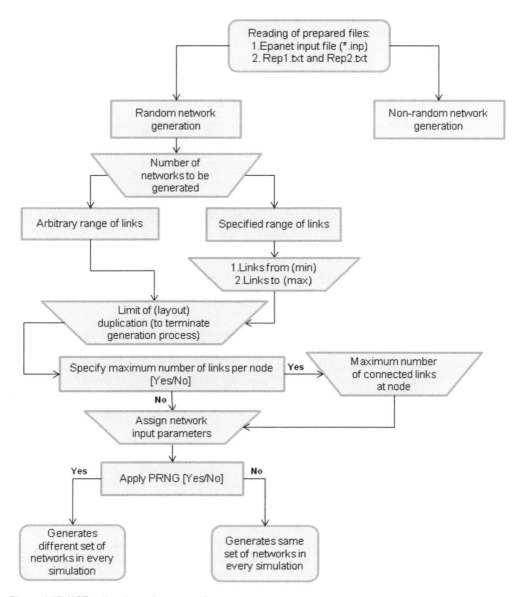

Figure 4.17 NGT options in random generation process

The options that are available while assigning the pipe and junction properties are shown in Figure 4.18.

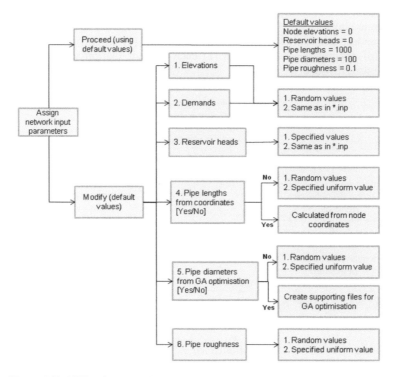

Figure 4.18 NGT options to assign the network properties

The default values can be assigned in NGT, taken over from EPANET defaults or defined as a range of random values. The exceptions are pipe lengths that can also be calculated from coordinates, and pipe diameters that can be GA-optimised. NGT does not optimise the pipe diameters directly but only prepares a number of supporting files that are further used by EO-optimiser in a separate programme, discussed more in details in Chapter 9.

4.6.2 Non-Random Generation

The list of input settings available in non-random generation process is given in Figure 4.19. The process of assigning the input values for nodes and links will be as in Figure 4.18.

Non-random network generation is potentially a time-consuming process that can be however influenced by selecting specific columns in the adjacency list that eventually influences the complexity of generated networks. The menu of available options includes:

- *Required number of networks*. Has the same meaning as in the random generation and is equally tentative, because its feasibility will depend on the number of nodes and other settings. With more nodes and columns, the process can become very long. Selecting a smaller number of networks can make it shorter but the variety in layouts becomes limited for the adjacency list has not been fully exhausted.
- *Range of links*. Has the same meaning as in the random generation.

- *Complexity.* The complexity of generated networks can be controlled in three ways: (1) by specifying one of the three categories ('Low', manipulating the three columns with single connections, 'Mid', manipulating the three columns with dual connections, and 'High', manipulating the column with triple connection), (2) by specifying any range of consecutive columns between 1 and 7, and (3) by skipping a number of columns. A number of equal column combinations can be created by all of these approaches, which is an overlap to be eliminated after the programme testing stage is over.
- *Maximum links at node.* Has the same meaning as in the random generation.

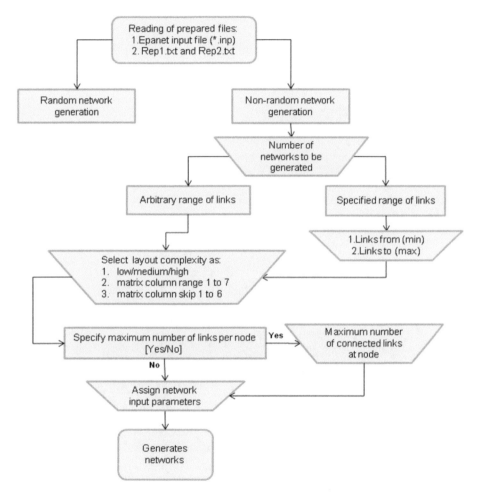

Figure 4.19 NGT options in non-random generation process

4.7 TEST CASES

Three nodal sets have been used to test the tool: of 20, 77 and 200 junctions. The layout of the first set has been already shown in Figure 4.16. The other two are shown in Figures 4.20 and 4.21, respectively.

Figure 4.20 Layout of set of 1 reservoir and 76 nodes

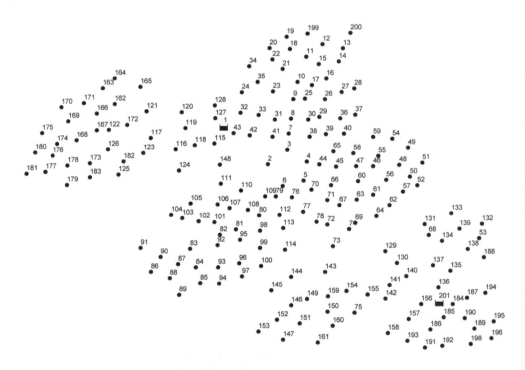

Figure 4.21 Layout of set of 2 reservoirs and 198 nodes

The networks have been generated applying the extensive range of options, random and non-random, and assigning the properties uniformly, randomly or applying coordinates and optimised diameters; one example of the settings is shown in Table 4.14.

Table 4.14 Assigned parameters for complex networks

Parameter	Assigned value
Elevation	Randomised: 0 to 20 msl
Demand	Randomised: 1 to 5 l/s
Reservoir head	Fixed: 60 msl
Length	Variable: calculated from coordinates
Diameter	Uniform: 300 mm
Roughness	Uniform: 0.5 mm

The selection of generated layouts has been shown in Figures 4.22 to 4.24, for each set of junctions, respectively.

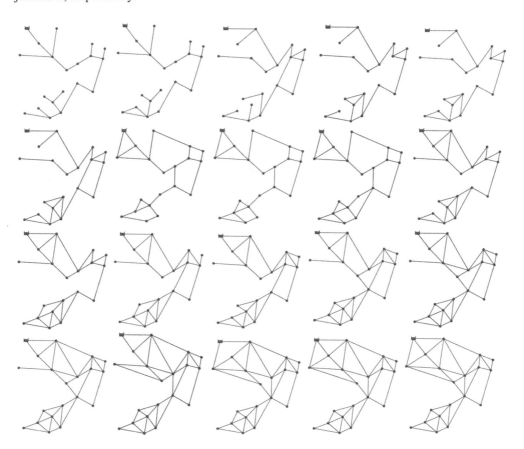

Figure 4.22 Selection of 20 layouts of generated networks for set of 20 junctions

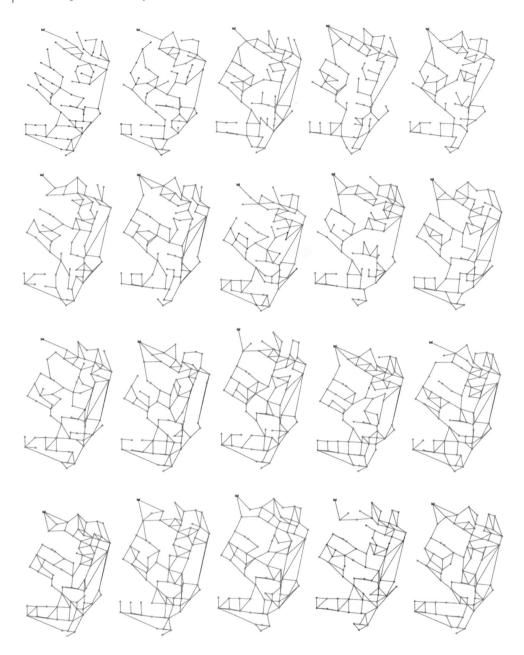

Figure 4.23 Selection of 20 layouts of generated networks for set of 77 junctions

Three examples of non-randomly generated networks of 20 junctions in different categories of complexity, have been shown in Figures 4.25 to 4.27. Finally, the results of EPANET hydraulic simulations of the two generated layouts: (1) of 77 junctions and 100 pipes, and (2) of 200 junctions and 271 pipe, both in non-optimised and optimised version, have been shown in Figures 4.28 to 4.30, respectively.

Figure 4.24 Selection of 21 layout of generated networks for set of 200 junctions

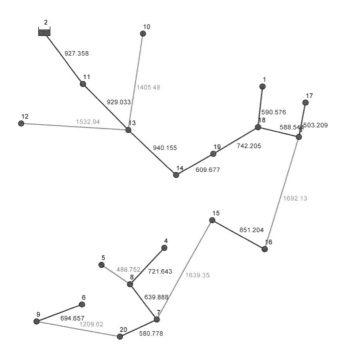

Figure 4.25 Example of non-random generated low-complex network layout of 20 junctions - L (m)

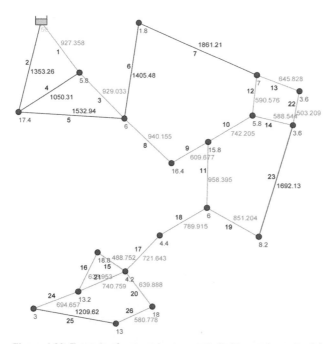

Figure 4.26 Example of non-random generated mid-complex network layout of 20 junctions - L (m)

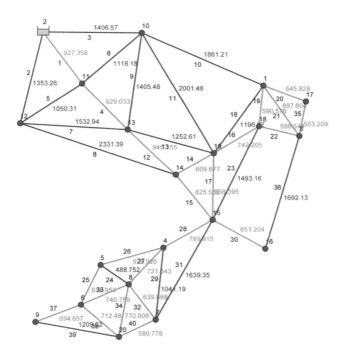

Figure 4.27 Example of non-random generated high-complex network layout of 20 junctions - L (m)

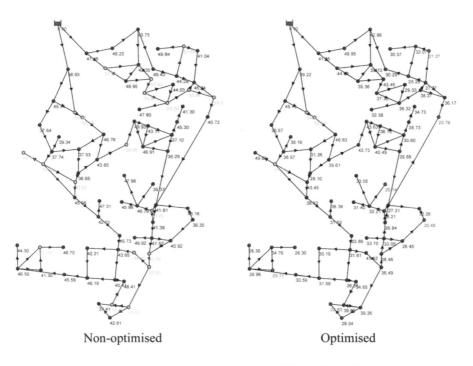

Non-optimised   Optimised

Figure 4.28 EPANET hydraulic simulation of sample network of 77 junctions (100 pipes) - p/pg (mwc)

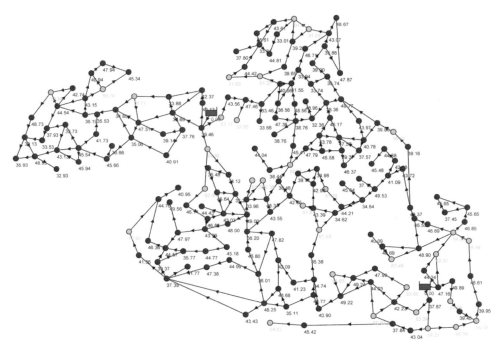

Figure 4.29 EPANET hydraulic simulation of sample network of 200 junctions (271 pipe) - p/pg (mwc)

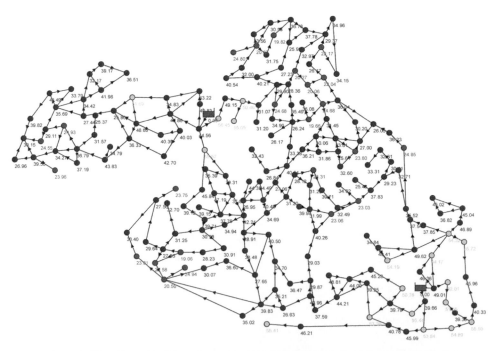

Figure 4.30 EPANET hydraulic simulation of optimised sample of 200 junctions (271 pipe) - p/pg (mwc)

All the generated networks show the layouts resembling what could be considered as a 'real water distribution outlook'. For the samples of ±50 networks, the generation time has been between a few seconds and a few minutes, depending on the selected settings. It is however not always that the generation time is proportional to the number of junctions, preferred number of networks or the level of complexity; the random generation process has been more erratic in this respect, which can also be attributed to the occasionally less logical combination of input settings. Tables 4.15 and 4.16 give indications about running times of NGT for various conditions in random and non-random mode of generation, respectively.

Table 4.15 NGT running times for various modes of random network generation (15 in each segment)

20 junctions			77 junctions			200 junctions		
No of links	Run time	Network type	No of links	Run time	Network type	No of links	Run time	Network type
19 - 20	5 sec	Branched	76 - 90	10 sec	Low complex	199 - 260	1 hrs, not complete	Low complex
21 - 22	3 sec	Low complex						
30 - 31	2 sec	Mid complex	106 - 120	4 sec	Mid complex	260 - 275	224 sec	Mid complex
						276 - 285	75 sec	
32 - 33	4 sec		121 - 135	7 sec		286 - 300	59 sec	
						301 - 315	33 sec	
34 - 35	218 sec		> 135	2 hrs, not complete		315 - 330	135 sec	
						330 - 345	35 min, not complete	

Table 4.16 NGT running times for various modes of non-random network generation

20 junctions			
No of Links	Column range	Network type	Run time
19 - 20	1 to 3	Branched	< 1 sec
21 - 22	1 to 3	Low complex	< 1 sec
30 - 31	1 to 6		2 sec
32 - 33	1 to 6	Mid complex	2 sec
34 - 35	1 to 6		3 sec
36 - 37	5 to 7		3 sec
38 - 39	6 to 7	High complex	3 sec
39 - 40	6 to 7		5 sec

Table 4.15 shows that the preferred number of networks could have not been reached for some segments of networks, next to the inconsistent running time, which is not necessarily a surprise. For instance, insisting on a network layout of 260 links and 200 junctions and setting the low maximum number of nodal connections, in order to maintain the preferred low complexity, is practically impossible assignment and random generator may not be able

to generate any such layout. Table 4.16 shows that the non-random generation has been more consistent in the segment of 20 junctions but the running times would certainly increase with the larger number of nodes and/or links, for the same reason as in case of randomly generated networks.

The network layouts shown in the previous figures are just a fraction of several hundreds of networks that have been generated in the process of the development and testing of NGT. Some of these have been used in the research presented in Chapters 6 and 7. The initial observations from running the tool in random and non-random mode are summarised in Table 4.17.

Table 4.17 Comparison of random and non-random network generation

Performance parameters	Random generation	Non-random generation
Effectiveness	Mostly effective for mid-complex networks.	Effective for various degree of network complexity.
Network type	Specific network configuration is not easy to control; difficult to preset branched or highly looped networks; more difficult to increase the complexity gradually.	Various configurations can be obtained with gradual increase in complexity.
Complexity	Random and difficult to control; only by defining the range of links, and maximum number of nodal connections.	Either defining the range of links and/or the complexity and/or the column range, or using all can fix this. Some practice is however needed to arrive at ideal combinations of input settings.
Running time	Sufficiently fast but inconsistent; not clearly linked to the network size. Sensitive on the ratio of the number of links and junctions, and the maximum number of nodal connections.	Fast for smaller networks (a few dozen junctions). Slower for full range of complexity due to large number of combinations; can be reduced by selecting specific column (range). Can be highly influenced by the (awkward) selection of input settings.
Seed value	Network generation is dependent on random generator seed value; may generate similar variety of networks regardless the selected number of layouts.	Not applicable.

4.8   CONCLUSIONS

The network generation tool (NGT) presented in this chapter has been developed based on the principles of graph theory by using the EPANET toolkit functions, with the idea of generating smaller, yet sufficiently large samples of synthetic networks with properties that resemble real water distribution networks. NGT uses the information about the set of junctions prepared in the EPANET input file (in INP-format) and generates the networks using the principle of connecting up to three closest junctions, starting from the source and avoiding pipe crossings and duplications. The pipe connections are created either by random or non-random generation. The screened networks will finally have the set of properties that are assigned in random or controlled way (fixed values, or calculated from coordinates, for pipe lengths, or in the process of GA-optimisation, for pipe diameters).

The tool has been tested on three sets of junctions (20, 77, and 200) and the conclusions about its functionality can be summarised in the following bullets:

1. NGT has proven to produce network layouts that fairly resemble those seen in water distribution practice.
2. The complexity of network layouts can be controlled (better in the non-random mode), as well as the running time to generate several dozens of networks up to a few hundred links is fast enough (between a few seconds and a few minutes). The random generation is showing faster running times for larger networks but with more unpredictable layouts, both in terms of complexity and number of networks. Various calculations will show different running times, which is highly sensitive to the selection of the number of links for given number of nodes and specified maximum nodal connectivity.
3. The library of case networks can be further diversified by the range of possibilities in definition of network properties.
4. During the development phase, the tool has not been tested for extreme conditions of several hundreds of networks (of several thousands of pipes). The large samples of a few thousand layouts have been generated in the research presented in Chapter 9, but the testing of upper limits of the tool is actually yet to be done. That has not been seen as immediate priority in this research, for two reasons. First, a few initial tests showed a huge increase in running times (of several hours) without any guarantee of completing the process; hence, the input settings are first to be refined, in fact simplified. Secondly, the improved robustness that would enable generation of large number of big networks may have little practical use: for design purposes very likely, and for research purposes, possibly. Analyses of extremely large samples ask much more time, whilst the additional benefit of the (large number of) results is yet to be proven compared to the analyses of smaller number of alternatives.
5. Nonetheless, it is believed that the NGT algorithm can be further improved to provide smooth generation of larger number of bigger networks. For particular purposes, allowing long running times of several hours or even days could be tolerated as the network generation process will mostly be a 'one time task'.
6. The room exists to improve the user friendliness of NGT by simplifying the programme options e.g. to integrate three ways of column/complexity definition into one, in the non-random generation mode.

The initial idea of NGT has been to serve the research in water distribution. With small addition of filtering the networks on particular pipe/street routes, NGT can easily be converted into a tool to be effectively used for design purposes, as well. This add-on has been developed within the decision support tool discussed in Chapter 9.

REFERENCES

1. Booth, K.S., and Lueker, G.S. (1976). *Testing for the consecutive ones property, interval graphs, and graph planarity using PQ-tree algorithms.* Journal of Computer and System Sciences, 13(3), 335-379.
2. de Fraysseix, H., de Mendez, P.O., and Rosenstiehl, P. (2006). *Trémaux Trees and Planarity.* International Journal of Foundations of Computer Science, 17(1017–1030).
3. *Evolving Objects (EO),* distributed under the GNU Lesser General Public License by SourceForge (http://eodev.sourceforge.net)
4. Ghosh, I. L., Hellweger, F., and Fritch, G.T. (2006). *Fractal Generation of Artificial Sewer Networks for Hydrologic Simulations.* ESRI International User Conference, San Diego, California, August 7-11.
5. Hopcroft, J., and Tarjan, R.E. (1974). *Efficient planarity testing.* Journal of the Association for Computing Machinery, 21(4), 549–568.
6. Möderl, M., Fetz, T., and Rauch, W. (2007). *Stochastic approach for performance evaluation regarding water distribution systems.* Water Sci Technol., 56(9), 29-36.
7. Möderl, M., Butler, D., and Rauch, W. (2009). *A stochastic approach for automatic generation of urban drainage systems.* Water Science and Technology 59(No. 6), 1137-1143.
8. Ráez, C. P. (2003). *Overview of Graph Theory: Ad Hoc Networking.* [Lecturer notes in Centre for Wireless Communications (ppt), University of Oulu, Finland].
9. Rodionov, A.S., & Choo, H. (2004). *On Generating Random Network Structures: Connected Graphs,* in H.-K. Kahng & S. Goto (Eds.), Information Networking (Vol. 3090, pp. 483-491): Springer Berlin / Heidelberg.
10. Sempewo, J., Pathirana, A., and Vairavamoorthy, K. (2008). *Spatial Analysis Tool for Development of Leakage Control Zones from the Analogy of Distributed Computing.* ASCE Conf. Proc., UNESCO-IHE Institute for Water Education, DA Delft, The Netherlands
11. Sitzenfrei, R., Fach, S., Kinzel, H., and Rauch, W. (2010). *A multi-layer cellular automata approach for algorithmic generation of virtual case studies: VIBe.* Water Sci. Tech., 61(1), 37-45
12. Urich, C., R, S., M, M., and W., R. (2010). *An agent based approach for generating virtual sewer systems in the software VIBe.* 8th International Conference on Urban Drainage Modeling, Tokyo, Japan, 7-11 Sep.

CHAPTER 5

Hydraulic Reliability Diagram and Network Buffer Index as Indicators of Water Distribution Network Resilience[1]

The research presented in this chapter aims at alternative way of expressing network reliability index, which has been derived from a diagram named the *hydraulic reliability diagram* (HRD). This diagram shows the correlation between the pipe flows under regular supply conditions and the loss of demand caused by the pipe failure. The reliability index calculated from the position of all the dots on the graph, and also adding proportional weighting to the pipes carrying variable flows, has been named the *network buffer index* (NBI). In this way, the NBI quantifies the HRD that clearly shows the pipes that are major contributors to the loss of demand. Moreover, the HRD hints the network buffer for particular level of demand. The concept has been illustrated on the networks of different degree of complexity and the NBI has been compared with two other indices from the literature: the resilience index of Todini (2000) and the network resilience defined by Prasad and Park (2004). The results show clear correlation between all three indices, but the NBI evaluates network reliability more realistically being more sensitive towards the change of network configuration than the resilience indices that are predominantly influenced by energy balance in the network.

[1] Paper submitted by Trifunović, N., Pathirana, A. and Vairavamoorthy, K., (2012) under the title *Network Buffer Index and Visual Representation of Water Distribution Network Resilience*, to the Journal of Water Research; under review.

5.1   INTRODUCTION

Hydraulic reliability of water distribution networks is commonly considered against a threshold pressure indicating sufficient service level. Below this value, the reduction of demand will occur to a certain degree and, combined with the probabilities of failure, an overall reliability index will be calculated that normally takes value between 0 and 1. Such approach is proposed by Ozger and Mays (2003) who introduced the term *available demand fraction* (ADF) to express the demand proportion still available in the network after a pipe failure event. Consequently, the loss of demand i.e. the drop in service level can be correlated to *1 – ADF*. If taken alone, this figure will reflect only the impact of the failure. Combined with the probability P_j that the failure of pipe j will happen, the reliability index R for the network of m pipes is calculated according to Equation 5.1:

$$R = 1 - \frac{1}{m}\sum_{j=1}^{m}(1 - ADF_j)P_j \qquad\qquad 5.1$$

Practical considerations assume a series of snapshot hydraulic simulations for particular demand scenario, starting with no-failure condition and then failing the pipes, one by one, and calculating the ADF in the network for each pipe, eventually obtaining the mean value for the entire network. To accomplish this analysis, the hydraulic simulation will switch into the pressure driven demand (PDD) mode and the outcome of the calculation will be the demand that is available in the network under stress conditions.

Having a PDD mode built into the hydraulic solver and adding computational loop that will break all the pipes in a sequence, makes the calculation of R relatively easy. Nevertheless, the obtained average value of ADF says much neither about the area affected by the failure i.e. the impact coverage, nor about the extent of the failure i.e. the impact intensity. The same value of ADF can be in theory a result of large network area affected to a lesser extent, or a small network portion that is affected severely. Furthermore, little can be grasped from the index about the buffer capacity of the network, and consequently no viable investment decision can be made purely based on the increase of ADF and overall reliability index, as long as it is not clear what it adds in practical terms i.e. to the improvement of the service level in irregular supply conditions.

The concept of resilience introduced by Todini (2000) and upgraded by Prasad and Park (2004) throws more light on the network buffer and its capacity to withstand a certain degree of stress but equally lacks insight about the impact coverage and intensity; the indices used to express the resilience will again be averaged values representing the entire network and possibly hiding the implications of some critical pipe failures.

5.2   HYDRAULIC RELIABILITY DIAGRAM

Serial- and branched networks have straight-forward calculation of ADF; the failure of any pipe j will mean the loss of all the demand downstream of that pipe, which equals the pipe flow Q_j under regular conditions. Hence:

$$Q_j = Q_{tot}(1 - ADF_j) \qquad\qquad\qquad 5.2$$

Q_{tot} in Equation 5.2 is the total network demand intended to be supplied from the source(s). In this respect, the drop of demand at no-failure condition, resulting from insufficient source head leading to the pressures below the threshold, would also be considered as a violation of the service level. Equally, the failure of any single pipe causing any loss of demand means that the buffer capacity of serial/branched networks that could be utilised during irregular supply scenarios is practically void.

Looped networks designed on the least-cost pipe diameters, by using genetic algorithms (GA), will show similar tendency. GA optimisers normally tend to develop a tree-like skeleton of secondary mains, towards the areas/nodes of higher demand, and then close the loops with the pipes of the least diameter available. From the perspective of reliability, these small pipes will serve little purpose as they carry low flows during regular supply and will therefore not be capable to convey surplus flows redirected after the failure of larger pipes, whatsoever. Figure 5.1 shows the layouts of 16 simple networks illustrating this point. The GA optimisation was conducted by *optiDesigner* software developed by OptiWater (http://www.optiwater.com) for the threshold pressure of 20 mwc and using the following list of available diameters (in mm): 50, 80, 100, 125, 150, 200, 250, 300, 350, 400, 500 and 600. Variable pipe lengths, nodal elevations and demands were assigned, as well as arbitrary head was set at the supplying point. The only fixed parameter was the absolute roughness set at 0.5 mm for all pipes/networks.

The results for flows and pressures are shown in Figure 5.2. The GA optimisation has been further repeated using two other programmes: *GANetXL* from Exeter University in UK (Savić et al., 2007), and *Evolving Objects* (EO) distributed under the GNU Lesser General Public License by SourceForge (http://eodev.sourceforge.net). The obtained results have been almost identical, which makes Equation 5.2 mostly valid for the least-cost designed looped networks in the sample.

The consequences of pipe failures in the four serial/branched networks from Figure 5.1a are shown in the diagram in Figure 5.3 where the loss of demand, $1 - ADF$, is plotted against the relative pipe flow Q_j/Q_{tot} under regular operation. Obviously, all the dots will be placed on the diagonal of the diagram, which results from the flow continuity i.e. the linearity of Equation 5.2.

The ADF of the 12 looped network configurations from Figures 5.1b-d has been calculated in the PDD mode. The results of ADF calculations at uniform PDD threshold of 20 mwc are shown in Figure 5.4. All of these dots are laid very close to the diagonal, matching the situation in Figure 5.3.

Creating additional buffer in the network, by adding extra pipes i.e. increasing the connectivity and/or the pipe diameters (also by reducing the demand, in theory) will result in migration of the dots towards the Y-axis; by increasing the connectivity significantly, all the dots will start converging towards the origin of the diagram. From the moment all of the dots have landed on the Y-axis, further increase of the diameters and/or connectivity makes no difference because the network has already achieved the level of buffer in which any failure causes no loss of the demand, whatsoever. This is the opposite extreme compared to the position of the dots on the diagonal of the diagram in Figures 5.3 and 5.4.

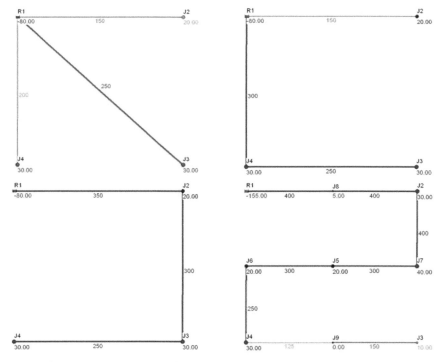

Figure 5.1a GA optimised nets 1 to 4 (optiDesigner) – pipes: D (mm), nodes: Q (l/s)

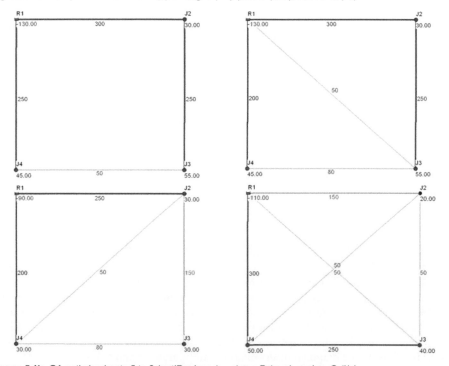

Figure 5.1b GA optimised nets 5 to 8 (optiDesigner) – pipes: D (mm), nodes: Q (l/s)

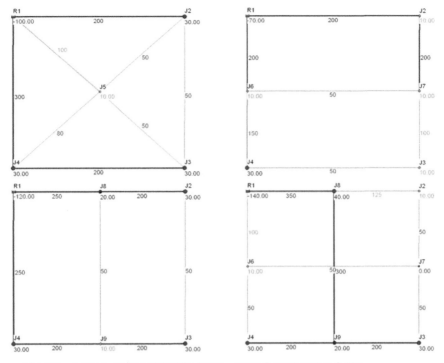

Figure 5.1c GA optimised nets 9 to 12 (optiDesigner) – pipes: D (mm), nodes: Q (l/s)

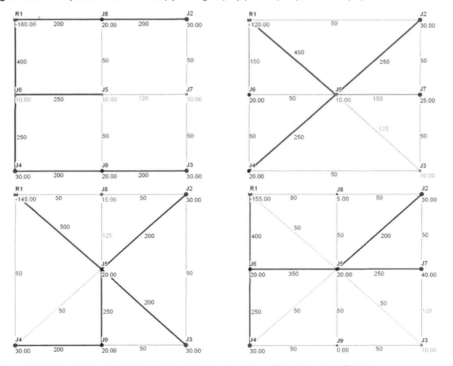

Figure 5.1d GA optimised nets 13 to 16 (optiDesigner) – pipes: D (mm), nodes: Q (l/s)

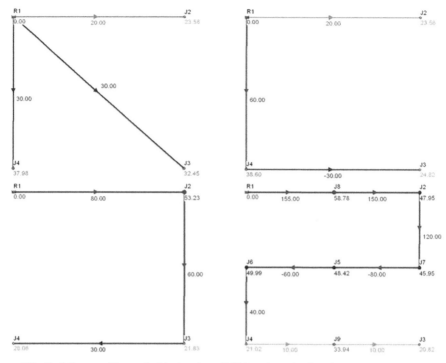

Figure 5.2a No-failure condition, nets 1 to 4 – pipes: Q (l/s), nodes: p/ρg (mwc)

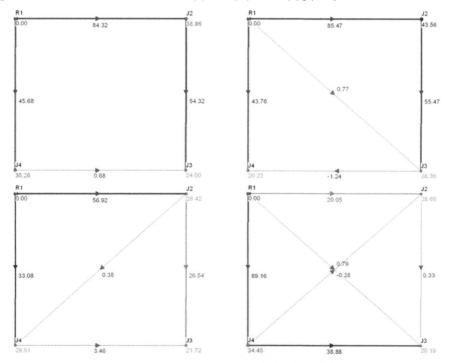

Figure 5.2b No-failure condition, nets 5 to 8 – pipes: Q (l/s), nodes: p/ρg (mwc)

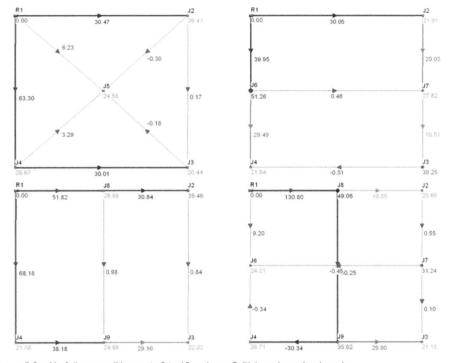

Figure 5.2c No-failure condition, nets 9 to 12 – pipes: Q (l/s), nodes: p/ρg (mwc)

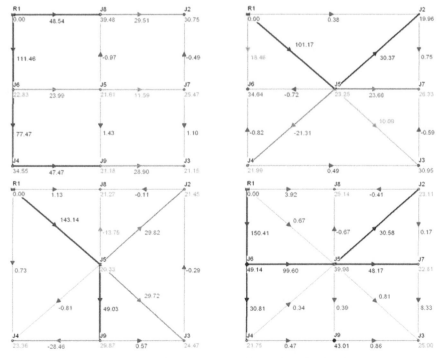

Figure 5.2d No-failure condition, nets 13 to 16 – pipes: Q (l/s), nodes: p/ρg (mwc)

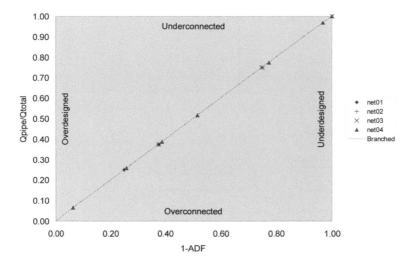

Figure 5.3 Loss of demand for serial/branched configurations

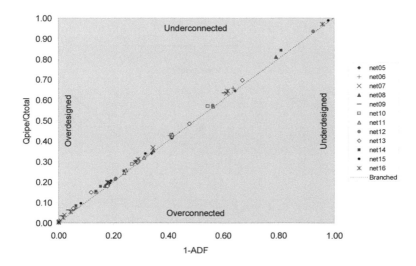

Figure 5.4 Loss of demand for GA optimised looped nets (PDD threshold = 20 mwc)

These trends are illustrated in the example in Figure 5.5, which shows the implications of the buffer increase for networks 10 and 16 (the upper-right in Figure 5.1c and the lower right in Figure 5.1d, respectively). Next to the optimised layouts (*net10* and *net16* on the graph), two additional layouts have been created by increasing each pipe diameter with the first higher value from the available range used for the GA optimisations (the layouts *a* and *b*).

The diagram as shown in Figures 5.3 to 5.5 is named here the *hydraulic reliability diagram* (HRD), and presents a footprint of network operation for given demand scenario, by showing the distribution of pipe flows under regular conditions against corresponding loss of demand in case of particular pipe failure.

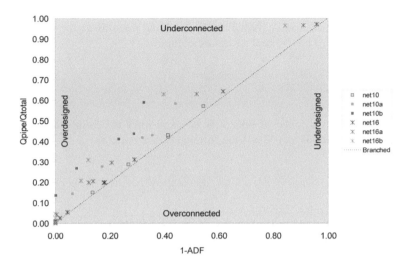

Figure 5.5 Loss of demand for nets 10 and 16 at various levels of buffer (D increase)

The following conclusions can be made by analysing position of the dots on such a diagram:
1. The closer the dots to the Y-axis are, the more overdesigned the network is (D >>).
2. The closer the dots to the diagonal of the diagram are, the closer the network is closer to the branched- or the looped least-cost design layout. The buffer for irregular supply scenarios is then also being reduced to a minimum (eventually becoming zero).
3. The network is under-designed if the dots are positioned below the diagonal of the graph (D <<); the minimum pressure i.e. design demand is not delivered even at no-pipe failure condition, because the head at the source(s) is too low for the selected pipe diameters.
4. The network has lower connectivity in case the dots are positioned closer to the upper X-axis (indicating small number of pipes).
5. The network has higher connectivity in case the dots are positioned closer to the lower X-axis i.e. the origin of the graph (indicating large number of pipes).

5.3   RELATION BETWEEN PIPE FLOW AND LOSS OF DEMAND

Mathematical relation between the pipe flow under regular supply and the loss of demand after the pipe fails is illustrated in simplified example of two pipes in parallel, shown in Figure 5.6.

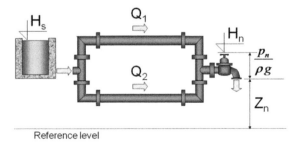

Figure 5.6 Supply through parallel pipes

In case of failure of any of the two pipes, the additional capacity will be taken over by the remaining pipe, increasing the head loss i.e. reducing the head and pressure in the discharge node (H_n and $p_n/\rho g$, respectively). Thus, in case of failure of pipe 1:

$$H_n - H_{n,f1} = \frac{p_n - p_{n,f1}}{\rho g} = R_{2,f1}Q_{2,f1}^2 - R_2Q_2^2 \qquad 5.3$$

Parameter $R_{2,(f1)}$ stands for the resistance of pipe 2 before and after the failure of pipe 1, respectively, which by applying the Darcy-Weisbach formula is calculated as:

$$R_{2,(f1)} = \frac{8\lambda_{2,(f1)}L_2}{g\pi^2 D_2^5} \qquad 5.4$$

taking into consideration the pipe length L_2, diameter D_2 and the friction factor $\lambda_{2,(f1)}$ before and after the failure of pipe 1. Furthermore, for $Q_{tot} = Q_1 + Q_2$ Equation 5.3 transforms into:

$$\frac{1}{R_2}\frac{\Delta p_{n,f1}}{\rho g} = \frac{R_{2,f1}}{R_2}Q_{2,f1}^2 - (Q_{tot} - Q_1)^2 \qquad 5.5$$

and finally, for:

$$\frac{1}{R_2}\frac{\Delta p_{n,f1}}{\rho g} = b \quad ; \quad \frac{R_{2,f1}}{R_2} = \frac{\lambda_{2,f1}}{\lambda_2} = a \qquad 5.6$$

Equation 5.5 can be rewritten in the format of Equation 5.2:

$$Q_1 = Q_{tot}\left(1 - \frac{\sqrt{aQ_{2,f1}^2 - b}}{Q_{tot}}\right) \qquad 5.7$$

Having the available demand fraction in case of the failure of pipe 1 as $ADF_1 = Q_{2,f1}/Q_{tot}$, Equations 5.2 and 5.7 become identical in case of parameters $a = 1$ and $b = 0$.

The condition $a = 1$ is satisfied in case of flow in the zone of developed (i.e. rough) turbulence where the condition $Q_{2,f1} > Q_2$ still yields $\lambda_{2,f1} = \lambda_2$. In the zone of transitional turbulence, factor a (actually its squared root) would significantly deviate from 1 only in case of significant pipe flow increase resulting from the failure. The Moody diagram in Figure 5.7 shows that in the most extreme cases, a doubling of pipe flow (the Reynolds number increase from, say, 10,000 to 20,000) for extraordinary low values of k/D (say, 0.000001 representing very big and smooth pipes) would result in the increase of λ up to 20 %; this is for instance a situation where a 1000 mm pipe would carry flow in order of only 10 l/s, which is not very common in practice. Much more likely is that significant flow increase will occur in pipes of smaller diameters (i.e. with higher k/D) which then affects the friction factor not more than a few percent. Therefore, $a \approx 1$ for flows in the zone of transitional turbulence in most of situations that normally occur in practice.

The condition $b = 0$ is more dependent on the pressure drop and the resistance of pipes that are to carry the surplus flow. In reality, the flow/demand reduction starts for the drop of pressures in lower range, which is in hydraulic simulations compared with the PDD threshold, $p_{min}/\rho g$. Hence, in the above example, $\Delta p_{n,f1} = p_{min} - p_{n,f1}$, and $Q_{2,f1}$ will equal Q_{tot} as long $\Delta p_{n,f1} < 0$.

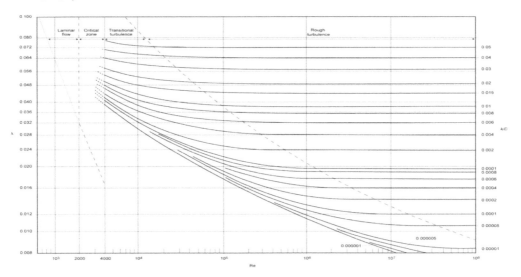

Figure 5.7 Moody diagram

In general, the order of value of the resistance R will always be (much) higher than the one of $\Delta p_{n,f1}/\rho g$, which makes the value of b initially low. Moreover, the smaller is the pipe, the closer the value of b will be to 0. For instance, a 1000 m pipe of 500 mm diameter with typical friction factor value of 0.025 makes R calculated from Equation 5.4 equal to 66.10 and with pressure drop of say 5 mwc, $b = 0.0756$. For shorter and smaller pipe, say, $L = 300$ m, $D = 150$ mm with $\lambda = 0.030$, R will equal 9792.81 and for the same pressure drop of 5 mwc, $b = 0.00051$. Hence, it can also be assumed that $b \approx 0$ in most of practical cases.

The similar analogy as presented in the case of two pipes in parallel can be assumed while analysing the results of the GA optimised looped networks, shown in Figure 5.4. It is the high resistance of small pipes in these networks, which is causing relatively high pressure drop once they are to take over the flows resulting from the burst of bigger pipes. Due to the relation between the nodal pressure and demand, the pipe friction loss will eventually be reduced as a result of the demand reduction, making the pipe resistance R far more dominant component than the pressure drop $\Delta p_{n,f}/\rho g$. Hence, the looped networks optimised on the least-cost pipe diameter are likely to perform hydraulically as branched networks, both in regular and irregular situations. The smaller is the minimum pipe diameter selected in the GA optimisation, the closer these hydraulic performances will be.

5.4   HYDRAULICS OF LOOPED NETWORKS UNDER STRESS CONDITIONS

Looped networks with certain degree of buffer capacity will be able to withstand the consequences of pipe failure to the extent that can be assessed by taking into consideration the same network parameters as discussed in the previous paragraph. Mathematical proof of correlation between the pipe flow and the loss of demand after the pipe failure is however more difficult to derive in case of fully looped 'non-optimised' networks. The resistance of pipes and the head at supply point(s) will be factors of influence i.e. whether there will be any loss of demand and how significant it will be in terms of the network coverage and intensity. The changes in governing equations after the pipe failure occurs are described in the following sections.

5.4.1 The Law of Continuity in Each Junction

After the failure of pipe f, the change of flow ($\Delta Q_{i,f}$) in k pipes connected to junction i will match the change of demand/supply ($\Delta Q_{i,f}$) in that junction:

$$Q_i + \Delta Q_{i,f} = \sum_{j=1}^{k}\left(Q_j + \Delta Q_{j,f}\right) \quad \Leftrightarrow \quad \Delta Q_{i,f} = \sum_{j=1}^{k}\Delta Q_{j,f} \qquad 5.8$$

If the demand of node i has been reduced, $\Delta Q_{i,f}$ will be negative. If the flow rate of pipe j has decreased and/or its direction has been reversed, $\Delta Q_{j,f}$ will also be negative. If the flow rate has increased and has kept the same direction, $\Delta Q_{j,f}$ becomes positive.

5.4.2 Total Loss of Demand from Failure of Pipe

For l sources and n nodes, the loss of demand as a result of the failure of pipe f will be:

$$1 - ADF_f = \frac{1}{Q_{tot}}\sum_{s=1}^{l}\Delta Q_{s,f} = -\frac{1}{Q_{tot}}\sum_{i=1}^{n}\Delta Q_{i,f} \qquad 5.9$$

$$Q_{tot} = -\sum_{s=1}^{l}Q_s = \sum_{i=1}^{n}Q_i \qquad 5.10$$

Having in mind that the intended- i.e. design demand can be reduced even at no-failure conditions, it will be more accurate to assume that Q_i is the target demand ($Q_{i,t}$), which is to be supplied with sufficient head at the sources.

5.4.3 Relation Between Nodal Demand and Pressure

For given range of pressures $0 < p_i < p_{min}$, the p_{min} being the PDD threshold, the relation between the nodal demand and pressure is usually assumed to be exponential, as follows:

$$Q_i = k_i\left(\frac{p_i}{\rho g}\right)^{0.5} \qquad 5.11$$

Above the threshold, the targeted demand ($Q_{i,t}$) will be supplied 100 %. Consequently, the loss of demand in node i caused by the failure of pipe f can be expressed as:

$$\Delta Q_{i,f} = k_i \left[\left(\frac{p_{i,f}}{\rho g} \right)^{0.5} - \left(\frac{p_{min}}{\rho g} \right)^{0.5} \right] = k_i \left(\frac{p_{i,f}}{\rho g} \right)^{0.5} - Q_{i,t} \qquad 5.12$$

5.4.4 Balance of head losses in loops

After the failure of pipe f, the balance of friction losses in each loop of a network changes as follows (respecting the loop orientation):

$$\sum_{j=1}^{k} h_{f,j,f} = \sum_{j=1}^{k} \left(h_{f,j} + \Delta h_{f,j,f} \right) = 0 \quad \Leftrightarrow \quad \sum_{j=1}^{k} \Delta h_{f,j,f} = 0 \qquad 5.13$$

h_{fj} is the initial friction loss of pipe j, and $h_{fj,f}$ the value after the failure. Additional impact of pumps and valves should be included in Equation 5.13 if they are part of the loop. Furthermore, Equation 5.13 will be invalid in case of the loop to which pipe f belongs. Nevertheless, Equation 5.13 will be valid for two adjacent loops sharing pipe f, and effectively merging into one bigger loop after that pipe has been closed.

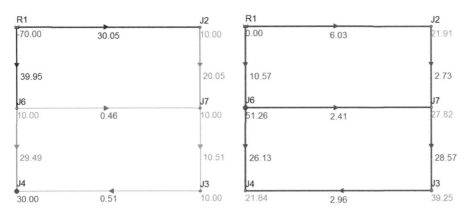

Figure 5.8 Net 10, no-failure – left: flow-demand (l/s), right: h_f/L-p/ρg (mwc)

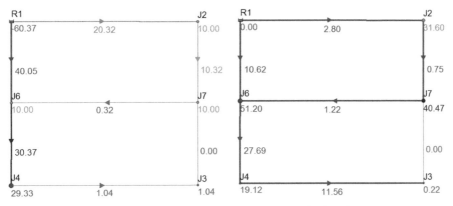

Figure 5.9 Net 10, failure J7-J3 – left: flow-demand (l/s), right: h_f/L-p/ρg (mwc)

Figure 5.8 and 5.9 show the above balances of flows and pressures/friction losses in the example of network 10. The results obtained for PDD threshold pressure of 20 mwc are further processed in Tables 5.1 and 5.2 for pipes and nodes, respectively.

Table 5.1 Net10, results for pipes

Geometry				No-failure			Failure J7-J3				Change		
ID_j	L_j	D_j	λ_j	R_j	Q_j	$h_{r,j}$	$\lambda_{j,r}$	$R_{j,r}$	$Q_{j,r}$	$h_{r,j,r}$	$\Delta Q_{j,r}$	$\Delta h_{r,j,r}$	→
(-)	(m)	(mm)	(-)	(s^2/m^5)	(l/s)	(mwc)	(-)	(s^2/m^5)	(l/s)	(mwc)	(l/s)	(mwc)	(Y)
R1-J2	3000	200	0.026	20140	30.05	18.19	0.026	20140	20.32	8.32	-9.73	-9.87	
R1-J6	1300	200	0.026	8727	39.95	13.93	0.026	8727	40.05	14.00	0.10	0.07	
J2-J7	1500	200	0.026	10070	20.05	4.05	0.027	10457	10.32	1.11	-9.73	-2.94	
J6-J7	3500	50	0.043	39793094	0.46	8.42	0.045	41643936	-0.32	-4.26	-0.78	-12.68	Y
J7-J3	1000	100	0.031	256143	10.51	28.29	0	0	0	0	-10.51	-28.29	
J6-J4	1700	150	0.028	51793	29.49	45.04	0.028	51793	30.37	47.77	0.88	2.73	
J3-J4	2500	50	0.043	28423639	0.51	7.39	0.041	27101609	-1.04	-29.31	-1.55	-36.70	Y

Table 5.2 Net10, results for nodes (PDD threshold, $p_{min}/\rho g$ = 20 mwc)

Topography			No-failure		Failure J7-J3		Change		
ID_i	z_i	k_i	$p_i/\rho g$	Q_i	$p_{i,r}/\rho g$	$Q_{i,r}$	$\Delta p_{i,r}/\rho g$	$\Delta Q_{i,r}$	↓
(-)	(msl)	$(l/s/mwc^{0.5})$	(mwc)	(l/s)	(mwc)	(l/s)	(mwc)	(l/s)	(Y)
R1	90	-	0	-70	0	-60.37	0	9.63	Y
J2	50	2.236	21.91	10	31.60	10	9.69	0	
J3	0	2.236	39.25	10	0.22	1.04	-39.03	-8.96	Y
J4	10	6.708	21.84	30	19.12	29.33	-2.72	-0.67	Y
J6	25	2.236	51.26	10	51.20	10	-0.06	0	
J7	40	2.236	27.82	10	40.47	10	12.65	0	

Resulting from the failure of the pipe connecting nodes J7 and J3, the pressure will drop below the threshold in nodes J3 and J4 causing the reduction of demand of 8.96 and 0.67 l/s respectively. The drop of pressure in node J6 has no impact on the demand because the pressure in that node after the pipe failure is still well above the threshold. Finally, the pressure in nodes J2 and J7 will grow as a result of overall demand i.e. head loss reduction. Also, the failure has caused the change of flow direction in pipes connecting nodes J6 and J7, and J3 and J4.

Two categories of junctions give possibility to correlate the flow in the failed pipe under regular supply with the loss of demand resulting from the calamity. In junctions that represent sources, the negative demand increase indicates the total demand loss. Hence, after combining Equation 5.8 and 5.9:

$$Q_{tot}\left(1 - ADF_f\right) = \sum_{s=1}^{l} \Delta Q_{s,f} = \sum_{s=1}^{l}\sum_{j=1}^{k} \Delta Q_{s,j,f} \qquad 5.14$$

Thus, each pipe j connected to the source s, carries after the failure of pipe f a flow increment/decrement that can be correlated to the total loss. In the above network, $\Delta Q_{R1,f}$ = 9.63 l/s, which balances the flow decrement of pipe connecting the source in R1 with J2 (of -9.73 l/s) and the flow increment of the pipe connection with J6 (of 0.10 l/s).

In the second category are the two junctions that connect the failed pipe f. The continuity equation in this case will look as:

$$Q_f = \Delta Q_{i,f} - \sum_{j=1}^{k-1} \Delta Q_{j,f} \qquad\qquad\qquad 5.15$$

Thus, each pipe j connected to the junction with the failed pipe f carries the flow increment/decrement that can be correlated to the original flow of pipe f. In the above network, $Q_f = Q_{J7-J3} = 10.51$ l/s during regular operation. After the closure of pipe $J7$-$J3$, in node $J7$ $\Delta Q_{J2-J7,f} = $ -9.73 l/s and $\Delta Q_{J6-J7,f} = $ -0.78 l/s; both flows balance the value of Q_f because $\Delta Q_{J7,f} = 0$. Equally, in node $J3$, $\Delta Q_{J3-J4,f} = $ -1.55 l/s and $\Delta Q_{J3,f} = $ -8.96 l/s, which also balance the Q_f.

Eventually, the changes in flow rates and nodal demands in all the pipes and nodes can be correlated to certain proportion p of the loss of demand and/or the flow of pipe f before the failure. For instance, after combining Equation 5.9 and 5.15, the continuity of flows in two junctions that connect the failed pipe yields:

$$Q_f = -p_i Q_{tot}\left(1 - ADF_f\right) - \sum_{j=1}^{k-1} p_j Q_f \quad \Leftrightarrow \quad Q_f = \alpha_i Q_{tot}\left(1 - ADF_f\right); \alpha_i = \frac{-p_i}{1 + \sum_{j=1}^{k-1} p_j} \qquad 5.16$$

which resembles the format of Equation 5.2 and 5.7. Similar relations can be developed for other junctions connected to these, or to the source(s), confirming the dependency of the loss of demand after particular pipe fails on the flow it carries under regular conditions.

Equation 5.16 will be valid in the nodes where the demand reduction takes place i.e. the nodal pressure is below the PDD threshold. Where $\Delta Q_{i,f} = 0$, the sum $\sum p_j = 0$; in this case the impact of the pipe closure is propagated towards surrounding nodes.

In the above network, the following values of p_i, p_j and α_i can be calculated by analysing the flow continuity in each junction (also shown in Figure 5.10):

$R1$: $p_{R1-J2} = $ -9.73/9.63 = -1.0104, $p_{R1-J6} = $ 0.10/9.63 = 0.0104

$J7$: $p_{J2-J7} = $ -9.73/10.51 = -0.9258, $p_{J6-J7} = $ -0.78/10.51 = -0.0742

$J2$: $p_{R1-J2} = $ -1.0104, $p_{J2-J7} = $ -0.9258, $\alpha_{J2} = $ -1.0104/-0.9258 = 1.09

$J3$: $p_{J3} = $ -8.96/9.63 = -0.9304, $p_{J3-J4} = $ -1.55/10.51 = -0.1475, $\alpha_{J3} = $ 0.9304/0.8525 = 1.09

After combining the continuity equations for nodes $J6$ and $J4$, the flow increment of 0.88 l/s registered in pipe $J6$-$J4$ will disappear. Hence:

$J6$: $p_{R1-J6} = $ 0.0104, $p_{J6-J7} = $ -0.0742   and $J4$: $p_{J4} = $ -0.67/9.63 = -0.0696, $p_{J3-J4} = $ -0.1475

and consequently, $\alpha_{J4} = $ (0.0104 + 0.0742)/(-0.0696 + 0.1475) = 1.09

In all the cases, the continuity equations in junctions lead to the value of α_i of 1.09, which is the ratio between the flow in pipe $J7$-$J3$ during regular operation and the loss of demand after that pipe has been closed. Thus, 10.51/9.63 = 1.09. Consequently, the bigger value of α_i

would indicate more buffer i.e. less of a demand reduction caused by the pipe failure, while α_i = 1 indicates serial/branched networks. Finally, $\alpha_i \approx 1$ in case of GA optimised networks (for the least-cost diameter); in this case, the pipe flows under regular conditions, also give indications about the loss of demand in case of pipe failures.

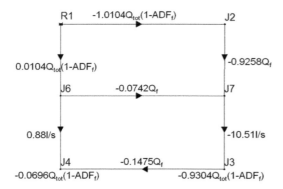

Figure 5.10 Net10, flow increments/decrements showing relation between Q_f and $1 - ADF_f$

5.5   NETWORK BUFFER INDEX

Adopting that the dots positioned on Y-axis of HRD reflect 100% of the network buffer i.e. an ADF index equal to one, while the dots positioned on the diagonal reflect no buffer in the pipes, all other footprints in between these two extremes characterise a certain reliability level that can be correlated to a unique index derived directly from the HRD. This index has been named the *network buffer index* (NBI) and is determined using the analogy between the network buffer and actual loss of demand shown in Figure 5.11.

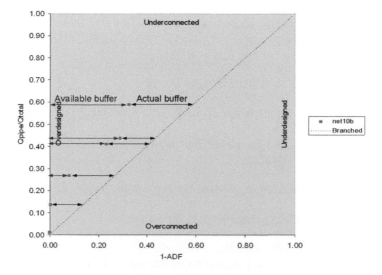

Figure 5.11 Determination of the Network Buffer Index

By adding the weighting proportional to the pipe flows under regular supply conditions, the NBI will be calculated as in Equation 5.17.

$$NBI = \sum_{j=1}^{m} \frac{Q_j/Q_{tot} - (1 - ADF_j)}{Q_j/Q_{tot}} \cdot \frac{Q_j}{\sum_{j=1}^{m} Q_j} = \sum_{j=1}^{m} \left(\frac{Q_j}{Q_{tot}} + ADF_j - 1 \right) \frac{Q_{tot}}{\sum_{j=1}^{m} Q_j} \qquad 5.17$$

For serial- and branched networks, Equation 5.17 yields the NBI value of 0 (after plugging Equation 5.1), while in case of $1 - ADF_j = 0$ for all the pipes, the NBI will equal 1.

Applying Equation 5.17, the networks from Figure 5.5 have the following values of *NBI*: *net10* – 0.050, *net10a* – 0.270, *net10b* – 0.499, *net16* – 0.060, *net16a* – 0.216 and *net16b* – 0.370. The trend of increase is obvious as the diameters have also been increased in *a-* and *b* alternative, whilst the index values for optimised networks are very low: 0.05 and 0.06, respectively, reflecting very low buffer capacity. Furthermore, the indices for *a-* and *b* alternative of *net16* are lower than those of *net10*; based on the pipe connectivity alone, this might look peculiar, but *net16* is supplying much higher demand with lower supplying head than *net10* (155 l/s and 60 msl, against 70 l/s and 90 msl, respectively). Moreover, the GA-optimisation has minimized the connectivity effect on the reliability by forming branched skeletons in both networks. Alternative *a* and *b* also maintain this skeleton as all pipe diameters have been uniformly increased. In fact, the position of the dots in Figure 5.5 suggests the poorer connectivity of *net16* compared to *net10*, which based on the selection of diameters is actually a valid conclusion. All this makes the NBI values fairly logical reflection of the layout and hydraulic performance of both networks.

With additional derivation, Equation 5.17 transforms into a simpler form:

$$NBI = 1 - \sum_{j=1}^{m} (1 - ADF_j) \frac{Q_{tot}}{\sum_{j=1}^{m} Q_j} = 1 - \frac{\sum_{j=1}^{m} (Q_{tot} - Q_{tot,j})}{\sum_{j=1}^{m} Q_j} = 1 - \frac{\sum_{j=1}^{m} \Delta Q_{tot,j}}{\sum_{j=1}^{m} Q_j} \qquad 5.18$$

where $Q_{tot,j}$ is the total demand in the network that is available after the failure of pipe j and $\Delta Q_{tot,j}$ is the corresponding loss of demand.

5.6   COMPARISON OF NBI WITH THE RESILIENCE INDICES

The network buffer index has been compared with two other indices from the literature: the *resilience index*, I_r, of Todini (2000) and its upgrade named the *network resilience*, I_n, which was defined by Prasad and Park (2004). The initial layouts of the networks *net10* and *net16* have been modified by introducing more looped skeletons of secondary mains i.e. reducing the range of used diameters. Furthermore, the supply heads and nodal elevations have been set for both networks at 75 msl (in *R1*) and 45 msl (in all other nodes), respectively. The remaining input data, namely the nodal demands, pipe lengths and k-values are kept the same as in the initial network. Figure 5.12 shows the layout and the hydraulic performance of the adapted *net10* and *net16*.

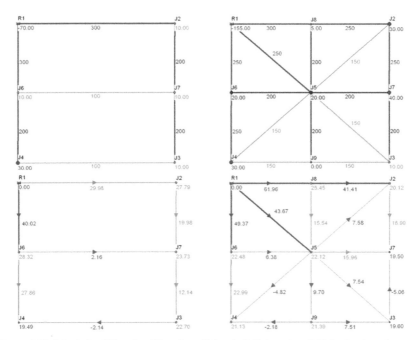

Figure 5.12 Adapted net10 and net16 – pipes: D (mm), Q (l/s), nodes: Q (l/s), p/ρg (mwc)

In order to describe the network resilience, Todini (2000) uses the concept of power balance between (1) the sources of supply, (2) the dissipated amount due to the network hydraulic losses, and (3) the remainder at discharge points. It is valid in all situations that:

$$\sum_{s=1}^{l} Q_s H_s + \sum_{p=1}^{k} Q_p h_p = \sum_{i=1}^{n} Q_i H_i + \sum_{j=1}^{m} Q_j h_{f,j} + \sum_{v=1}^{t} Q_v h_{m.v} \qquad 5.19$$

$H_{s/i}$ in Equation 5.19 indicate the piezometric heads at l sources (which includes all the reservoirs, and tanks that supply the network), and n nodes (which includes all the junctions with discharge, and tanks that are supplied from the network), respectively. The values of $h_{p/f/m}$ include the heads of k pumps, the pipe friction losses of m pipes and the minor losses of t major valves. In all cases, $Q_{s/p/i/j/v}$ stand for corresponding flows supplying the network (s), being conveyed through it (p,j and v), or withdrawn from it (i). Todini's resilience index, I_r, is a measurement of the system's vulnerability of letting some nodes without service in the occurrence of failure, expressed as:

$$I_r = 1 - \frac{P^*_{int}}{P^*_{max}} \qquad 5.20$$

P^*_{int} in Equation 5.20 is the amount of power dissipated in the network to satisfy the total demand, ΣQ_i, under regular operation i.e. at the minimum required pressure, and P^*_{max} is the maximum power that would be dissipated internally during the pipe failure in order to satisfy the constraints of the demand and pressure in the nodes. By substituting Equation 5.19 into 5.20, the I_r can be expressed as:

$$I_r = \frac{\sum_{i=1}^{n} Q_i (H_i - H_i^*)}{\sum_{s=1}^{k} Q_s H_s + \sum_{p=1}^{n} Q_p h_p - \sum_{i=1}^{n} Q_i H_i^* - \sum_{v=1}^{t} Q_v h_{m.v}}$$

5.21

where H^* is the minimum piezometric head required to satisfy the demand in node i, equal to the sum of nodal elevation and PDD threshold pressure .

Prasad and Park (2004) propose more conservative value of the index by introducing the weighting of the nodal power, based on possibly larger discrepancies in diameters of connecting pipes. The multiplier, c_i, is expressed as:

$$c_i = \frac{\sum_{j=1}^{m,i} D_j}{m_i \max\{D_j\}}$$

5.22

where D_j are the diameters of m pipes connected in node i. The corresponding network resilience, I_n, will be calculated as in Equation 5.23.

$$I_n = \frac{\sum_{i=1}^{n} c_i Q_i (H_i - H_i^*)}{\sum_{s=1}^{k} Q_s H_s + \sum_{p=1}^{n} Q_p h_p - \sum_{i=1}^{n} Q_i H_i^* - \sum_{v=1}^{t} Q_v h_{m.v}}$$

5.23

After running the hydraulic simulations, the HRD for the two networks from Figure 5.12 will look as shown in Figure 5.13, with the NBI of 0.234 for *net10*, and 0.660 for *net16*. Unlike was the situation with the layouts *a* and *b* of these networks, the layouts *c* show clearly more buffer and better effect of connectivity from *net16* than in case of *net10*.

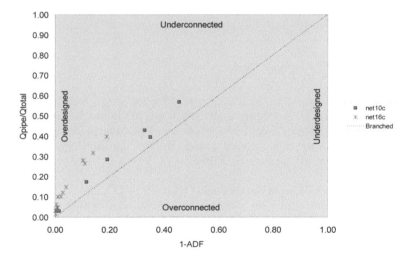

Figure 5.13 HRD for adapted nets 10 and 16 (PDD threshold = 20 mwc)

The values of I_r, I_n, and NBI were further calculated for both networks increasing the pipe diameters gradually and uniformly, for 2 %; fourteen simulation runs have been conducted applying the same increments and the results showing the change of three indices for the *net10* and *net16*, for total diameter increase of 32 %, are given in Figures 5.14 and 5.15.

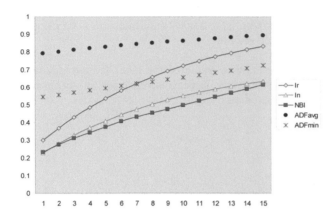

Figure 5.14 Comparison of I_r, I_n and NBI for increased diameters of net10

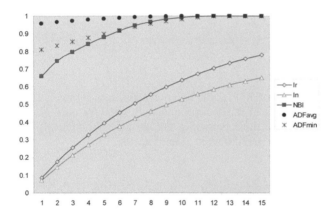

Figure 5.15 Comparison of I_r, I_n and NBI for increased diameters of net16

Another series of simulation runs was done in the same way for the single increase of the demand multiplier of 2 % (total 32 %) and those results are shown in Figures 5.16 and 5.17. Finally, Figures 5.18 and 5.19 show the effect of 2 % increase of PDD threshold pressure. In all the figures, the corresponding average network values of ADF are plot together with ADF_{min}, which is the value after the worst-case pipe failure.

There are four major observations looking at the results in Figures 5.14 to 5.19:
1. All three indices follow the same trend. Nevertheless, the correlation of NBI to the values of ADF is better. This is not a surprise knowing that the ADF is the input parameter for calculation of NBI.
2. All three indices show negative value if the level of service is (significantly) disturbed during regular supply i.e. at no-failure conditions; this can occur in case of insufficient

head at supply points or extreme head loss in the network. Here as well, the I_r, and I_n indicate more radical problem than is the case with NBI (and with ADF_{avg} and ADF_{min}). One clear reason is that the resilience indices are products of demand driven simulations while the NBI is calculated from PDD simulations.

3. The resilience indices evaluate *net16* less favourably than NBI (in Figure 5.15). At the 12th simulation (the total pipe diameter increase of 24 %) the NBI, ADF_{avg}, and ADF_{min} all get value 1.0 i.e. the PDD simulation indicating failures without any loss of demand, whilst the I_r, and I_n for the same situation have values of 0.704 and 0.585, respectively; this is considered low in view of the fact that the system operates without substantial loss of pressure during any pipe failure.

4. Unlike is the case with NBI, the values of I_r, and I_n, still indicate the lower reliability of *net16* compared to *net10*. The likely reason for this is the bigger surplus head of *net10* resulting from lower demand than in the case of *net16*, which can be envisaged comparing the hydraulic performances shown in Figure 5.17. Still, the network performance is better described by considering the value of NBI; the resilience indices seem to be less sensitive to the change of the network layout.

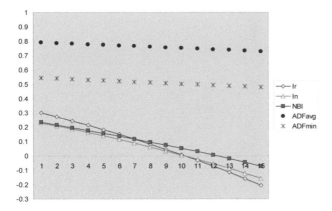

Figure 5.16 Comparison of I_r, I_n and NBI for increased demand of net10

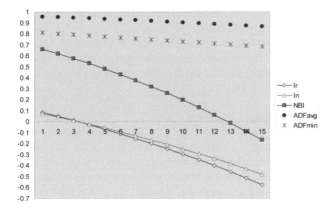

Figure 5.17 Comparison of I_r, I_n and NBI for increased demand of net16

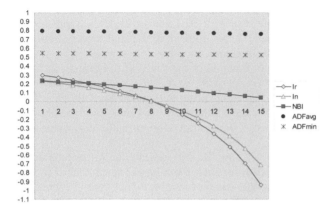

Figure 5.18 Comparison of I_r, I_n and NBI for increased PDD threshold pressure of net10

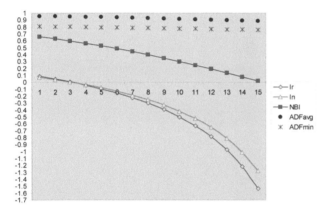

Figure 5.19 Comparison of I_r, I_n and NBI for increased PDD threshold pressure of net16

To be more confident with the last conclusion, another simulation has been run for all 16 optimised networks shown in Figure 5.1a-d and repeated by equalising the geometrical properties of all the networks in the following way: the supply head of 70 msl in R1, the total demand of 90 l/s allocated at *J2*, *J3* and *J4* (30 l/s each), all nodal elevations of 5 msl, all pipe diameters of 250 mm, all distances to the demand nodes of 3000 m (the intermediate nodes being located at the half of that distance, and all the k-values set at 0.5 mm.The PDD threshold pressure was in all cases kept at 20 mwc.

The results of these two series of calculations shown in Figures 5.20 and 5.21 confirm the observations from the previous figures. NBI follows the trend of ADF_{avg} but to a lesser extent in the cases of larger difference between the ADF_{avg} and ADF_{min} values, as is the case of *net7* and *net8* in Figure 5.20. While the lower value of NBI for *net8* signals the problem caused by the failure of major pipe, both resilience indices show higher resilience of this network. At the same time, *net9* has lower values of I_r, and I_n, compared to *net8*, while the NBI (ADF_{avg} and ADF_{min}) are clearly higher in case of *net8*. Both observations lead to a conclusion that the NBI values reflect the implications of the pipe failures more closely than the resilience indices, in this example.

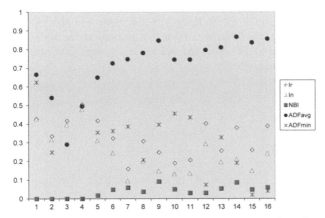

Figure 5.20 Comparison of I_r, I_n and NBI for 16 optimised nets from Figure 5.1a-d

Figure 5.21 shows the result for networks of equalised diameters where these are clearly too small in case of *net3* and *net4*, resulting in negative values of all three indices, and pretty sufficient, actually too big for the looped configurations. As all the nets have the same pipe diameters, there will obviously be no difference between the I_r, and I_n values. In case of networks 5, 7 and 10 to 13, both indices will range between 0.7 and 0.8 for otherwise highly reliable networks; *net12* and *net13* would have the NBI (ADF_{avg} and ADF_{min}) values of 1.0.

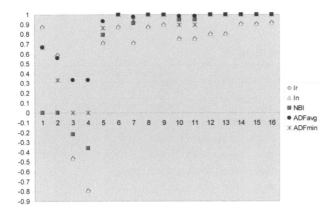

Figure 5.21 Comparison of I_r, I_n and NBI for 16 equalised nets from Figure 5.1a-d

5.7   TEST NETWORK FROM LITERATURE

The discrepancy between the values for NBI, and I_r and I_n, which is visible in Figures 5.15, 5.17 and 5.19, raised some concerns about the correctness of calculation of the two resilience indices in view of somewhat unclear definition of the parameter H^* in Equations 5.20 and 5.22 in the papers of Todini (2000) and Prasad and Park (2004). To be sure, the comparison of all three indices was also done on the two-loop case network, shown in Figure 5.22, which was used in both of these references. Assuming the PDD threshold of 20 mwc, the minimum head found in Table 2 of Prasad and Park (2004) was used to calculate the nodal elevations, which became (in msl): node 1 = 190, node 2 = 160, node 3 = 170, node 4 = 165, node 5 = 160, node 6 = 175, node 7 = 170. The same head of 210 msl was used at the source, and uniform k-value of 0.5 mm was assigned to all the pipes, which is an approximation of the originally used Hazen-Williams roughness factor.

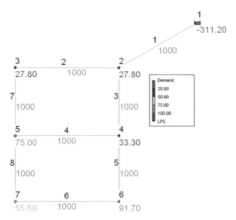

Figure 5.22 Case network from Todini (2000) – pipes: L(m), nodes: Q(l/s)

Eight different configurations of diameters were randomly taken over from Table 3 of the same reference and the results were compared with the calculations done in this research; these are shown in Table 5.3.

Table 5.3 Results comparison with Prasad and Park (2004)

Network Number	Pipe Diameter (mm), from Prasad and Park (2004)								Prasad and Park		Trifunović		
	1	2	3	4	5	6	7	8	I_r	I_n	I_r	I_n	NBI
1	609.6	609.6	609.6	25.4	609.6	25.4	609.6	609.6	0.9002	0.6223	0.8919	0.6168	0.0029
2	609.6	609.6	609.6	25.4	609.6	25.4	609.6	558.8	0.8999	0.6141	0.8917	0.6087	0.0029
6	609.6	609.6	609.6	609.6	609.6	203.2	609.6	609.6	0.9038	0.8007	0.8955	0.7938	0.6110
9	609.6	609.6	609.6	609.6	609.6	101.6	609.6	609.6	0.9036	0.7749	0.8955	0.7682	0.5030
11	609.6	609.6	609.6	609.6	609.6	609.6	609.6	609.6	0.9038	0.9038	0.8957	0.8955	0.6520
15	609.6	609.6	609.6	609.6	609.6	406.4	609.6	609.6	0.9037	0.8523	0.8956	0.8446	0.6506
19	609.6	609.6	558.8	609.6	609.6	609.6	609.6	609.6	0.8989	0.8927	0.8904	0.8842	0.6541
20	609.6	609.6	609.6	558.8	609.6	609.6	609.6	609.6	0.9037	0.8923	0.8956	0.8841	0.6503

The table confirms the assumption of taking the minimum required head H^* as the sum of nodal ground elevation and PDD threshold pressure. Minor differences with the results of Prasad and Park come likely from the conversion of the roughness factor. The table also shows that the NBI values are consistently lower than those of I_r and I_n, which is easy to explain; the source of supply is connected to the rest of the system with single pipe. NBI is significantly lower in case of the network configurations 1 and 2, for the same reason as in case of the networks in Figure 5.1b-d: pipes 4 and 6 are too small and the entire system functions actually as branched network. This again shows that NBI captures the impact of network connectivity better than the resilience indices.

5.8 CASE: WATER DISTRIBUTION NETWORK AMSTERDAM NORTH

To demonstrate the concept on a real case network, the HRD diagram and the corresponding NBI have been determined for the portion of water distribution network of the city of Amsterdam, NL, shown in Figure 5.23.

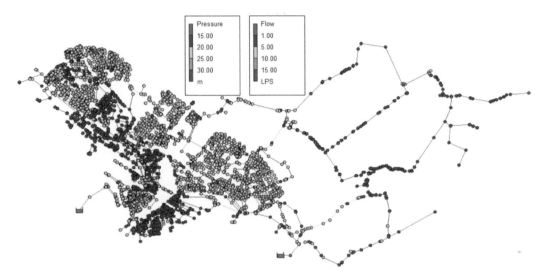

Figure 5.23 Network of Amsterdam North (Courtesy: Waternet, Amsterdam)

The constructed EPANET model consists of 4552 nodes and 5044 pipes and the connection to the rest of the network is simulated by two dummy reservoirs at fixed heads of 25 msl (on the left side) and 21 msl (on the right side). The colours in the figure depict the hydraulic performance of the network for snapshot simulation at the demand of 468 l/s. Most of the pressures under normal operating conditions fall within the range of 15 to 25 mwc, with two nodes below the PDD threshold pressure of 15 mwc (13.83 and 12.30 mwc, which is the lowest pressure in the network, in the node connected to the right dummy reservoir). The pipe flows are relatively low; quite some pipes have the flow rate below 1.0 l/s suggesting quite oversized i.e. over-connected network. This is confirmed looking at the corresponding HRD shown in Figure 5.24.

Therefore, the failure of the majority of pipes shows no loss of demand, whatsoever. The NBI for this network has been calculated at 0.970. The average ADF for the network equals 0.9996, which indicates a highly reliable network but is in fact the result influenced by huge amount of pipes conveying mostly small flows in a network of nearly 500 loops. The value of ADF_{min} is 0.761 with only 28 pipes causing the loss of demand in the network above 1 %. Seven pipes of diameter 762 to 800 mm, connected in series in total length of 429 m create the ADF_{min} after any of them fails. Hence, the network is over-designed, actually. At the same time, the calculation of resilience indices, I_r, and I_n, will yield the values of 0.605 and 0.560, respectively, which is not an order of value immediately suggesting a high reliability.

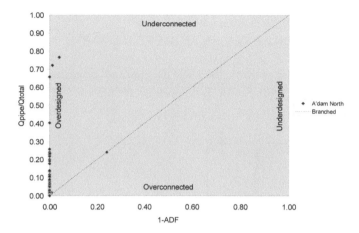

Figure 5.24 HRD of Amsterdam North (Q_{tot} = 467.83 l/s, PDD threshold = 15 mwc)

It has been therefore assumed that the network is capable of conveying higher flows and its limits have been tested in the same way as in case of *net10* and *net16*, but at the PDD threshold pressure of 15 mwc. The results for 15 simulations at the increment of the demand multiplier of 4 % (total 73%) are shown in Figure 5.25. At the same time the HRD of the network has been plot for three levels of demand increase: 32 %, 73 % and 98 %, reflecting the 15 simulations at increment of 2, 4 and 5 %, respectively, which is shown in Figure 5.26. The zoomed-in origin of the diagram in this figure (the results for both Q_{pipe}/Q_{total} and *1 – ADF* in the range of 0 to 0.05) is shown in Figure 5.27.

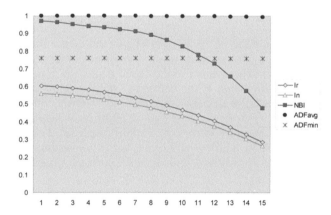

Figure 5.25 Comparison of I_r, I_n and NBI for demand increase of 4 % (total 73%)

The results in Figure 5.25 show again more conservative values of the two resilience indices than is the case with the NBI. It is also visible that all three indices start dropping more substantially in the last few simulation runs. At the same time, an insignificant drop of the values for ADF_{avg} and ADF_{min} is registered for rather large increase in demand. The following are the values in the final simulation run, nr. 15: I_r = 0.286, I_n = 0.263, NBI = 0.479, ADF_{avg} = 0.9938 and ADF_{min} = 0.758. The explanation for this discrepancy can be found looking at the results in Figures 5.26 and 5.27. The applied uniform increase of demand, by changing the demand multiplier alone, creates uniform consequences after the pipe failures. Thus, the ranking of pipes based on their criticality will remain the same except that the demand loss will grow. The visual representation of this phenomenon is that the dots on the HRD start to migrate towards the diagonal of the graph in uniform pattern, as shown in Figure 5.27. With further demand increase, they will gradually enter the area underneath the diagonal indicating the condition in which the pressure in the network is below the PDD threshold i.e. the demand reduction takes place even at no-pipe failure condition. The dots on the other side of the diagonal introduce negative values in the calculation of NBI, which starts affecting, i.e. reducing the index more significantly. This however has no big impact on the reduction of ADF values for the network as the dots located below the diagonal represent in this particular case the pipes conveying marginal flows and therefore inflicting marginal loss of demand, if any, in case of their failure. Moreover, these pipes influence the value of ADF_{avg} in negative way, showing the network more reliable than it really is.

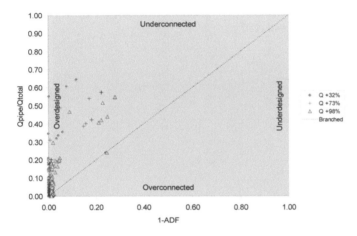

Figure 5.26 HRD of Amsterdam North for increased level of demand

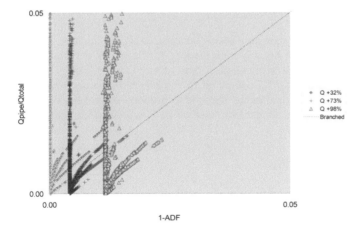

Figure 5.27 Zoom-in of Figure 5.26

Drawing the conclusion about the reliability of the Amsterdam North network, based on the outlook of the corresponding HRD, the network is generally over designed and could function properly with smaller diameters for majority of the pipes. To increase the reliability on the other hand, one should concentrate on the couple of major pipes, alone. Seven of these are the main reason for 24 % of the demand loss, which is the consequence of the worst pipe failure scenario. How far this should be done, is a decision for the water company.

The seven most critical pipes identified by the developed network diagnostics software are (D in mm/L in m): 15003 (800/18), 15004 (800/20), 15005 (800/19), 15006 (762/35), 15007 (769/320), 17824 (800/9) and 17825 (800/8). To demonstrate the implications of the network renovation on the improvement of the three indices, a pipe of 800 mm diameter and 430 m length has been connected between the node 14226, which is the first one (upstream) in the series of seven pipes, and 12444 being the last one (downstream) and the major discharge node of Q = 112.14 l/s. In this case, a parallel pipe has been added to the network strengthening one of its branches towards the area of high demand. The results in Figure 5.28 show the improvement, with following values in the final simulation run, nr. 15: $I_r = 0.288$, $I_n = 0.264$, NBI = 0.500, $ADF_{avg} = 0.9941$ and $ADF_{min} = 0.777$.

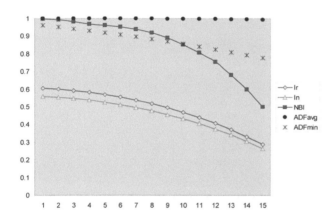

Figure 5.28 Implication of network renovation on I_r, I_n and NBI (total demand incr. 73%)

While the values of I_r and I_n do not change significantly compared to the situation in Figure 5.25, the effect of having parallel pipe on a branch leading to the major demand node is clearly captured in higher value of NBI in the first couple of simulations, being indirectly influenced by much higher value of ADF_{min} compared to the original situation with single branch. This effect is obviously diminished as the demand is growing so that the NBI value towards the final simulation showing the demand increase of 73 % is less different than in the first case.

5.9   CONCLUSIONS

The research presented in this chapter aims at alternative way of expressing the reliability index being derived from a diagram showing the correlation between the pipe flows under the regular supply conditions and the loss of demand caused by their failure. The proposed diagram, called the hydraulic reliability diagram (HRD), considers the reliability of zero in the events of no buffer in the network i.e. the situation when the failure of any pipe will cause the loss of demand, which is the case with serial/branched networks, or GA-optimised looped networks reflecting the least-cost design. The other extreme, the reliability index having the value of one, depicts the situation where no single pipe failure will cause the loss of demand in the network, whatsoever. The reliability index derived from the position of all the dots on the graph, including proportional weighting to the pipes carrying more flow under normal condition, has been proposed as the network buffer index (NBI), which takes only into consideration the consequences of the failure and not the chance that it would happen.

The conducted analysis and the comparisons done between the NBI and two resilience indices from the literature make clear that no easy conclusion can be drawn based on a single index whatever is the value it takes. Moreover, no direct conclusion could have been drawn from the pipe connectivity alone; the pipe diameters and supplying heads play important role for network reliability, too. Hence, one number can hardly describe the complexity of hydraulic performance in irregular supply scenarios, which is determined by both geometric and hydraulic parameters.

Nevertheless, giving a visual i.e. graphical interpretation of the index, in this case by having the NBI value 'explained' with the HRD, gives better idea how this index has been influenced i.e. what is the source of the loss of demand. Moreover, the HRD gives impression about the buffer in the network, possible loss of service level resulting for insufficient supply heads (no pipe failure), and the source of reliability achieved either by increased connectivity, and/or increased pipe diameters, and as such can be seen as useful diagnostics tool.

As far the computational tool, the PDD software developed in C++ using the EPANET toolkit function library for the purpose of network diagnostics and calculation of ADF has proven to be pretty robust and fast. For example, the calculation of ADF values for all the pipes of the Amsterdam North network was taking anything between five and 10 minutes for one demand scenario, depending on the PC that was used for this purpse.

Finally, all the results showed clear correlation between the three indices. Nevertheless, the NBI index appears to be more responsive towards the changes of network configuration i.e. the pipe connectivity and diameters, compared to the resilience indices that are mostly driven by the change of piezometric heads. Those will therefore occasionally show unreasonably low or high values compared to NBI.

REFERENCES

1. Ozger, S.S. (2003). *A semi-pressure-driven approach to reliability assessment of water distribution network.* PhD dissertation, Arizona State University, Tempe, Arizona, USA.
2. Mays, L.W. (2003). *Water supply systems security.* McGraw-Hill.
3. Todini, E. (2000). *Looped water distribution networks design using a resilience index based heuristic approach.* Urban Water, Elsevier, Article No 63, p 1–8.
4. Prasad, T. D., and Park, N.S. (2004). *Multiobjective genetic algorithms for design of water distribution networks.* J. Water Resources Planning and Management. ASCE, 130(1), p 73–82.
5. Savić, D.A., Bicik, J., Morley, M.S., and Keedwell, E.C. (2007). *GANetXL (Evolutionary Optimisation for Microsoft Excel)*, University of Exeter, UK.
6. *optiDesigner software*, developed by OptiWater (http://www.optiwater.com).
7. *Evolving Objects (EO)*, distributed under the GNU Lesser General Public License by SourceForge (http://eodev.sourceforge.net).

CHAPTER 6

Impacts of Node Connectivity on Reliability of Water Distribution Networks[1]

The research presented in this chapter analyses impacts of node connectivity on network reliability by comparing two approaches: a simplified one, which evaluates connectivity in relative terms by taking into consideration two most dominant node categories (in three different ways), and the other one that is based on graph theory. The results of the first approach have been processed by the *network diagnostics tool* (NDT), while NodeXL, an MS Excel add-on template in public domain, has been used for the second approach. Several hydraulic simulations have been conducted on various sets of networks partly generated by the tool discussed in Chapter 4, with fixed or variable pipe diameters and demands. The mutual correlation of the connectivity measures has been analysed, and compared separately for each approach with the reliability measures discussed in Chapter 5, namely the ADF_{avg}, NBI and I_n. The results processed statistically by the Spearman test show limited correlations existing mostly in those layouts that vary significantly in the degree of complexity. This points the fact that node connectivity can hardly be analysed independently from other network parameters, although the concept of network connectivity index as a parameter of more complex reliability formula is recommended for further research.

---

[1] Paper submitted by Trifunović, N., Chobhe, S. and Vairavamoorthy, K., (2012) under the title *Analysis of Network Connectivity and its Impact on Water Distribution Network Resilience*, to the Journal of Water Resources Planning and Management; under review.

6.1   INTRODUCTION

Graph theory is widely used to describe the degree of connectivity in networks of various types: for transport of electricity, vehicles, or water. Node connectivity in those networks will be presented in the form of matrices; the application of these in generation of water distribution networks has been discussed in Chapter 4. In addition, the node connectivity plays a role in network reliability. Obviously, having more pipes means more alternative routes in the event of particular pipe failure. Next to sufficient head at the source(s) combined with sufficient conveying capacity of the pipes, increased node connectivity helps to maintain the guaranteed service levels during irregular supply conditions.

The research related to the node connectivity as an element of water distribution network reliability is mostly dealing with the concepts of *spanning trees* and *minimum cut-sets*. Jain and Gopal (1988) have proposed an algorithm for generation of mutually disjoint spanning trees of the network graph, named as *appended spanning trees* (AST). Each AST represents a probability term in the final global reliability expression. The algorithm calculates the global reliability of the network directly, which can also be terminated at an appropriate stage for an approximate value of global reliability.

Kansal et al. (1995) use the concept of AST to calculate the global network connectivity, which is defined as the probability of the source node being connected with all the demand nodes simultaneously. Since a water distribution network is a 'repairable system', a general expression for pipeline availability using the failure/repair rate is considered. Furthermore, the sensitivity of global reliability estimates due to likely error in the estimation of failure/repair rates of various pipelines is also studied in this research.

More recently, an efficient algorithm for connectivity analysis of moderately sized distribution networks has been suggested by Kansal and Devi (2007). This algorithm, based on generation of all possible minimum system cut-sets, identifies the necessary and sufficient conditions of system failure conditions and is demonstrated with the help of *saturated* and *unsaturated* distribution networks. The computational efficiency of the algorithm is compared to those of AST having the added advantage in generation of system inequalities, which is useful in reliability estimation of *capacitated* networks.

Applications of graph theory and complex network principles in the analysis of vulnerability and robustness of water distribution networks are also investigated by Yazdani and Jeffrey (2010). Several benchmark water networks of different size and configuration, including their vulnerability-related structural properties have been studied in this research. The metrics, grouped as *basic connectivity*, *spectral metrics* and *statistical measurements*, are used to correlate the network structure to the resilience against failures or targeted removal of the nodes and links.

Network- i.e. graph structures are extensively studied by the researchers in other fields. For example, Jamaković and Uhlig (2007) analyse in their work, serving predominantly electrical networks, a relationship between the algebraic connectivity and graph's robustness to the node and link failures. Furthermore, they have studied how the algebraic connectivity is affected by topological changes caused by random node/link removal. The conclusion is that the random node or link removal increases the value of the algebraic connectivity only if the resulting sub-graphs have approximately equal number of nodes and links. On the other hand, random node or link removal will decrease the value of the algebraic connectivity if the resulting sub-graphs have a larger number of nodes than links. In their further topological analyses (2008), the authors compare the relationships among several topological measures such as: the *clustering coefficient*, the *assortativity coefficient* and the *rich-club coefficient*, which is analysed in relation to the average *node coreness*, *distance*, *eccentricity*, *degree* and *betweenness*. The results show various degree of correlation implying redundancy between topological measures. Consequently, a significantly smaller set of topological measures has been proposed to characterize real-world network's structures. The concepts elaborated in this research look potentially applicable to water distribution networks, as well.

6.2 TOOLS FOR ANALYSIS OF NETWORK CONNECTIVITY

Several tools based on graph theory are available for generic analysis of network connectivity, directly or indirectly. The interest for research in network connectivity has significantly grown with the increase in numbers of web users and exponential growth in memberships of various social networks on the Internet. While there are relatively few freely available programs for graph-based analysis, these programs are of generally high quality, and allow wide range of analyses and visualization tasks. In situations where such programs are not sufficient, either because they cannot perform the required analysis or they are not fast enough or user-friendly, there are freely available libraries for several common programming languages. These libraries provide a wide range of graph-based data structures and algorithms that can significantly reduce the coding process. Brief overview of a few tools coming out of the Internet search is given in Table 6.1.

None of these tools has direct application in the field of water distribution. Yet, the parameters they calculate aiming to assess the network connectivity can potentially be used in the assessment of water distribution network reliability, too. The programmes developed for social network analyses enable relatively easy processing of connectivity parameters and visual representation of large networks, which makes them in any case initially attractive for water distribution network analyses. The assumption of applicability is made in spite of the fact that social networks have configurations that can substantially differ from physical networks. For instance, they are predominantly branched and may possess loops formed by a self-connected node.

NodeXL was selected for this study being easily accessible and transparent for use. This MS Excel add-on, developed by Social Media Research Foundation with contributions from the group of scientists is primarily a support to social network analyses; more information on that can be found in Smith et al. (2009). The workbook template includes multiple worksheets where all the information needed to represent a network/graph can be stored. Network connections (i.e. graph edges) are represented as an 'edge list'. Other worksheets contain information about each vertex (i.e. node) and cluster. MS Excel graphical interface is used to represent network layouts with various map data attributes; this also includes filtering, clustering, and customized mapping of vertex and edge-level data. The tool can easily handle networks of several thousand vertices/edges. Figure 6.1 gives a screen snap-shot of it.

Table 6.1 Tools for analysis of network connectivity (adapted from http://www.itee.uq.edu.au/)

No.	Name	Type	License	Details
1	Node XL	MS Excel 2007 template	Free	Open source add-on to the widely used spreadsheet programme; provides a range of basic network analysis and visualization features; used mostly for social network analysis; user-friendly and easy to install; no specific computer skills required.
	http://nodexl.codeplex.com/			
2	Pajek	Programme	Free	General graph analysis tool providing excellent range of metrics including tools using clusters, cliques, components, various partitioning schemes, general graph properties such as diameter and radius, node degrees and many other features; can visualise large graphs; has a large user community in the field of social network analysis; can be unintuitive to use.
	http://vlado.fmf.uni-lj.si/pub/networks/pajek/ http://vlado.fmf.uni-lj.si/pub/networks/pajek/apply.htm			
3	UCINET	Programme	Commercial	Social network analysis tool providing a variety of metrics in a user-friendly manner; more user-friendly and less capable than Pajek; makes heavy use of terminology from social network analysis; free licence for demo version available for 30 days.
	http://www.analytictech.com/ucinet.htm			
4	Matlab	Programme /Library	Commercial	Matrix-based development tool; not designed to work directly with graph-based representations but capable of efficiently and simply dealing with connectivity matrix representations of graphs; tool boxes provided by user community contain algorithms such as minimum spanning tree, minimum vertex cover and all-pairs shortest path, as well as those for manipulation and plotting of graphs; recommended only for already familiar Matlab users.
	http://www.mathworks.com/ http://www.mathworks.com/matlabcentral/			
5	Boost Graph Libraries	Library	Free and open source (BSD-like)	C++ graph library providing graph data structures and many common graph algorithms; developed by extensive use of C++ template functionality; extending the libraries possible but difficult for programmers without significant experience in C++ templates.
	http://www.boost.org/			
6	Graph Template Libraries	Library	Free only for academic use	C++ graph library providing graph data structures and some common graph algorithms; graph data structures integrate well with the C++ Standard Template Libraries; limited functionality compared to the Boost Graph Libraries; commercial use of the library is prohibited.
	http://www.infosun.fmi.uni-passau.de/GTL/index.html			
7	JUNG	Library	Free and open source (BSD)	Java graph library providing multiple graph data structures, many common graph algorithms and visualisation tools; easily extensible through standard Java mechanisms; imports and exports Pajek data files.
	http://jung.sourceforge.net/			

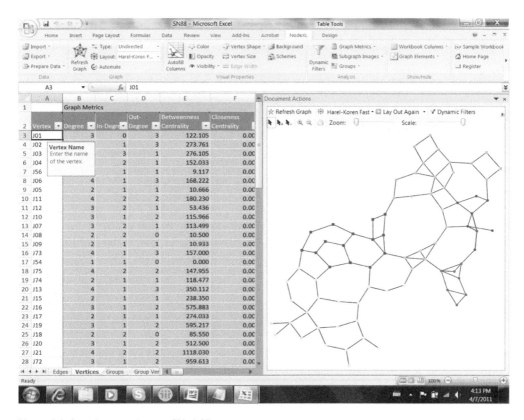

Figure 6.1 Sample screen layout of NodeXL

6.3   GEOMETRIC PROPERTIES AS INDICATORS OF NETWORK RELIABILITY

Hydraulic performance of water distribution networks results from combination of factors emerging from their geometric properties on one side, and supplying heads and distribution of demands, on the other side. The required level of service defined by the minimum pressure that can guarantee 100 % of the design demand determines the room for loss of energy inflicted by the network resistance. This resistance will mostly be controlled by appropriate selection of pipe diameters; the lengths and routes of pipes will in reality be predefined by the greed of streets and roads. Nevertheless, the connectivity of network nodes will play a role, especially in irregular supply scenarios. More pipes connected to the discharge points will cause the reduction of flow rates increasing in that way the buffer in the system.

6.3.1 Network Configuration Assessment

Water distribution network is configured by selected number of nodes and links. In branched configurations, the number of branches is equal to the number of nodes in which three or more links are connected. In case of loops made of n nodes and m links, the total number of loops l will be calculated as:

$$l = m - n + 1 \qquad\qquad\qquad 6.1$$

These are basic loops which, if mutually connected, also merge into larger loops, increasing the number of alternative routes for water flow in case of pipe failure. For adjacent loops, the total number of network loops L_n, which includes both the basic loops and complex loops combined from the basic loops, will be:

$$L_n = l(l-1) + 1 = (m-n)^2 + (m-n) + 1 \qquad\qquad 6.2$$

Equations 6.1 and 6.2 are also valid if pipes cross each other without having a connection, which is rare but possible case in reality.

Thus, the number of loops will be unique for specified value of $a = m - n$. A *network grid index* (NGI) can be defined taking into consideration the ratio between the number of basic loops and the total number of loops, assuming that the basic loops are adjacent or are at least sharing one node.

$$NGI = 1 - \frac{l}{L_n} = \frac{a^2}{a^2 + a + 1} \qquad\qquad 6.3$$

Equation 6.1 is correct for any network configuration. In case of serial- and branched configurations, $a = -1$. This makes the value of L_n in Equation 6.2 equal to 1, and consequently the *NGI* in Equation 6.3 is also equal to 1. For looped configurations, $l > 0$, and *NGI* takes the values greater than 0 and lower than 1. To arrive at practical range, the *NGI* for serial- and branched configurations is set at 0 instead of 1, to indicate the lowest connectivity; this is also done in view of the fact that the result $L_n = 1$ leads to a false conclusion because those configurations have no loops. Hence, the higher value of *NGI* should indicate the networks of higher degree of connectivity, in general.

Parameter a denotes a unique value of *NGI* that can be achieved for various combinations of m and n. For example, either for $n = 999$ and $m = 1001$, or for $n = 5$ and $m = 7$, there will be three basic loops, although not necessarily adjacent in case of larger networks. Due to the difference in size, these two networks will never be compared for any viable reason. Still, there is a problem that selected combination of m and n offers more than one possible layout evaluated by the same value of *NGI*. Four networks shown in Figure 6.2, illustrate this point.

In all four cases, $m = 13$, $n = 11$, $l = 3$, $a = 2$, $L_n = 7$, and $NGI = 0.571$. Also, all four networks have total 26 connections to the nodes, which are distributed in different way. In case of network L1, 11 nodes are connected in the following scheme: two nodes with one connection (R1 and N11), three nodes with two connections (N4, N5 and N10) and the remaining six nodes with three connections (N2, N3, N6, N7, N8 and N9).

Reconnecting pipe P7 to node N11, instead of N8, yields configuration L2 (upper-right in the figure); as a result, the supply reliability of N11 is likely to improve being now connected from two sides. Thus, the connectivity scheme has changed, with only the source in R1 having one connection, five nodes having two connections and five nodes having three connections. Alternatively, reconnecting P7 with N6 would result in the connectivity scheme L3 (lower-left in the figure) consisting of two nodes with one connection, four with two, four with three and one with four connections (node N6). Although in any fairly developed network in practice, especially in urban setup, the number of loops will overwhelm the number of branches, describing the network shape with *NGI* value alone will not be entirely correct approach to express the network connectivity.

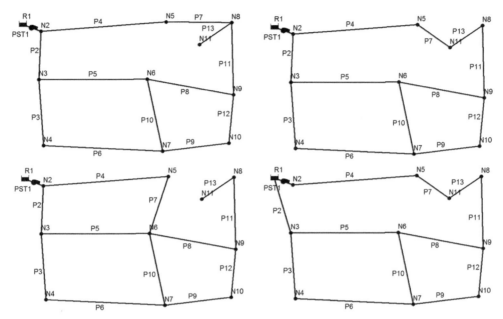

Figure 6.2 Network layouts L1 (upper-left) to L4 (bottom-right) of 13 links and 11 nodes

The following can be written for network layout of *m* links and *n* nodes:

$$\sum_{i=1}^{max\,con} in_i = 2m \quad ; \quad \sum_{i=1}^{max\,con} n_i = n \qquad\qquad 6.4$$

where *i* is the number of connections to a node, n_i the number of nodes with *i* connections and max *con* is the maximum number of connections to at least one node in the system. This value is influenced by the network connectivity itself; for fixed values of *m* and *n*, more nodes with lower connectivity will mean fewer nodes with higher connectivity i.e. a wider range of nodal connectivity. In the same way, aiming at higher value of max *con* will not increase the reliability but rather the number of lower connected nodes.

Hence, all four layouts in Figure 6.2 have three basic loops but L4 is the only fully looped layout; L2 is also fully looped but connected to the source with single link, while L1 and L3 have one dead-end node, on top of the single connection to the source. The network reliability will of course depend on the selection of pipe diameters, lengths and roughness values, but purely on the pipe connectivity i.e. by keeping all other network parameters fixed, the initial guess about the network reliability would possibly lead towards the following ranking: L3 (the lowest reliability), L1, L2 and L4 (the highest reliability). The above described nodal connectivity is summarised in Table 6.2.

Table 6.2 Connectivity of network layouts from Figure 6.2

No. conn.	1	2	3	4	2m	n
L1	2	3	6	0	26	11
L2	1	5	5	0	26	11
L3	2	4	4	1	26	11
L4	0	7	4	0	26	11

The table shows the dominant number of node connections to be two and three; the number of corresponding nodes is shown in the shaded cells. At the same time, the average connectivity of all four layouts is 2.36 connections per node $(2m/n)$. As both the figure and the table show, the fully looped network (L4) will consist only of the nodes with two or three connections, the average value of 2.36 suggesting more of those with two than of those with three. Reconnecting the nodes in a way that would create nodes with less than two or more than three connections will inevitably lead to a configuration combined of loops and branches, as is the case of L1 to L3.

6.3.2 Measures of Network Connectivity

Very rarely in reality, the nodal connectivity will result in the max *con* value above 4. It can therefore be assumed that the network connectivity is sufficiently reflected by analysing the two most dominant connectivity categories. Based on this hypothesis, the measure named the *network connectivity factor* (NCF) is formulated to arrive at the description of network shape:

$$NCF = \frac{1}{n}\left(i_{max,1}n_{max,1} + i_{max,2}n_{max,2}\right)$$

6.5

There are three possible ways to look at the values of the two connectivity categories represented by indices *max,1* and *max,2* in Equation 6.5:
1. by taking into consideration the two largest numbers of nodes *n* with connectivity *i*. In case of two categories with equal number of nodes, being the second largest, the one with higher *i* would qualify. This approach bases the value of NCF_1 on dominant connectivity categories in terms of maximum number of nodes per category.
2. by taking into consideration the largest product between the number of nodes *n* and their corresponding connectivity *i*; this approach bases the value of NCF_2 on dominant connectivity categories in terms of maximum number of connections per category.
3. by taking into consideration the number of nodes *n* in the two connectivity categories *i* closest to the average network connectivity $(2m/n)$; this approach bases the value of NCF_3 on dominant connectivity categories in terms of average number of connections.

Independent from the approach, the results of calculations for the layouts L1 to L4 will be the same; they are shown in Table 6.3. The obtained NCF values show the same ranking of the networks coming from the visual perception of their reliability.

Table 6.3 NCF values for network layouts from Figure 6.2

No. conn.	1	2	3	4	NCF
L1	0.18	0.55	1.64	0.00	2.18
L2	0.09	0.91	1.36	0.00	2.27
L3	0.18	0.73	1.09	0.36	1.82
L4	0.00	1.27	1.09	0.00	2.36

The other extreme to the fully looped layout L4 is to have a node with ten connections, six nodes with two connections and four nodes with one connection; this layout (L5) is shown in Figure 6.3, on the left. Reconnecting P8 of this layout to N9, instead of N8, creates layout L6 and making the connection between N7 and N10 with P8, instead of N8 and N11, creates layout L7. The connectivity of layouts L5 to L7 using the *NCF* values is summarised in Tables 6.4 and 6.5, respectively. Based on the applied approach, the results will differ but show the same trend in all three cases.

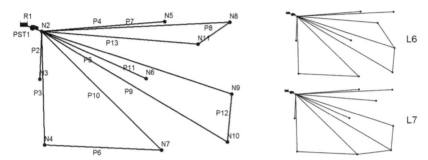

Figure 6.3 Network layouts L5-L7 of 13 links and 11 nodes

Table 6.4 Connectivity of network layouts L5 to L7

No. conn.	1	2	3	10	2m	n
L5	4	6	0	1	26	11
L6	5	4	1	1	26	11
L7	6	2	2	1	26	11

Table 6.5a NCF$_1$ values, based on two largest numbers of nodes

No. conn.	1	2	3	4	NCF
L5	0.36	1.09	0.00	0.91	1.45
L6	0.45	0.73	0.27	0.91	1.18
L7	0.55	0.36	0.55	0.91	1.09

Table 6.5b NCF$_2$ values, based on two largest values of *in*

No. conn.	1	2	3	4	NCF
L5	0.36	1.09	0.00	0.91	2.00
L6	0.45	0.73	0.27	0.91	1.64
L7	0.55	0.36	0.55	0.91	1.46

Table 6.5c NCF$_3$ values, based on average nodal connectivity

No. conn.	1	2	3	4	NCF
L5	0.36	1.09	0.00	0.91	1.09
L6	0.45	0.73	0.27	0.91	1.00
L7	0.55	0.36	0.55	0.91	0.91

The *NCF* values in Tables 6.5a-c indicate the layout L7 with three adjacent loops to be the least connected, which is a fair assumption in view of the fact that (a) this layout has the largest number of nodes with single connection (6), and (b) the node N2 connecting all three loops of L5, enables combination of these loops into complex loops i.e. the connectivity of the nodes is not affected by the fact that these loops are not adjacent because they share the same connecting node. Theoretical limit of max *con* for the worst connected network will be $n - 1$. In that case, for l adjacent loops, $l - 1$ nodes will have three connections, two nodes will have two connections, one node will have max *con* connections and $(n - (l - 1) - 2 - 1)$ node will be with single connection. Hence:

$$\max con = 2m - (l-1) \times 3 - 2 \times 2 - (n-l+1-3) \times 1 = n-1 \qquad 6.6$$

All serial networks have two nodes with single connection and the rest with two connections, leading to the *NCF* value calculated as in Equation 6.7.

$$NCF = \frac{1}{n}[2 + 2(n-2)] = 2 - \frac{2}{n} \qquad 6.7$$

The lowest possible *NCF* = 1, for two nodes and one link, will be increasing for increased number of nodes/pipes, actually resembling the increase of reliability, too.

Branched networks will have the value of max *con* ≥ 2 while its theoretical maximum value is also $n - 1$, as for the looped networks. The values of *NCF* for branched networks will regularly be lower than 2, as is the case with serial networks. Looped and combined networks will have the *NCF* value regularly higher than 2, except for the extreme cases of too high value of max *con* not typical in practice.

Taking the average value of *NCF* as:

$$NCF_{avg} = \frac{2m}{n} \qquad 6.8$$

makes possible to express the *network connectivity index* (NCI) within the range of 0 to 1, by combining Equations 6.5 and 6.8:

$$NCI = \frac{NCF}{NCF_{avg}} = \frac{1}{2m}\left(i_{max,1}n_{max,1} + i_{max,2}n_{max,2}\right) \qquad 6.9$$

Equation 6.9 calculates the degree of connectivity for fixed m and n yielding the value of 1 for optimal connectivity. For instance, by comparing Equations 6.7 and 6.8, *NCF* will equal NCF_{avg}, giving the $NCI = 1$ for all serial networks; this result is justified because there is only one possible connectivity scheme. Fully looped networks will also have the *NCI* value of 1, indicating the optimal connectivity, while branched and combined networks will have the value between 0 and 1, based on the value of max *con* and selected connectivity scheme.

Finally, combining the *NGI* and *NCI*, the *network shape index* (NSI) expresses the network connectivity taking into consideration combined effect of loops and branches:

$$NSI = NGI \times NCI \qquad 6.10$$

Based on the assumed *NGI* = 0, the *NSI* value for serial and branched networks will be 0 in all cases and will be approaching the value of 1 for increasingly complex looped networks.

The results of all calculations assessing the configuration of networks L1 to L7 are summarised in Table 6.6 ($NCF_{avg} = 2.36$).

Table 6.6 Configuration assessment of network layouts L1 to L7

No. conn.	NGI	NCF1	NCF2	NCF3	NCI1	NCI2	NCI3	NSI1	NSI2	NSI3
L1	0.571	2.18	2.18	2.18	0.923	0.923	0.923	0.527	0.527	0.527
L2	0.571	2.27	2.27	2.27	0.962	0.962	0.962	0.549	0.549	0.549
L3	0.571	1.82	1.82	1.82	0.769	0.769	0.769	0.440	0.440	0.440
L4	0.571	2.36	2.36	2.36	1.000	1.000	1.000	0.571	0.571	0.571
L5	0.571	1.45	2	1.09	0.613	0.846	0.461	0.351	0.484	0.264
L6	0.571	1.18	1.64	1	0.499	0.694	0.423	0.285	0.396	0.242
L7	0.571	1.09	1.46	0.91	0.461	0.618	0.385	0.264	0.353	0.220

6.3.3 Network Diagnostics Tool

To be able to check the correlations of the proposed indices and factors with network reliability, the *network diagnostics tool* (NDT) has been developed in C++ programming language by using EPANET toolkit functions, which can process unlimited number of large networks (each up to 10,000 nodes/links) in single simulation run, assess their geometry and hydraulic performance, eventually draw the energy/pressure balance and correlate these to the loss of demand under failure conditions.

The input for the programme is a network file or group of files created in *.inp* format in EPANET, the latter including a text file listing the total number of INP-files and their names. In addition, a uniform PDD threshold pressure and exponent are specified for PDD calculations in which the loss of demand is calculated with or without pipe failure. Furthermore, each network is assessed on the following list of parameters:
1. number of nodes and links, (n and m, respectively),
2. number of basic loops and complex loops (l and L_n, respectively),
3. the network grid index (NGI),
4. the network connectivity factor and index (NCF, NCI), in all three versions,
5. he network shape index (NSI), in all three versions,
6. total pipe length and volume (L_{tot} and V_{tot}, respectively),
7. total demand in demand driven mode of hydraulic simulation (DD),
8. minimum pressure and corresponding node ID (in DD mode),
9. maximum pressure and corresponding node ID (in DD mode),
10. the resilience index of Todini (I_r) in DD mode,
11. the network resilience of Prasad and Park (I_n) in DD mode,
12. the network buffer index proposed in Chapter 5 (NBI) in PDD mode,
13. the average available demand fraction (ADF_{avg}) calculated by failing the pipes consecutively,
14. the minimum available demand fraction (ADF_{min}) and corresponding pipe ID causing the highest loss of demand.

In the single network calculation, additional table will include complete results of PDD hydraulic analysis per pipe, which includes the calculation of ADF_j. The pipes will be counted based on the loss of demand they cause, in 10 categories, at increments of 10%. Furthermore, it is possible to do multiple calculations on a single network using fixed or randomized (range of) multipliers for major hydraulic parameters such as nodal discharge, pipe diameters, or PDD threshold. All results are saved in the text file readable by MS Excel, which enables easy graphical presentation.

6.4 ANALYSIS OF NETWORK CONNECTIVITY BASED ON GRAPH THEORY

As mentioned in Chapter 4, in the terminology of graph theory, water distribution networks consisting of nodes and links are actually *graphs* composed of *vertices* and *edges*, respectively. Equally, the degree of each vertex, showing the number of corresponding edges, will mean the number of pipe connections in each node. Graph will be *directed* if the edges have fixed direction (for instance, pipes of fixed flow/velocity direction). In this case the *degree* of vertex can be in- or out-degree, showing the number of edges directed in or out of the vertex, respectively (i.e. the pipes with inflow or outflow to the node). If the edges have arbitrary direction, the graph is *undirected*.

Besides the vertex degree, several other parameters exist in the analysis of graph structures i.e. network configurations, which can be used to investigate particular structural characteristics. As most of these can be measured in different ways, the final decision on appropriate parameters may depend on the specific domain of the networks and can even be a result of trial and error process. The list of the most common connectivity terminology and related aspects of network configuration is given in Table 6.7, followed by short description of the parameters specifically analysed in this research. The terminology typical for water distribution field is further extensively used.

Table 6.7 Connectivity terminology and network configuration (adapted from http://www.itee.uq.edu.au/)

Parameter	Aspects of network configuration
Degree (in- and out-)	Shows importance of a node, based on how connected it is.
Degree distribution	Set of related properties, such as average shortest path length; indicates probability of creating disconnected components through node/link removal.
Shortest path length	Distance between two nodes; shows degree of mutual influence.
Clique	Identifies highly interconnected sub-networks.
Cluster	Identifies highly connected nodes.
Pivots, Cut-points	Identifies nodes crucial to keep the network connected.
Bridges	Identifies links crucial to keep the network connected.
Node-connectivity, Link-connectivity	Indicates buffer the network layout has before becoming disconnected.
Centralization	Indicates how much the network centres on a single node or group of nodes.
Betweenness centrality, Closeness centrality	Shows importance of a node based on its relationship to other nodes in the network.

6.4.1 Node Degree

In network of n nodes, the node degree, Deg_i, indicates the number of connections to node i. If the network links are directed, in- and out-degree will show the number of connections to- and from the node, respectively. The average nodal degree has the same meaning as the

NCF_{avg} in Equation 6.8. In normal situations, all nodes are connected and the minimum value of Deg_i is 1 while the maximum one is $n-1$ (i.e. equal to max con in Equation 6.6). Nodes can however have the values of $Deg_{In,i}$ and $Deg_{Out,i}$ eqaul to 0. Finally, the *degree centrality* of node i, $C_{D,i}$, will be calculated as:

$$C_{D,i} = \frac{Deg_i}{n-1} \qquad\qquad 6.11$$

Figure 6.4 shows a simple network and the corresponding values of above-discussed parameters. The average node degree of this network is 2.

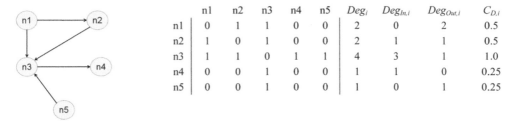

	n1	n2	n3	n4	n5	Deg_i	$Deg_{In,i}$	$Deg_{Out,i}$	$C_{D,i}$
n1	0	1	1	0	0	2	0	2	0.5
n2	1	0	1	0	0	2	1	1	0.5
n3	1	1	0	1	1	4	3	1	1.0
n4	0	0	1	0	0	1	1	0	0.25
n5	0	0	1	0	0	1	0	1	0.25

Figure 6.4 Node connectivity matrix, degree and degree centrality

6.4.2 Graph Density

In an undirected network of n nodes that are all connected to each other, the (maximum) number of links, m_{max} will be calculated as:

$$m_{max} = \frac{n(n-1)}{2} \qquad\qquad 6.12$$

The graph density, GD, as the ratio between the actual number- and maximum number of links, indicates the level of interconnectivity amongst the nodes. Hence:

$$GD = \frac{2m}{n(n-1)} \qquad\qquad 6.13$$

For the network from Figure 6.4, m_{max} equals 10 and consequently, $GD = 0.5$. The value of m_{max} for directed network will double, in which case $GD = 0.25$.

6.4.3 Geodesic Distance and Diameter

The geodesic distance, $d_{i,j}$, between any two nodes i and j is determined by the number of links in a shortest possible path connecting these nodes. The pair of nodes with maximum geodesic distance (MGD) defines the *geodesic diameter*. The *average geodesic distance* (AGD) for n nodes is determined as:

$$AGD = \frac{\sum_{i,j} d_{i,j}}{2n}$$

6.14

The network in Figure 6.5 shows the geodesic diameter of undirected network of five nodes. The total number of shortest paths between any node pair i and j, $P_{i,j} = 1$, in this case. There are total ten unique pairs of nodes and consequently ten shortest paths, namely: $n1n2$, $n1n3$, $n1n3n4$, $n1n3n5$, $n2n3$, $n2n3n4$, $n2n3n5$, $n3n4$, $n3n5$, $n4n3n5$. Five of these paths have one link while the other five are composed of two links. Hence, the value of $AGD = 1.5$, and $MGD = 2$.

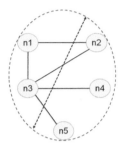

Figure 6.5 Geodesic distances and diameter

If two nodes cannot be connected, their geodesic distance is conventionally assumed to be infinite.

6.4.4 Betweenness Centrality

The betweenness centrality (BC) of a particular node analyses in how many shortest paths the node is present, being a conveyor of (important) information from/to other nodes. For node k, the BC_k will be calculated as:

$$BC_k = \sum_{i,j} \frac{P_{i,k,j}}{P_{i,j}} \quad ; \quad i \neq j \neq k$$

6.15

In the network in Figure 6.5, five of the ten shortest paths pass though node $n3$. As already mentioned, all ten pairs of nodes have one shortest path each, and also $P_{i,k,j} = 1$ for k being node $n3$. Thus, $BC_{n3} = 5$. No other node is on any other shortest path; hence, their betweenness centrality will equal 0 (also in case of nodes $n1$ and $n2$, because $n2n1n3$ and $n1n2n3$ are not the shortest paths between $n2$ and $n3$, and $n1$ and $n3$, respectively).

6.4.5 Closeness Centrality

The closeness centrality, Cc_i, of particular node i is a measure of its average distance (along the shortest path) to all other nodes. Hence:

$$Cc_i = \frac{n-1}{\sum_j d_{i,j}}$$ 6.16

The results for the network in Figure 6.5 are shown in Figure 6.6. $\sum d_{i,j}$ indicates the total number of links on the shortest path between node i and all other nodes.

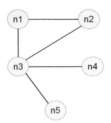

	n1	n2	n3	n4	n5	$\sum d_{i,j}$	Cc_i
n1	0	1	1	0	0	6	0.67
n2	1	0	1	0	0	6	0.67
n3	1	1	0	1	1	4	1.0
n4	0	0	1	0	0	7	0.57
n5	0	0	1	0	0	7	0.57

Figure 6.6 Node connectivity matrix and closeness centrality

6.4.6 Clustering Coefficient

The clustering coefficient, Cp_i, measures how the neighboring nodes of node i are connected amongst themselves. It is expressed as the ratio of the actual number of their interconnecting links and the total possible number of these links that can be calculated in the same was as in Equation 6.12. Hence, for k_i neighboring nodes of node i, which are mutually connected with m_k links in an undirected network:

$$Cp_i = \frac{2m_k}{k_i(k_i-1)}$$ 6.17

The results for the network in Figure 6.6 are shown in Figure 6.7. Node $n3$ has four neighboring nodes mutually connected with only connection, whilst the maximum number of their connections is six.

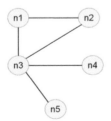

	n1	n2	n3	n4	n5	k_i	m_k	Cp_i
n1	0	1	1	0	0	2	1	1.0
n2	1	0	1	0	0	2	1	1.0
n3	1	1	0	1	1	4	1	0.17
n4	0	0	1	0	0	1	0	0
n5	0	0	1	0	0	1	0	0

Figure 6.7 Node connectivity matrix and clustering coefficient

In fact, the clustering coefficient defines the probability that two neighbouring nodes of node i are connected to each other, which enhances the connectivity in case the node i is removed from the system. This is related to the concept of cut-points and bridges. A node is a *cut-point* if its removal disintegrates the network i.e. creates sub-networks without mutual connection.

A link that creates the same effect if being removed from the system is called a *bridge*. Both cases are illustrated in Figure 6.8, in red colour. The figure also shows that removal of any other node or link will maintain the network integrity.

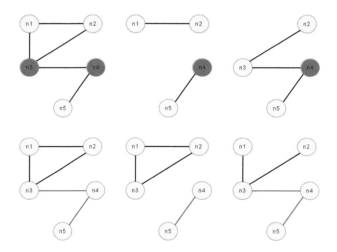

Figure 6.8 Cut-points and bridges

6.5   SIMULATION RUNS, CASE 16 NETWORKS

The first run of simulations was done with the set of 16 networks discussed in Chapter 5, schematised here in Figure 6.9. The layouts with gradual increase of complexity have been constructed to enable easier monitoring of correlation with the connectivity measures.

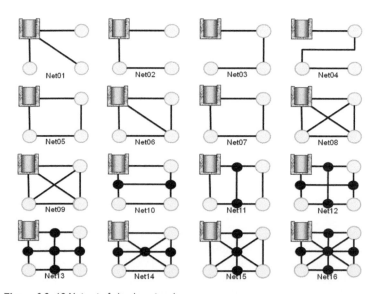

Figure 6.9 16 Net set of simple networks

The results of calculation using NDT are shown in Table 6.8 and those from NodeXL in Table 6.9.

Table 6.8 Connectivity indices calculated by NDT - case 16 nets

Epanet	n	m	l	Ln	NGI	NCF1	NCF2	NCF3	NCF$_{avg}$	NCI1	NCI2	NCI3	NSI1	NSI2	NSI3
Net01	4	3	0	0	0.00	1.50	1.50	0.75	1.50	1.00	1.00	0.50	0.00	0.00	0.00
Net02	4	3	0	0	0.00	1.50	1.50	1.50	1.50	1.00	1.00	1.00	0.00	0.00	0.00
Net03	4	3	0	0	0.00	1.50	1.50	1.50	1.50	1.00	1.00	1.00	0.00	0.00	0.00
Net04	9	8	0	0	0.00	1.78	1.78	1.78	1.78	1.00	1.00	1.00	0.00	0.00	0.00
Net05	4	4	1	1	0.00	2.00	2.00	2.00	2.00	1.00	1.00	1.00	0.00	0.00	0.00
Net06	4	5	2	3	0.33	2.50	2.50	2.50	2.50	1.00	1.00	1.00	0.33	0.33	0.33
Net07	4	5	2	3	0.33	2.50	2.50	2.50	2.50	1.00	1.00	1.00	0.33	0.33	0.33
Net08	4	6	3	7	0.57	3.00	3.00	3.00	3.00	1.00	1.00	1.00	0.57	0.57	0.57
Net09	5	8	4	13	0.69	3.20	3.20	3.20	3.20	1.00	1.00	1.00	0.69	0.69	0.69
Net10	6	7	2	3	0.33	2.33	2.33	2.33	2.33	1.00	1.00	1.00	0.33	0.33	0.33
Net11	6	7	2	3	0.33	2.33	2.33	2.33	2.33	1.00	1.00	1.00	0.33	0.33	0.33
Net12	8	10	3	7	0.57	2.50	2.50	2.50	2.50	1.00	1.00	1.00	0.57	0.57	0.57
Net13	9	12	4	13	0.69	2.22	2.22	2.22	2.67	0.83	0.83	0.83	0.58	0.58	0.58
Net14	7	12	6	31	0.81	3.43	3.43	2.57	3.43	1.00	1.00	0.75	0.81	0.81	0.60
Net15	7	12	6	31	0.81	3.43	3.43	2.57	3.43	1.00	1.00	0.75	0.81	0.81	0.60
Net16	9	16	8	57	0.86	3.56	3.56	2.67	3.56	1.00	1.00	0.75	0.86	0.86	0.64

Table 6.9 Connectivity indices calculated by NodeXL - case 16 nets

Epanet	n	m	AGD	GD	Deg$_{min}$	Deg$_{max}$	Deg$_{avg}$	BC$_{avg}$	Cc$_{avg}$	Cp$_{avg}$
Net02	4	3	1.25	0.5	1	2	1.5	1	0.21	0
Net03	4	3	1.25	0.5	1	2	1.5	1	0.21	0
Net01	4	3	1.13	0.5	1	3	1.5	0.75	0.23	0
Net05	4	4	1	0.67	2	2	2	0.5	0.25	0
Net06	4	5	0.88	0.83	2	3	2.5	0.25	0.29	0.83
Net07	4	5	0.88	0.83	2	3	2.5	0.25	0.29	0.83
Net08	4	6	0.75	1	3	3	3	0	0.33	1
Net09	5	8	0.96	0.8	3	4	3.2	0.4	0.21	0.67
Net10	6	7	1.39	0.47	2	3	2.33	1.67	0.12	0
Net11	6	7	1.39	0.47	2	3	2.33	1.67	0.12	0
Net14	7	12	1.22	0.57	3	6	3.43	1.29	0.12	0.63
Net15	7	12	1.22	0.57	3	6	3.43	1.29	0.12	0.63
Net12	8	10	1.5	0.36	2	3	2.5	2.5	0.08	0
Net04	9	8	2.96	0.22	1	2	1.78	9.33	0.04	0
Net13	9	12	1.78	0.33	2	4	2.67	4	0.06	0
Net16	9	16	1.38	0.44	3	8	3.56	2.22	0.08	0.62

The first conclusion analysing the results is that in neither of the two tables the indices follow the trend corresponding to the order of the networks i.e. their growth of complexity anticipated after observing Figure 6.9. Those are actually related to the clusters of networks with fixed number of nodes following the trend of increase in number of links. For this reason are the results in Table 6.9 sorted per number of nodes and then links, while in Table 6.8, the clusters of networks of 4 and 9 nodes are shown in different shades. Looking at the results from this perspective, the increased node connectivity within a cluster of networks with fixed number of nodes will be reflected in the following trends of the indices shown in Table 6.8:

1. Network grid index (*NGI*): growing for looped networks, and (forcefully) equal to 0 for the branched/ serial networks.
2. Network connectivity factor (*NCFx*): growing for all the networks.
3. Network connectivity index (*NCIx*): equal to 1 for nearly all the networks except Net13 to Net16; this gives distorted impression because the networks are too simple to be diversified by this index.
4. Network shape index (*NSIx*): growing for looped networks, and (forcefully) equal to 0 for the branched/ serial networks.

The results in Table 6.9 show the following trends for the increased number of links within a cluster of fixed number of nodes, regardless the type of network configuration:

1. Average geodesic distance (*AGD*): reducing in the value.
2. Graph density (*GD*): increasing in the value.
3. Average node degree (Deg_{avg}): increasing in the value.
4. Average betweeness centrality (BC_{avg}): reducing in the value.
5. Average closeness centrality (Cc_{avg}): increasing in the value.
6. Average clustering coefficient (Cp_{avg}): no clear trend: equal to 0 for all the branched/serial networks, but also for five looped networks.

6.6   SIMULATION RUNS, CASE 30 NETWORKS

The above set of networks could barely serve for more than to understand the concept behind the indices and give the first indication of their trends. To get more insight, another set of simulation runs has been executed, including the calculation of reliability indices, on a sample of synthetic gravity networks all consisting of a reservoir and 76 discharge nodes, connected in 30 random configurations ranging between 79 and 109 pipes (shown in Figure 6.10).

All the networks have configuration combined of loops and branches with variable degree of connectivity. Some layouts are almost fully looped (such as *SN8* or *SN53*) while the others are having more branches (e.g. *SN13*, or *SN73*). The connectivity to the source is provided with one to four pipes (the latter done in *SN34*). Table 6.10 gives more information about the geometry of each configuration and the connectivity factors and indices calculated by NDT. The results for indices that have been obtained from NodeXL are shown in Table 6.11. The order of results in both tables follows the order networks in Figure 6.10.

The correlation between the results in the tables and the number of links in each network is shown in Figures 6.11 and 6.12, respectively. For the purpose of presenting all the NodeXL results in one graph, the GD- and Cc-values have been multiplied by 100 and 1000, respectively, while the BC-values are divided by 100. All the NodeXL results have been calculated assuming undirected graphs i.e. networks. The connectivity calculations for directed graph have been done after running the hydraulic simulations in EPANET, and reversing the order of nodes for each link with negative flow/velocity. As a result, the assumption of directed graph has only influenced the BC-values; the correlation with the number of pipes has improved slightly, with the R^2-value increasing from 0.6429 to 0.6754.

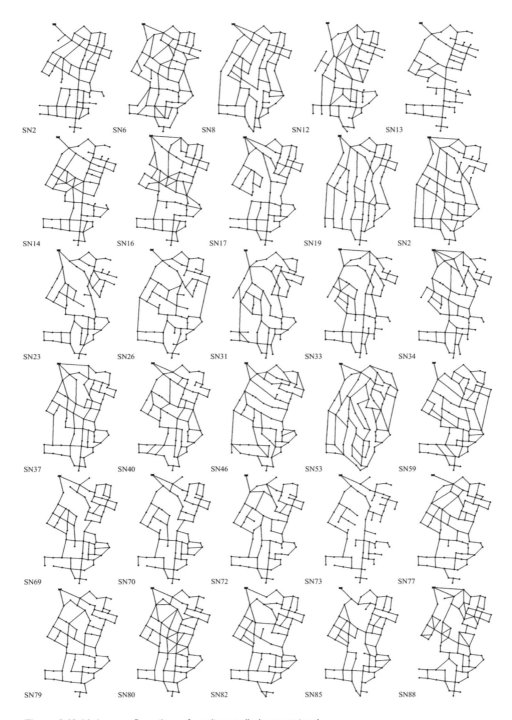

Figure 6.10 Various configurations of gravity supplied case network

Table 6.10 Connectivity indices calculated by NDT - case 30 nets

Epanet	n	m	l	Ln	NGI	NCF1	NCF2	NCF3	NCF$_{avg}$	NCI1	NCI2	NCI3	NSI1	NSI2	NSI3
SN2	77	105	29	813	0.96	2.09	2.09	1.55	2.73	0.77	0.77	0.57	0.74	0.74	0.55
SN6	77	108	32	993	0.97	1.90	1.90	1.90	2.81	0.68	0.68	0.68	0.65	0.65	0.65
SN8	77	103	27	703	0.96	2.14	2.14	2.14	2.68	0.80	0.80	0.80	0.77	0.77	0.77
SN12	77	99	23	507	0.95	1.73	1.73	1.73	2.57	0.67	0.67	0.67	0.64	0.64	0.64
SN13	77	89	13	157	0.92	1.49	1.49	1.49	2.31	0.65	0.65	0.65	0.59	0.59	0.59
SN14	77	105	29	813	0.96	1.57	1.73	1.57	2.73	0.58	0.63	0.58	0.56	0.61	0.56
SN16	77	107	31	931	0.97	1.77	1.77	1.77	2.78	0.64	0.64	0.64	0.61	0.61	0.61
SN17	77	99	23	507	0.95	2.05	2.05	2.05	2.57	0.80	0.80	0.80	0.76	0.76	0.76
SN19	77	99	23	507	0.95	2.01	2.01	2.01	2.57	0.78	0.78	0.78	0.75	0.75	0.75
SN21	77	105	29	813	0.96	1.96	1.99	1.96	2.73	0.72	0.73	0.72	0.69	0.70	0.69
SN23	77	93	17	273	0.94	1.99	1.99	1.99	2.42	0.82	0.82	0.82	0.77	0.77	0.77
SN26	77	92	16	241	0.93	2.05	2.05	2.05	2.39	0.86	0.86	0.86	0.80	0.80	0.80
SN31	77	95	19	343	0.94	1.92	1.92	1.92	2.47	0.78	0.78	0.78	0.74	0.74	0.74
SN33	77	96	20	381	0.95	1.68	1.68	1.68	2.49	0.67	0.67	0.67	0.64	0.64	0.64
SN34	77	99	23	507	0.95	1.77	1.77	1.77	2.57	0.69	0.69	0.69	0.66	0.66	0.66
SN37	77	105	29	813	0.96	1.94	1.94	1.94	2.73	0.71	0.71	0.71	0.68	0.68	0.68
SN40	77	100	24	553	0.96	1.95	1.95	1.95	2.60	0.75	0.75	0.75	0.72	0.72	0.72
SN46	77	99	23	507	0.95	2.06	2.06	2.06	2.57	0.80	0.80	0.80	0.77	0.77	0.77
SN53	77	101	25	601	0.96	2.12	2.12	2.12	2.62	0.81	0.81	0.81	0.77	0.77	0.77
SN59	77	107	31	931	0.97	2.01	2.09	2.01	2.78	0.72	0.75	0.72	0.70	0.73	0.70
SN69	77	100	24	553	0.96	1.94	1.94	1.94	2.60	0.75	0.75	0.75	0.71	0.71	0.71
SN70	77	94	18	307	0.94	1.92	1.92	1.92	2.44	0.79	0.79	0.79	0.74	0.74	0.74
SN72	77	93	17	273	0.94	1.95	1.95	1.95	2.42	0.81	0.81	0.81	0.76	0.76	0.76
SN73	77	79	3	7	0.57	1.09	1.48	1.48	2.05	0.53	0.72	0.72	0.30	0.41	0.41
SN77	77	101	25	601	0.96	2.08	2.08	2.08	2.62	0.79	0.79	0.79	0.76	0.76	0.76
SN79	77	96	20	381	0.95	2.19	2.19	2.19	2.49	0.88	0.88	0.88	0.83	0.83	0.83
SN80	77	109	33	1057	0.97	1.94	1.94	1.94	2.83	0.68	0.68	0.68	0.66	0.66	0.66
SN82	77	103	27	703	0.96	2.23	2.23	2.23	2.68	0.83	0.83	0.83	0.80	0.80	0.80
SN85	77	105	29	813	0.96	2.16	2.16	2.16	2.73	0.79	0.79	0.79	0.76	0.76	0.76
SN88	77	109	33	1057	0.97	1.84	1.97	1.84	2.83	0.65	0.70	0.65	0.63	0.68	0.63

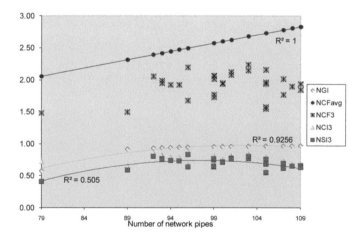

Figure 6.11 Correlation between number of pipes and connectivity indices calculated by NDT

Table 6.11 Connectivity indices calculated by NodeXL - case 30 nets (undirected)

Epanet	n	m	AGD	GD	Deg$_{min}$	Deg$_{max}$	Deg$_{avg}$	BC$_{avg}$	Cc$_{avg}$	Cp$_{avg}$
SN2	77	105	6.30	0.036	1	4	2.73	205	0.0021	0.00
SN6	77	108	6.00	0.037	1	5	2.81	193	0.0022	0.03
SN8	77	103	6.84	0.035	1	4	2.68	225	0.0019	0.00
SN12	77	99	6.22	0.034	1	5	2.57	201	0.0022	0.03
SN13	77	89	6.90	0.030	1	4	2.31	228	0.0019	0.00
SN14	77	105	6.40	0.036	1	6	2.73	208	0.0021	0.03
SN16	77	107	6.70	0.037	1	6	2.78	220	0.0020	0.06
SN17	77	99	7.34	0.034	1	4	2.57	245	0.0018	0.01
SN19	77	99	6.49	0.034	1	4	2.57	212	0.0020	0.00
SN21	77	105	6.14	0.036	1	5	2.73	199	0.0022	0.00
SN23	77	93	7.85	0.032	1	4	2.42	264	0.0017	0.00
SN26	77	92	6.67	0.031	1	4	2.39	219	0.0020	0.00
SN31	77	95	6.84	0.032	1	5	2.47	225	0.0019	0.01
SN33	77	96	6.67	0.033	1	5	2.49	219	0.0020	0.01
SN34	77	99	6.54	0.034	1	4	2.57	214	0.0020	0.02
SN37	77	105	6.11	0.036	1	4	2.73	197	0.0022	0.01
SN40	77	100	6.76	0.034	1	4	2.60	222	0.0020	0.02
SN46	77	99	6.70	0.034	1	4	2.57	220	0.0020	0.05
SN53	77	101	6.21	0.035	1	5	2.62	201	0.0021	0.01
SN59	77	107	6.30	0.037	1	4	2.78	204	0.0021	0.00
SN69	77	100	6.87	0.034	1	4	2.60	227	0.0019	0.00
SN70	77	94	7.14	0.032	1	4	2.44	237	0.0019	0.00
SN72	77	93	7.05	0.032	1	4	2.42	234	0.0019	0.00
SN73	77	79	8.64	0.027	1	4	2.05	295	0.0015	0.00
SN77	77	101	6.66	0.035	1	5	2.62	219	0.0020	0.00
SN79	77	96	7.06	0.033	1	4	2.49	234	0.0019	0.00
SN80	77	109	6.04	0.037	1	6	2.83	195	0.0022	0.06
SN82	77	103	6.58	0.035	1	4	2.68	215	0.0020	0.02
SN85	77	105	6.69	0.036	1	4	2.73	219	0.0020	0.00
SN88	77	109	6.82	0.037	1	5	2.83	225	0.0020	0.10

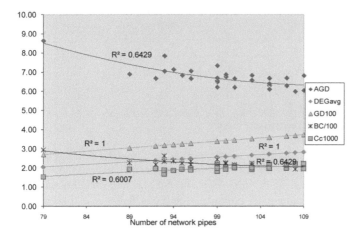

Figure 6.12 Correlation between number of pipes and connectivity indices calculated by NodeXL

Having the results between the three alternatives of *NCF*, *NCI* and *NSI* similar, only the values of the 3rd alternative, the one based on the two nodal categories with the closest number of connections to the average number of connections, have been further compared to the other parameters. Table 6.12 shows the network ranking for each of these, 1 being the most connected, and 30 the least connected network. The shaded figures indicate equal network ranking for three or more measures in either of the approaches, or four or more measures in both approaches together.

Table 6.12 Connectivity ranking (1-highest, 30-lowest) per factor/index - case 30 nets (undirected)

Epanet	NGI	NCF3	NCF$_{avg}$	NCI3	NSI3	AGD	GD	Deg$_{avg}$	BC$_{avg}$	Cc$_{avg}$
SN2	6	28	10	30	29	8	8	10	8	6
SN6	3	21	3	23	22	1	3	3	1	2
SN8	11	4	12	8	6	22	12	12	22	24
SN12	17	25	21	25	23	6	17	21	6	5
SN13	29	29	29	27	27	24	29	29	24	21
SN14	7	27	9	29	28	9	9	9	9	9
SN16	4	23	5	28	26	18	5	5	18	15
SN17	18	8	18	9	9	28	21	18	28	28
SN19	19	11	19	13	12	10	18	19	10	11
SN21	8	13	7	19	18	4	7	7	4	4
SN23	26	12	26	4	5	29	27	26	29	29
SN26	28	9	28	2	3	15	28	28	15	19
SN31	24	19	24	14	14	21	24	24	21	22
SN33	22	26	23	24	24	14	22	23	14	14
SN34	20	24	20	21	21	11	19	20	11	10
SN37	9	17	8	20	19	3	6	8	3	3
SN40	15	14	15	15	15	19	15	15	19	16
SN46	21	7	17	7	7	17	20	17	17	18
SN53	13	5	13	5	4	5	13	13	5	8
SN59	5	10	4	17	17	7	4	4	7	7
SN69	16	18	16	16	16	23	16	16	23	23
SN70	25	20	25	12	13	27	25	25	27	27
SN72	27	15	27	6	11	25	26	27	25	26
SN73	30	30	30	18	30	30	30	30	30	30
SN77	14	6	14	10	10	13	14	14	13	13
SN79	23	2	22	1	1	26	23	22	26	25
SN80	1	16	1	22	20	2	1	1	2	1
SN82	12	1	11	3	2	12	11	11	12	12
SN85	10	3	6	11	8	16	10	6	16	17
SN88	2	22	2	26	25	20	2	2	20	20

4 or more in both groups, or 3 or more in one group

The results show pretty much of inconsistency in both approaches, however to a lesser extent of the Node XL results. The high correlation between the number of links and the values of *NGI*, *NCF$_{avg}$*, *DEG$_{avg}$* and *GD*, shown in Figures 6.11 and 6.12 is expected knowing that these measures directly depend on number of links, as Equations 6.3, 6.8 and 6.13 show, respectively. The correlation of other measures is (much) weaker, although the general trend of increase/decrease is visible in all the cases. Moreover, the results in Table 6.12 show somewhat erratic ranking, after comparing the two approaches. Nevertheless, a couple of networks give impression that these measures describe the connectivity fairly well, but may be too coarse to distinguish between the networks that actually look pretty similar. The example of good correlation is network *SN73* that is visibly the network of the lowest connectivity, which has been captured by all the measures except the *NCI3*. Furthermore, networks *SN13*, *SN23*, *SN26*, *SN31*, *SN33*, *SN34*, *SN70*, *SN72* and *SN79* all qualify for the ranking amongst the last 10, which is hinted by comparing Table 6.12 and Figure 6.10 although not unanimously. On the other hand, *SN80* and *SN6* appear to be ranked as well connected networks, which can also be anticipated for *SN37*, *SN53* and *SN59*. Finally, a couple of networks in the medium range of connectivity also shows reasonably uniform ranking, such as *SN14*, *SN40* or *SN77*, but there are networks where some measures rank them highly while the others low. The example is *SN88*, where possible reason is in more heterogeneous layout (some nodes with one connection, but also several of those with four or five connections).

All the results have been further compared with the reliability measures, discussed in Chapter 5, namely the network buffer index (*NBI*), the network resilience of Prasad and Park (I_n) and the average available demand fraction (ADF_{avg}). For this purpose, a series of hydraulic simulations has been conducted in EPANET, for the head at the source of 160 msl. All nodal elevations have been set to 50 msl, creating elevation difference with the source of 110 m. The uniform k-value of 0.5 mm has been applied for all the pipes. Two rounds of calculations have been executed to assess the correlation with the reliability measures: the first one for all the configurations having uniform pipe diameter of 500 mm and variable demand multiplier adjusted by NDT to reach the selected pressure threshold; this has resulted in the range of demands between 1995 l/s (in network *SN2*) and 5837 l/s (SN34). The second run has been conducted for GA optimised diameters at minimum pressure of 20 mwc, which was in both runs also taken as the threshold used in PDD calculations (with the exponent of 0.5). The range of diameters used for GA optimisation is between 100 and 750 mm, in increments of 50 mm; the EO-optimiser has been used for the calculations. The total demand of 2280 l/s has been kept fixed for all the networks.

The results in Figures 6.13 to 6.16 show insufficiently strong correlation between the connectivity and reliability measures, which is slightly better and more consistent in case of the networks with fixed D/variable Q and NodeXL results.

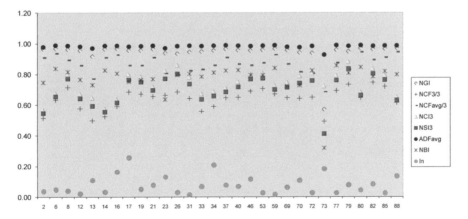

Figure 6.13 Reliability measures and connectivity indices calculated by NDT (fixed D, variable Q)

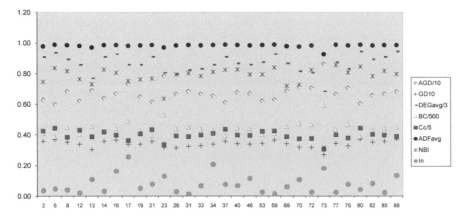

Figure 6.14 Reliability measures and connectivity indices calculated by NodeXL (fixed D, variable Q)

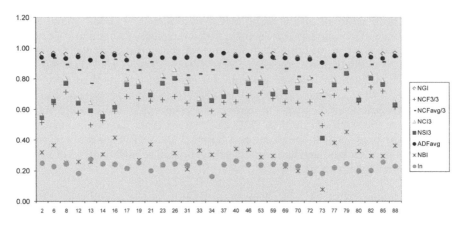

Figure 6.15 Reliability measures and connectivity indices calculated by NDT (optimised D, fixed Q)

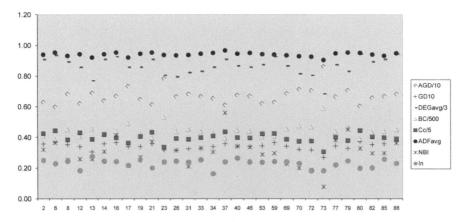

Figure 6.16 Reliability measures and connectivity indices calculated by NodeXL (optimised D, fixed Q)

That is not quite a surprise in view of the fact that pipe resistance has not been considered in this analysis. The Pearson correlation test gives the results shown in Tables 6.13 and 6.14, with shaded figures showing better correlation than the rest. The low values of I_n reflecting low-resilient layouts, may have contributed to the low correlation of this index.

Table 6.13 Pearson correlation - case 30 nets (undirected, fixed D, variable Q)

Measures	NGI	NCF3	NCF$_{avg}$	NCI3	NSI3	AGD	GD	Deg$_{avg}$	BC$_{avg}$	Cc$_{avg}$
ADFavg	0.94	0.47	0.75	0.01	0.54	-0.81	0.75	0.75	-0.81	0.74
NBI	0.91	0.43	0.71	0.00	0.51	-0.79	0.71	0.71	-0.79	0.72
In	-0.33	-0.20	-0.25	-0.05	-0.22	0.52	-0.25	-0.25	0.52	-0.49

Table 6.14 Pearson correlation - case 30 nets (undirected, optimised D, fixed Q)

Measures	NGI	NCF3	NCF$_{avg}$	NCI3	NSI3	AGD	GD	Deg$_{avg}$	BC$_{avg}$	Cc$_{avg}$
ADFavg	0.61	0.22	0.68	-0.18	0.19	-0.70	0.68	0.68	-0.70	0.70
NBI	0.53	0.21	0.60	-0.15	0.17	-0.60	0.60	0.60	-0.60	0.61
In	0.29	0.06	0.11	-0.01	0.16	-0.10	0.11	0.11	-0.10	0.03

A few remarks specific to the NDT results: these evaluate a couple of networks against the impression obtained from visual observation. For instance, $SN53$ ranks as the 4th with the value of $NSI3$ (0.77), the 5th with the value of $NCI3$ (0.81) and only the 13th for the value of NCF_{avg} (2.62). On the other hand, while looking as less connected network configuration, $SN79$ has the highest value of $NSI3$ (0.83) and $NCI3$ (0.88) but ranks only 22 for the value of NCF_{avg} (2.49). A few reasons possibly explaining this situation are:
- Unlike is the case with the measures of NodeXL, NCI and NSI both attempt to describe any network, regardless its size, with the connectivity factor that can never be greater than 1, i.e. in relative terms.
- With variable number of links, networks with fewer loops can still be well connected i.e. with fewer branches; this causes the higher value of NCI.
- The length of pipes gives visual impression of higher or lower connectivity. The total pipe length of $SN79$ is 13,231 m, while $SN53$ is 17,510 m long.
- Few configurations having non-adjacent basic loops, such as $SN17$ or $SN69$ have been evaluated unrealistically high due to too high value of L_n. This results in higher values of NGI (and consequently NSI).
- In networks with larger number of pipes, the selection of (only) two major connectivity categories for determination of NCF may yield too low values; hereby the highly connected nodes have been cut-off being in low numbers.

6.7   SIMULATION RUNS, CASE THREE CLUSTERS OF 10 NETWORKS

To further investigate the correlations, the 30 networks have been recreated into three groups of 10 networks with fixed number of pipes 79, 99, and 109, respectively. This has been achieved by adding/removing/reconnecting pipes in the networks of the previous batch. The layout of the new networks is shown in Figure 6.17. The rational for this analysis has been twofold: (a) to investigate the sensitivity of connectivity measures with fixed number of nodes and pipes and (b) to check the correlations in case of more radical differences in number of pipes/loops of the three groups.

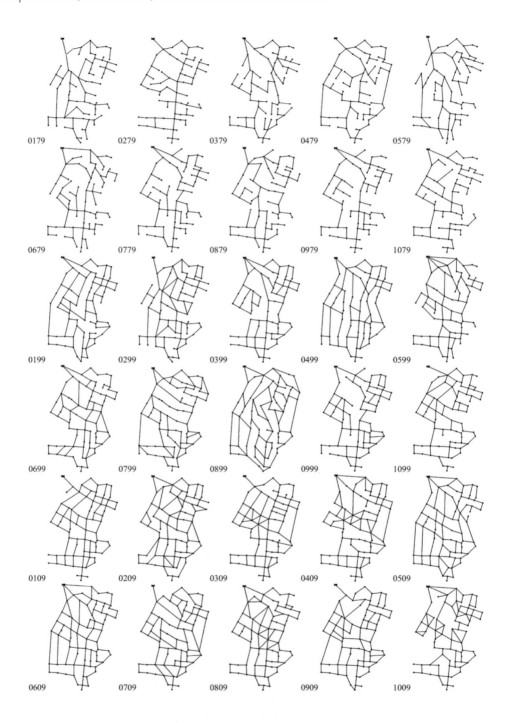

Figure 6.17 Configurations of three network clusters (77 nodes and 79, 99 and 109 links, each)

Table 6.15 Connectivity indices calculated by NDT - case three clusters

Epanet	n	m	l	Ln	NGI	NCF1	NCF2	NCF3	NCF$_{avg}$	NCI1	NCI2	NCI3	NSI1	NSI2	NSI3
SN0179	77	79	3	7	0.57	0.97	1.44	1.44	2.05	0.47	0.70	0.70	0.27	0.40	0.40
SN0279	77	79	3	7	0.57	1.05	1.29	1.29	2.05	0.51	0.63	0.63	0.29	0.36	0.36
SN0379	77	79	3	7	0.57	1.03	1.52	1.52	2.05	0.50	0.74	0.74	0.29	0.42	0.42
SN0479	77	79	3	7	0.57	1.64	1.64	1.64	2.05	0.80	0.80	0.80	0.46	0.46	0.46
SN0579	77	79	3	7	0.57	1.13	1.53	1.53	2.05	0.55	0.75	0.75	0.31	0.43	0.43
SN0679	77	79	3	7	0.57	1.08	1.42	1.42	2.05	0.53	0.69	0.69	0.30	0.39	0.39
SN0779	77	79	3	7	0.57	1.13	1.49	1.49	2.05	0.55	0.73	0.73	0.31	0.42	0.42
SN0879	77	79	3	7	0.57	1.29	1.55	1.55	2.05	0.63	0.75	0.75	0.36	0.43	0.43
SN0979	77	79	3	7	0.57	1.09	1.48	1.48	2.05	0.53	0.72	0.72	0.30	0.41	0.41
SN1079	77	79	3	7	0.57	1.70	1.70	1.70	2.05	0.83	0.83	0.83	0.47	0.47	0.47
SN0199	77	99	23	507	0.95	2.17	2.17	2.17	2.57	0.84	0.84	0.84	0.81	0.81	0.81
SN0299	77	99	23	507	0.95	1.73	1.73	1.73	2.57	0.67	0.67	0.67	0.64	0.64	0.64
SN0399	77	99	23	507	0.95	2.05	2.05	2.05	2.57	0.80	0.80	0.80	0.76	0.76	0.76
SN0499	77	99	23	507	0.95	2.01	2.01	2.01	2.57	0.78	0.78	0.78	0.75	0.75	0.75
SN0599	77	99	23	507	0.95	1.77	1.77	1.77	2.57	0.69	0.69	0.69	0.66	0.66	0.66
SN0699	77	99	23	507	0.95	2.03	2.03	2.03	2.57	0.79	0.79	0.79	0.75	0.75	0.75
SN0799	77	99	23	507	0.95	2.06	2.06	2.06	2.57	0.80	0.80	0.80	0.77	0.77	0.77
SN0899	77	99	23	507	0.95	2.17	2.17	2.17	2.57	0.84	0.84	0.84	0.81	0.81	0.81
SN0999	77	99	23	507	0.95	1.90	1.90	1.90	2.57	0.74	0.74	0.74	0.70	0.70	0.70
SN1099	77	99	23	507	0.95	2.01	2.01	2.01	2.57	0.78	0.78	0.78	0.75	0.75	0.75
SN0109	77	109	33	1057	0.97	2.25	2.25	1.70	2.83	0.79	0.79	0.60	0.77	0.77	0.58
SN0209	77	109	33	1057	0.97	1.92	1.92	1.92	2.83	0.68	0.68	0.68	0.66	0.66	0.66
SN0309	77	109	33	1057	0.97	1.70	1.83	1.70	2.83	0.60	0.65	0.60	0.58	0.63	0.58
SN0409	77	109	33	1057	0.97	1.83	1.83	1.83	2.83	0.65	0.65	0.65	0.63	0.63	0.63
SN0509	77	109	33	1057	0.97	2.10	2.18	2.10	2.83	0.74	0.77	0.74	0.72	0.75	0.72
SN0609	77	109	33	1057	0.97	1.99	2.12	1.99	2.83	0.70	0.75	0.70	0.68	0.72	0.68
SN0709	77	109	33	1057	0.97	2.01	2.22	2.01	2.83	0.71	0.78	0.71	0.69	0.76	0.69
SN0809	77	109	33	1057	0.97	1.94	1.94	1.94	2.83	0.68	0.68	0.68	0.66	0.66	0.66
SN0909	77	109	33	1057	0.97	2.12	2.14	2.12	2.83	0.75	0.76	0.75	0.72	0.73	0.72
SN1009	77	109	33	1057	0.97	1.84	1.97	1.84	2.83	0.65	0.70	0.65	0.63	0.68	0.63

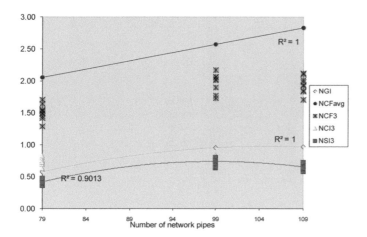

Figure 6.18 Correlation between the clusters and connectivity indices calculated by NDT

Table 6.16 Connectivity indices calculated by NodeXL - case three clusters (undirected)

Epanet	n	m	AGD	GD	Deg$_{min}$	Deg$_{max}$	Deg$_{avg}$	BC$_{avg}$	Cc$_{avg}$	Cp$_{avg}$
SN0179	77	79	7.72	0.027	1	4	2.05	259	0.0017	0.00
SN0279	77	79	7.85	0.027	1	4	2.05	264	0.0017	0.00
SN0379	77	79	10.28	0.027	1	4	2.05	358	0.0013	0.00
SN0479	77	79	8.66	0.027	1	4	2.05	295	0.0015	0.00
SN0579	77	79	8.62	0.027	1	5	2.05	294	0.0016	0.00
SN0679	77	79	8.23	0.027	1	4	2.05	279	0.0016	0.00
SN0779	77	79	8.26	0.027	1	4	2.05	280	0.0016	0.00
SN0879	77	79	9.84	0.027	1	4	2.05	341	0.0014	0.00
SN0979	77	79	8.64	0.027	1	4	2.05	295	0.0015	0.00
SN1079	77	79	9.77	0.027	1	4	2.05	338	0.0014	0.00
SN0199	77	99	7.23	0.034	1	4	2.57	240	0.0018	0.00
SN0299	77	99	6.22	0.034	1	5	2.57	201	0.0022	0.03
SN0399	77	99	7.34	0.034	1	4	2.57	245	0.0018	0.01
SN0499	77	99	6.49	0.034	1	4	2.57	212	0.0020	0.00
SN0599	77	99	6.54	0.034	1	4	2.57	214	0.0020	0.02
SN0699	77	99	6.85	0.034	1	4	2.57	226	0.0020	0.00
SN0799	77	99	6.70	0.034	1	4	2.57	220	0.0020	0.05
SN0899	77	99	6.42	0.034	1	5	2.57	209	0.0020	0.01
SN0999	77	99	6.90	0.034	1	4	2.57	228	0.0019	0.00
SN1099	77	99	7.04	0.034	1	5	2.57	233	0.0019	0.00
SN0109	77	109	6.03	0.037	1	4	2.83	194	0.0022	0.00
SN0209	77	109	5.93	0.037	1	5	2.83	190	0.0022	0.03
SN0309	77	109	5.88	0.037	1	6	2.83	188	0.0023	0.03
SN0409	77	109	6.33	0.037	1	6	2.83	206	0.0021	0.06
SN0509	77	109	5.95	0.037	1	5	2.83	191	0.0022	0.00
SN0609	77	109	5.93	0.037	1	4	2.83	190	0.0023	0.01
SN0709	77	109	6.09	0.037	1	4	2.83	196	0.0022	0.00
SN0809	77	109	6.04	0.037	1	6	2.83	195	0.0022	0.06
SN0909	77	109	6.09	0.037	1	5	2.83	196	0.0022	0.00
SN1009	77	109	6.82	0.037	1	5	2.83	225	0.0020	0.10

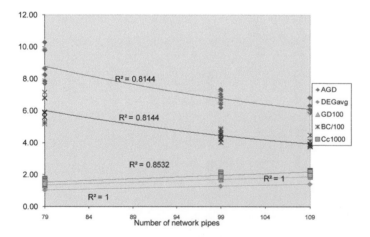

Figure 6.19 Correlation between the clusters and connectivity indices calculated by NodeXL

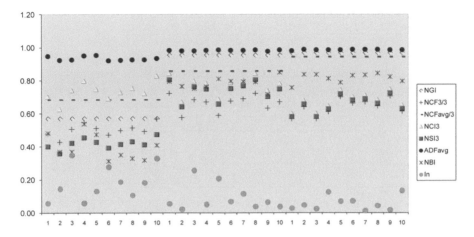

Figure 6.20 Reliability measures and connectivity indices calculated by NDT (fixed D, variable Q)

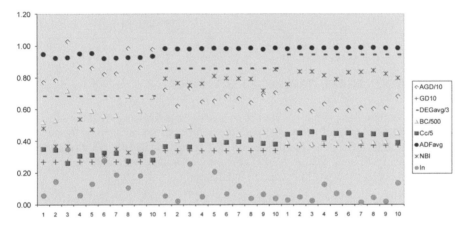

Figure 6.21 Reliability measures and connectivity indices calculated by NodeXL (fixed D, variable Q)

The hydraulic simulations have been conducted for all the networks having fixed diameter of 500 mm and variable demand multiplier, resulting in the demand range between 1949 l/s (in network 0279) and 5858 l/s (in 0609). Other network parameters have been the same as in the previous set of networks.

The results in Tables 6.15 to 6.18 and Figures 6.18 to 6.21 show the similar picture with slightly better correlations. It is also becoming clearer that NodeXL measures describe the networks' connectivity more consistently than the NDT ones. Significant improvement of correlation shown in Table 6.17, compared to Table 6.13, is also the result of clustering of the networks. The results split per cluster, in Table 6.18, show much weaker correlation although it should be taken into consideration that they are based on rather short series.

Table 6.17 Pearson correlation - case three clusters (undirected, fixed D, variable Q)

Measures	NGI	NCF3	NCF$_{avg}$	NCI3	NSI3	AGD	GD	Deg$_{avg}$	BC$_{avg}$	Cc$_{avg}$
ADFavg	0.96	0.84	0.93	-0.03	0.89	-0.88	0.93	0.93	-0.88	0.87
NBI	0.97	0.84	0.93	-0.04	0.89	-0.88	0.93	0.93	-0.88	0.88
In	-0.56	-0.45	-0.57	0.13	-0.48	0.69	-0.57	-0.57	0.69	-0.68

Table 6.18 Pearson correlation - results per cluster (undirected, fixed D, variable Q)

Measures	NGI	NCF3	NCF$_{avg}$	NCI3	NSI3	AGD	GD	Deg$_{avg}$	BC$_{avg}$	Cc$_{avg}$
cluster 1 - 77 nodes, 79 links										
ADFavg		0.38		0.38	0.38	-0.19			-0.19	0.16
NBI		0.37		0.37	0.37	-0.19			-0.19	0.17
In		0.17		0.17	0.17	0.54			0.54	-0.51
cluster 2 - 77 nodes, 99 links										
ADFavg		0.15		0.15	0.15	-0.24			-0.24	0.23
NBI		0.19		0.19	0.19	0.02			0.02	0.00
In		-0.11		-0.11	-0.11	0.38			0.38	-0.32
cluster 3 - 77 nodes, 109 links										
ADFavg		0.51		0.51	0.51	-0.28			-0.28	0.29
NBI		0.27		0.27	0.27	-0.26			-0.26	0.29
In		-0.12		-0.12	-0.12	0.73			0.73	-0.75

6.8 SIMULATION RUNS, CASE NGT NETWORKS

The final analysis has been done on the set of networks consisting of three clusters of 15 networks with fixed number of nodes and variable number of pipes, namely: 50 nodes and 66 to 76 pipes, 151 node and 217 to 246 pipes, and 200 nodes and 285 to 313 pipes; the network layouts are shown in Figures 6.22 to 6.24. These networks have been randomly generated by the network generation tool (NGT) discussed in Chapter 4.

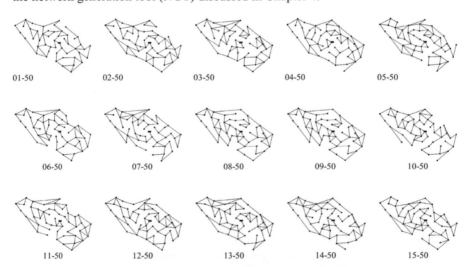

01-50 02-50 03-50 04-50 05-50

06-50 07-50 08-50 09-50 10-50

11-50 12-50 13-50 14-50 15-50

Figure 6.22 Configurations of the network cluster NGT50 (50 nodes, 66-76 links)

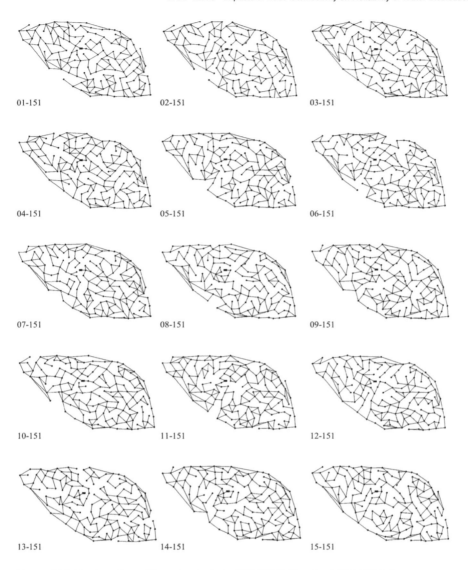

Figure 6.23 Configurations of the network cluster NGT151 (151 nodes, 217-246 links)

The source has been located in the middle area, and all the pipe diameters have been set at 100 mm; this much smaller diameter is chosen to test the correlations with increased resistance of the pipes, compared to the previous sets. To avoid low pressures, the demand multipliers have been adjusted by NDT to satisfy the minimum pressure condition of 20 mwc, used also for PDD simulations and calculations of ADF_{avg} and NBI, like in the previous simulation runs with fixed diameters. This has resulted in the demand ranges between (a) 5.8 l/s (in network *04-50*) and 20.2 l/s (in network *09-50*), for the cluster of networks with 50 nodes, (b) 6.1 l/s (in network *08-151*) and 20.8 l/s (in network *04-151*), for the cluster of networks with 151 nodes, and (c) 7.4 l/s (in network *06-200*) and 20.0 l/s (in network *13-200*), for the cluster of networks with 200 nodes. The baseline nodal demands and elevations have been randomised by NGT per cluster of networks, the nodal elevations resembling mildly hilly area with elevation differences up to 20 meters. Finally, the uniform roughness of 0.5 mm was used in all the cases.

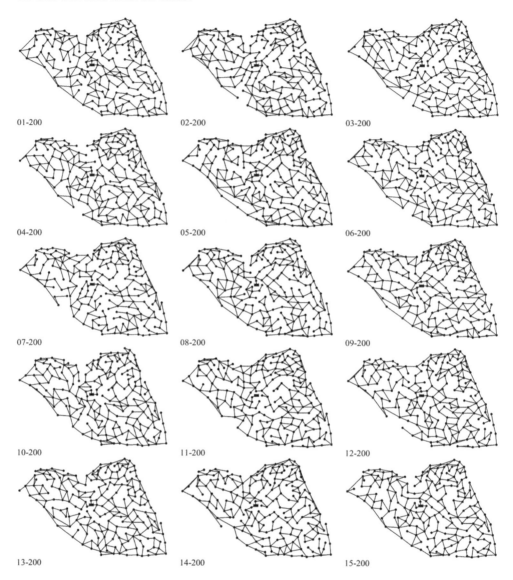

01-200   02-200   03-200

04-200   05-200   06-200

07-200   08-200   09-200

10-200   11-200   12-200

13-200   14-200   15-200

Figure 6.24 Configurations of the network cluster NGT200 (200 nodes, 285-313 links)

The results of calculations are shown in Tables 6.21 to 6.25 and in Figures 6.25 to 6.28, in the same formats as in the previous paragraphs. No substantial difference in the patterns compared to the previous sets of networks could have been observed. Again, while the correlations between the connectivity measures and number of pipes are visible in Figures 25 and 26, they are not too strong except for the values of NGI, NCF_{avg} (Deg_{avg}) and GD. The values of $NCIx$ and $NSIx$ are still rather similar, pointing with the low range of values relatively equally connected networks; this complies with the impression from the Figures 6.22 to 6.24. Unlike is the case with the NDT measures, the NodeXL measures are distinctively different between the three clusters.

Table 6.19 Connectivity indices calculated by NDT - NGT clusters

Epanet	n	m	l	Ln	NGI	NCF1	NCF2	NCF3	NCF$_{avg}$	NCI1	NCI2	NCI3	NSI1	NSI2	NSI3
01-50	50	66	17	273	0.94	2.12	2.12	2.12	2.64	0.80	0.80	0.80	0.75	0.75	0.75
02-50	50	72	23	507	0.95	1.70	2.06	1.70	2.88	0.59	0.72	0.59	0.56	0.68	0.56
03-50	50	71	22	463	0.95	1.64	1.88	1.64	2.84	0.58	0.66	0.58	0.55	0.63	0.55
04-50	50	69	20	381	0.95	1.56	1.56	1.56	2.76	0.57	0.57	0.57	0.54	0.54	0.54
05-50	50	76	27	703	0.96	1.60	1.76	1.76	3.04	0.53	0.58	0.58	0.51	0.56	0.56
06-50	50	70	21	421	0.95	1.64	1.84	1.64	2.80	0.59	0.66	0.59	0.56	0.62	0.56
07-50	50	66	17	273	0.94	1.66	1.66	1.66	2.64	0.63	0.63	0.63	0.59	0.59	0.59
08-50	50	71	22	463	0.95	1.88	1.88	1.88	2.84	0.66	0.66	0.66	0.63	0.63	0.63
09-50	50	72	23	507	0.95	1.88	1.88	1.48	2.88	0.65	0.65	0.51	0.62	0.62	0.49
10-50	50	68	19	343	0.94	1.74	2.02	1.74	2.72	0.64	0.74	0.64	0.60	0.70	0.60
11-50	50	72	23	507	0.95	1.86	2.02	1.86	2.88	0.65	0.70	0.65	0.62	0.67	0.62
12-50	50	75	26	651	0.96	1.64	2.00	2.00	3.00	0.55	0.67	0.67	0.52	0.64	0.64
13-50	50	76	27	703	0.96	1.66	1.82	1.82	3.04	0.55	0.60	0.60	0.53	0.58	0.58
14-50	50	73	24	553	0.96	1.58	1.62	1.58	2.92	0.54	0.55	0.54	0.52	0.53	0.52
15-50	50	73	24	553	0.96	1.78	1.78	1.78	2.92	0.61	0.61	0.61	0.58	0.58	0.58
01-151	151	233	83	6807	0.99	2.07	2.07	2.07	3.09	0.67	0.67	0.67	0.66	0.66	0.66
02-151	151	219	69	4693	0.99	1.62	1.88	1.62	2.90	0.56	0.65	0.56	0.55	0.64	0.55
03-151	151	229	79	6163	0.99	1.56	1.93	1.93	3.03	0.52	0.64	0.64	0.51	0.63	0.63
04-151	151	236	86	7311	0.99	2.09	2.09	2.09	3.13	0.67	0.67	0.67	0.66	0.66	0.66
05-151	151	223	73	5257	0.99	1.48	1.72	1.48	2.95	0.50	0.58	0.50	0.50	0.57	0.50
06-151	151	224	74	5403	0.99	1.58	1.67	1.58	2.97	0.53	0.56	0.53	0.52	0.55	0.52
07-151	151	236	86	7311	0.99	2.12	2.12	2.12	3.13	0.68	0.68	0.68	0.67	0.67	0.67
08-151	151	217	67	4423	0.98	1.64	1.82	1.64	2.87	0.57	0.63	0.57	0.56	0.62	0.56
09-151	151	228	78	6007	0.99	1.68	1.89	1.89	3.02	0.55	0.63	0.63	0.55	0.62	0.62
10-151	151	231	81	6481	0.99	1.58	1.87	1.87	3.06	0.52	0.61	0.61	0.51	0.60	0.60
11-151	151	246	96	9121	0.99	1.53	1.95	1.95	3.26	0.47	0.60	0.60	0.46	0.59	0.59
12-151	151	221	71	4971	0.99	1.53	1.94	1.53	2.93	0.52	0.66	0.52	0.52	0.65	0.52
13-151	151	223	73	5257	0.99	1.51	1.93	1.51	2.95	0.51	0.65	0.51	0.50	0.65	0.50
14-151	151	236	86	7311	0.99	2.01	2.01	2.01	3.13	0.64	0.64	0.64	0.64	0.64	0.64
15-151	151	223	73	5257	0.99	1.57	1.86	1.57	2.95	0.53	0.63	0.53	0.52	0.62	0.52
01-200	200	299	100	9901	0.99	1.54	2.01	1.54	2.99	0.51	0.67	0.51	0.51	0.66	0.51
02-200	200	302	103	10507	0.99	1.55	1.92	1.92	3.02	0.51	0.63	0.63	0.51	0.63	0.63
03-200	200	303	104	10713	0.99	1.54	1.82	1.82	3.03	0.51	0.60	0.60	0.50	0.59	0.59
04-200	200	311	112	12433	0.99	2.23	2.23	2.23	3.11	0.72	0.72	0.72	0.71	0.71	0.71
05-200	200	309	110	11991	0.99	2.06	2.06	2.06	3.09	0.67	0.67	0.67	0.66	0.66	0.66
06-200	200	304	105	10921	0.99	1.55	1.89	1.89	3.04	0.51	0.62	0.62	0.50	0.62	0.62
07-200	200	285	86	7311	0.99	1.66	1.66	1.66	2.85	0.58	0.58	0.58	0.58	0.58	0.58
08-200	200	296	97	9313	0.99	1.41	1.75	1.41	2.96	0.47	0.59	0.47	0.47	0.58	0.47
09-200	200	313	114	12883	0.99	2.15	2.15	2.15	3.13	0.69	0.69	0.69	0.68	0.68	0.68
10-200	200	300	101	10101	0.99	1.65	1.98	1.98	3.00	0.55	0.66	0.66	0.54	0.65	0.65
11-200	200	305	106	11131	0.99	2.04	2.04	2.04	3.05	0.67	0.67	0.67	0.66	0.66	0.66
12-200	200	311	112	12433	0.99	2.08	2.08	2.08	3.11	0.67	0.67	0.67	0.66	0.66	0.66
13-200	200	307	108	11557	0.99	1.64	1.99	1.99	3.07	0.53	0.65	0.65	0.53	0.64	0.64
14-200	200	291	92	8373	0.99	1.58	1.73	1.58	2.91	0.54	0.59	0.54	0.54	0.59	0.54
15-200	200	307	108	11557	0.99	1.59	1.91	1.91	3.07	0.52	0.62	0.62	0.51	0.61	0.61

Table 6.20 Connectivity indices calculated by NodeXL - NGT clusters (undirected)

Epanet	n	m	AGD	GD	Deg$_{min}$	Deg$_{max}$	Deg$_{avg}$	BC$_{avg}$	Cc$_{avg}$	Cp$_{avg}$
01-50	50	66	5.72	0.054	1	5	2.64	118	0.0036	0.10
02-50	50	72	5.03	0.059	1	5	2.88	101	0.0040	0.18
03-50	50	71	5.69	0.058	1	5	2.84	118	0.0036	0.18
04-50	50	69	4.76	0.056	1	5	2.76	94	0.0043	0.18
05-50	50	76	5.53	0.062	1	5	3.04	114	0.0038	0.24
06-50	50	70	5.47	0.057	1	5	2.80	112	0.0038	0.16
07-50	50	66	5.42	0.054	1	6	2.64	111	0.0038	0.19
08-50	50	71	5.81	0.058	1	5	2.84	121	0.0035	0.17
09-50	50	72	4.86	0.059	1	6	2.88	97	0.0042	0.22
10-50	50	68	6.71	0.056	1	4	2.72	143	0.0031	0.19
11-50	50	72	5.72	0.059	1	5	2.88	118	0.0036	0.23
12-50	50	75	4.76	0.061	1	5	3.00	94	0.0042	0.24
13-50	50	76	5.07	0.062	1	6	3.04	102	0.0040	0.33
14-50	50	73	4.78	0.060	1	6	2.92	95	0.0042	0.20
15-50	50	73	5.98	0.060	1	6	2.92	125	0.0035	0.22
01-151	151	233	8.32	0.021	1	6	3.09	553	0.00080	0.17
02-151	151	219	9.39	0.019	1	6	2.90	634	0.00072	0.17
03-151	151	229	8.38	0.020	1	6	3.03	558	0.00080	0.20
04-151	151	236	8.03	0.021	1	6	3.13	531	0.00084	0.21
05-151	151	223	8.41	0.020	1	6	2.95	560	0.00080	0.19
06-151	151	224	8.58	0.020	1	6	2.97	573	0.00079	0.20
07-151	151	236	8.39	0.021	1	6	3.13	558	0.00080	0.18
08-151	151	217	8.54	0.019	1	6	2.87	570	0.00078	0.16
09-151	151	228	7.84	0.020	1	6	3.02	517	0.00085	0.17
10-151	151	231	7.66	0.020	1	5	3.06	503	0.00088	0.16
11-151	151	246	8.28	0.022	1	6	3.26	550	0.00081	0.24
12-151	151	221	8.16	0.020	1	6	2.93	541	0.00082	0.17
13-151	151	223	10.19	0.020	1	6	2.95	694	0.00068	0.15
14-151	151	236	7.89	0.021	1	6	3.13	521	0.00084	0.18
15-151	151	223	8.28	0.020	1	6	2.95	550	0.00081	0.15
01-200	200	299	9.60	0.015	1	6	2.99	861	0.00053	0.19
02-200	200	302	9.60	0.015	1	6	2.99	861	0.00053	0.19
03-200	200	303	9.49	0.015	1	6	3.03	850	0.00053	0.16
04-200	200	311	9.96	0.016	1	6	3.11	897	0.00051	0.18
05-200	200	309	9.43	0.016	1	6	3.09	843	0.00054	0.23
06-200	200	304	9.84	0.015	1	6	3.04	884	0.00052	0.21
07-200	200	285	10.49	0.014	1	6	2.85	950	0.00048	0.13
08-200	200	296	9.84	0.015	1	6	2.96	884	0.00052	0.15
09-200	200	313	9.46	0.016	1	6	3.13	847	0.00054	0.21
10-200	200	300	10.07	0.015	1	6	3.00	907	0.00050	0.18
11-200	200	305	9.40	0.015	1	6	3.05	840	0.00054	0.16
12-200	200	311	9.58	0.016	1	6	3.11	858	0.00053	0.19
13-200	200	307	9.56	0.015	1	6	3.07	857	0.00053	0.15
14-200	200	291	9.74	0.015	1	6	2.91	875	0.00052	0.16
15-200	200	307	9.48	0.015	1	6	3.07	849	0.00054	0.18

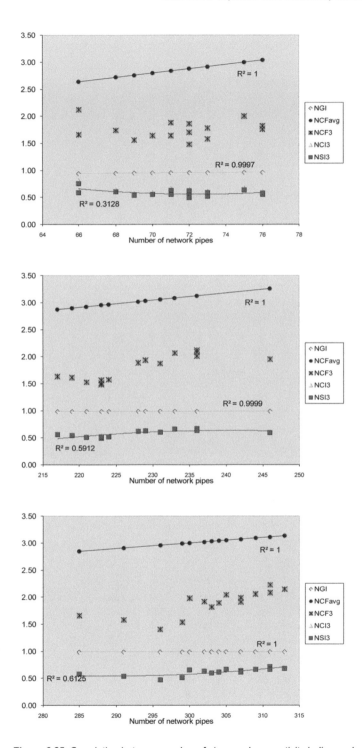

Figure 6.25 Correlation between number of pipes and connectivity indices calculated by NDT

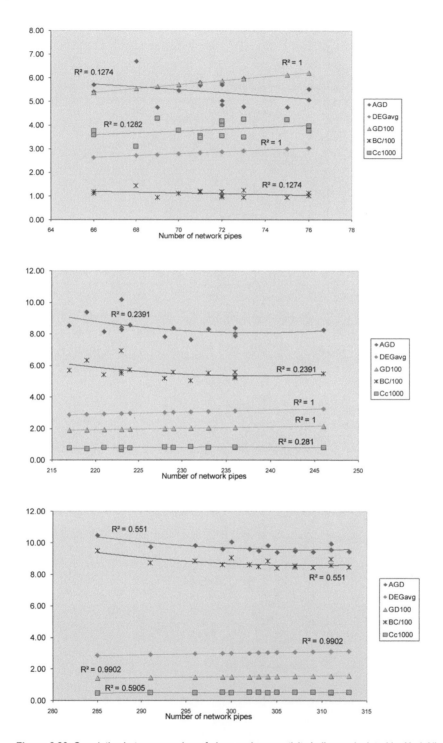

Figure 6.26 Correlation between number of pipes and connectivity indices calculated by NodeXL

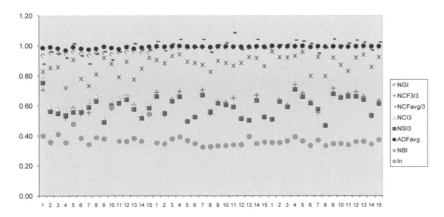

Figure 6.27 Reliability measures and connectivity indices calculated by NDT (fixed D, variable Q)

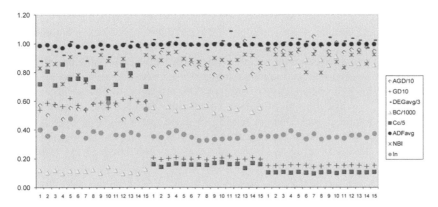

Figure 6.28 Reliability measures and connectivity indices calculated by NodeXL (fixed D, variable Q)

As can be seen on Figure 6.28, the values of *AGD* and BC_{avg} generally grow for substantial increase in the number of pipes, while the values of *GD* and Cc_{avg} will reduce, which are the patterns not entirely followed within the clusters. The values of Deg_{avg} fluctuate with less prediction, slightly increasing on average, but still with the extreme values in the networks not necessarily with the biggest number of nodes/pipes.

Table 6.21 Pearson correlation - case NGT clusters (undirected, fixed D, variable Q)

Measures	NGI	NCF3	NCF$_{avg}$	NCI3	NSI3	AGD	GD	Deg$_{avg}$	BC$_{avg}$	Cc$_{avg}$
ADFavg	0.82	0.27	0.68	0.00	0.15	0.73	-0.75	0.68	0.73	-0.77
NBI	0.56	0.28	0.54	0.08	0.17	0.48	-0.47	0.54	0.47	-0.50
In	-0.47	-0.03	-0.29	0.10	0.02	-0.32	0.48	-0.29	-0.41	0.41

Finally, the results of the Pearson correlation given in Tables 6.21 and 6.22 show relatively weak correlations with the reliability measures, somewhat better between the ADF_{avg} and NodeXL results and particularly weak in case of I_n. It shows again that making significant correlation without taking both pipe resistance and supply heads into consideration is hardly possible even after fixing some of these parameters for all the networks.

Table 6.22 Pearson correlation - results per cluster (undirected, fixed D, variable Q)

Measures	NGI	NCF3	NCFavg	NCI3	NSI3	AGD	GD	Degavg	BCavg	Ccavg
cluster 1 - 50 nodes, 66-76 links										
ADFavg	0.13			-0.07	-0.04	0.02			0.02	0.03
NBI	0.07			-0.11	-0.08	0.18			0.18	-0.12
In	0.09			0.08	0.08	0.77			0.77	-0.68
cluster 2 - 151 nodes, 217-246 links										
ADFavg	0.45			0.40	0.40	-0.29			-0.29	0.34
NBI	0.32			0.30	0.30	0.05			0.05	0.02
In	-0.12			-0.14	-0.14	0.42			0.42	-0.38
cluster 3 - 200 nodes, 285-313 links										
ADFavg	0.28			0.26	0.26	-0.19			-0.19	0.23
NBI	0.31			0.30	0.30	-0.17			-0.17	0.19
In	0.40			0.41	0.41	0.13			0.13	-0.07

6.9 STATISTICAL ANALYSIS

To verify the above conclusions, more complete statistical analysis has been conducted of all the results. The Spearman test has been used for this purpose being an alternative for the Pearson test that is known to be more suitable for linear correlations of normalised data. Despite large number of calculations, the number of analysed networks in each set/cluster has been relatively small from the perspective of statistical significance, to be sure about the data normalisation. Also, the assumption of linear correlation between the connectivity and reliability measures is highly questionable, having in mind quadratic head/flow relation in pipes. Hence, the idea of using also the Spearman test was to compare the results with the Pearson test and also having in mind its resistance towards isolated high or low values. The Spearman's Rank Correlation Coefficient, R, is calculated as:

$$R = 1 - \frac{6\sum_n d^2}{n(n^2 - 1)} \qquad \qquad 6.18$$

In Equation 6.18, d stands for difference in the ranking between n pairs of data. This ranking is done for each of the two data sets that are to be correlated, as it is shown in Table 6.12. The bandwidth of correlation is given in Table 6.23. The closer is the value of R to +1 or -1, the stronger is the likely correlation. The R-value of zero assumes no correlation, whatsoever.

Table 6.23 R-value bandwidth

-1	-0.8	-0.6	-0.4	-0.2	0	0.2	0.4	0.6	0.8	1
Strong positive correlation		Weak positive correlation	Little correlation		No correlation	Little correlation		Weak positive correlation	Strong positive correlation	

The correlation level, P, of probability distribution to determine whether two data sets have different degrees of diversity, is calculated as in Equation 6.19; parameter d_f, in the equation stands for the *degree of freedom* (equals $n-2$). The bandwidth of the significance level is shown in Table 6.24. The final conclusion about the correlation will depend on the comparison of the R and P against the value of d_f.

$$P = \frac{d_f R^2}{1 - R^2}$$

6.19

Table 6.24 *P*-value bandwidth

P-value	< 0.0001	< 0.001	0.001 >	< 0.001 & 0.01 >	< 0.01 & 0.05 >
P-value (%)	< 0.01%	< 0.1%	0.1% >	< 0.1% & 1% >	< 1% & 5% >
Level of correlation	100%	99.99%	99.9%	99%	95%
	Perfect	Strong	Fair	Moderate	Statistical; 5% likelihood the results occurred by chance.

The results of correlation are shown in the Tables 6.25 to 6.30. The tables show the correlation of the NodeXL measures with *NCIx* and *NSIx* calculated by NDT, and the reliability measures. This has been selected in view of more consistent correlation of the NodeXL measures indicated by the Pearson test. The results look better than with the Pearson's correlation but still, not good enough to talk about strong and/or generalised connectivity patterns. The best correlation has again been shown in case of the three clusters of the networks with fixed number of pipes and fixed pipe diameters, which is shown in Table 6.27. This case also shows fairly good correlation between the NodeXL and NDT measures. For the rest of the cases, these two comply mostly in the values of *GD* and *Deg$_{avg}$*, and to much more limited extent in case of the other measures. The clustering coefficient shows no correlation whosoever, although it is also to be observed that the tested networks are hardly of clustered configuration.

Table 6.25 Statistical analysis, case 30 networks (fixed D, variable Q)

R-value	NCI1	NCI2	NCI3	NSI1	NSI2	NSI3	ADF$_{avg}$	NBI	I$_n$
AGD-dia	0.24	0.34	0.36	0.19	0.20	0.21	-0.61	-0.44	0.42
GD	-0.28	-0.33	-0.40	-0.20	-0.15	-0.23	0.65	0.54	-0.16
Deg$_{avg}$	-0.28	-0.33	-0.40	-0.20	-0.15	-0.23	0.65	0.54	-0.16
BC$_{avg}$	0.24	0.34	0.36	0.19	0.20	0.21	-0.61	-0.44	0.42
CC$_{avg}$	-0.32	-0.43	-0.46	-0.27	-0.29	-0.31	0.59	0.44	-0.35
Cp$_{avg}$	-0.29	-0.34	-0.28	-0.27	-0.25	-0.19	0.42	0.36	0.26
P-value	NCI1	NCI2	NCI3	NSI1	NSI2	NSI3	ADF$_{avg}$	NBI	I$_n$
AGD-dia	0.210	0.068	0.048	0.327	0.300	0.259	0.000	0.014	0.022
GD	0.135	0.079	0.028	0.294	0.418	0.229	0.000	0.002	0.412
Deg$_{avg}$	0.135	0.079	0.028	0.294	0.418	0.229	0.000	0.002	0.412
BC$_{avg}$	0.210	0.068	0.048	0.327	0.300	0.259	0.000	0.014	0.022
CC$_{avg}$	0.081	0.017	0.010	0.144	0.124	0.096	0.001	0.015	0.057
Cp$_{avg}$	0.121	0.062	0.129	0.154	0.183	0.311	0.019	0.053	0.173
Correlation	NCI1	NCI2	NCI3	NSI1	NSI2	NSI3	ADF$_{avg}$	NBI	I$_n$
AGD-dia	R	R	95%	R	R	R	99.9%	95%	95%
GD	R	R	95%	R	R	R	99.9%	99%	R
Deg$_{avg}$	R	R	95%	R	R	R	99.9%	99%	R
BC$_{avg}$	R	R	95%	R	R	R	99.9%	95%	95%
CC$_{avg}$	R	95%	99%	R	R	R	99.9%	95%	R
Cp$_{avg}$	R	R	R	R	R	R	95%	R	R

Table 6.26 Statistical analysis, case 30 networks (optimised D, fixed Q)

R-value	NCI1	NCI2	NCI3	NSI1	NSI2	NSI3	ADF_{avg}	NBI	I_n
AGD-dia	0.24	0.34	0.36	0.19	0.20	0.21	-0.61	-0.52	0.12
GD	-0.28	-0.33	-0.40	-0.20	-0.15	-0.23	0.54	0.57	-0.04
Deg_{avg}	-0.28	-0.33	-0.40	-0.20	-0.15	-0.23	0.54	0.57	-0.04
BC_{avg}	0.24	0.34	0.36	0.19	0.20	0.21	-0.61	-0.52	0.12
CC_{avg}	-0.32	-0.43	-0.46	-0.27	-0.29	-0.31	0.65	0.56	-0.11
Cp_{avg}	-0.29	-0.34	-0.28	-0.27	-0.25	-0.19	0.51	0.38	-0.20
P-value	NCI1	NCI2	NCI3	NSI1	NSI2	NSI3	ADF_{avg}	NBI	I_n
AGD-dia	0.210	0.068	0.048	0.327	0.300	0.259	0.000	0.003	0.511
GD	0.135	0.079	0.028	0.294	0.418	0.229	0.002	0.001	0.819
Deg_{avg}	0.135	0.079	0.028	0.294	0.418	0.229	0.002	0.001	0.819
BC_{avg}	0.210	0.068	0.048	0.327	0.300	0.259	0.000	0.003	0.511
CC_{avg}	0.081	0.017	0.010	0.144	0.124	0.096	0.000	0.001	0.572
Cp_{avg}	0.121	0.062	0.129	0.154	0.183	0.311	0.004	0.038	0.291
Correlation	NCI1	NCI2	NCI3	NSI1	NSI2	NSI3	ADF_{avg}	NBI	I_n
AGD-dia	R	R	95%	R	R	R	99.9%	99%	R
GD	R	R	95%	R	R	R	99%	99%	R
Deg_{avg}	R	R	95%	R	R	R	99%	99%	R
BC_{avg}	R	R	95%	R	R	R	99.9%	99%	R
CC_{avg}	R	95%	99%	R	R	R	99.9%	99%	R
Cp_{avg}	R	R	R	R	R	R	99%	95%	R

Table 6.27 Statistical analysis, three clusters of 10 networks (fixed D, variable Q)

R-value	NCI1	NCI2	NCI3	NSI1	NSI2	NSI3	ADF_{avg}	NBI	I_n
AGD-dia	-0.22	0.17	0.42	-0.52	-0.52	-0.47	-0.86	-0.78	0.64
GD	0.33	-0.02	-0.28	0.60	0.63	0.56	0.89	0.84	-0.47
Deg_{avg}	0.33	-0.02	-0.28	0.60	0.63	0.56	0.89	0.84	-0.47
BC_{avg}	-0.22	0.17	0.42	-0.52	-0.52	-0.47	-0.86	-0.78	0.64
CC_{avg}	0.21	-0.19	-0.42	0.51	0.51	0.47	0.86	0.79	-0.63
Cp_{avg}	0.15	-0.18	-0.18	0.27	0.28	0.34	0.58	0.60	0.06
P-value	NCI1	NCI2	NCI3	NSI1	NSI2	NSI3	ADF_{avg}	NBI	I_n
AGD-dia	0.245	0.364	0.022	0.003	0.003	0.009	0.000	0.000	0.000
GD	0.074	0.915	0.136	0.000	0.000	0.001	0.000	0.000	0.008
Deg_{avg}	0.074	0.915	0.136	0.000	0.000	0.001	0.000	0.000	0.008
BC_{avg}	0.246	0.367	0.021	0.003	0.003	0.009	0.000	0.000	0.000
CC_{avg}	0.260	0.315	0.021	0.004	0.004	0.009	0.000	0.000	0.000
Cp_{avg}	0.427	0.329	0.336	0.146	0.129	0.069	0.001	0.001	0.734
Correlation	NCI1	NCI2	NCI3	NSI1	NSI2	NSI3	ADF_{avg}	NBI	I_n
AGD-dia	R	R	95%	99%	99%	99%	100%	100%	99.9%
GD	R	R	R	99.9%	99.9%	99%	100%	100%	99%
Deg_{avg}	R	R	R	99.9%	99.9%	99%	100%	100%	99%
BC_{avg}	R	R	95%	99%	99%	99%	100%	100%	99.9%
CC_{avg}	R	R	95%	99%	99%	99%	100%	100%	99.9%
Cp_{avg}	R	R	R	R	R	R	99.9%	99.9%	R

Table 6.28 Statistical analysis, 15 NGT networks of 50 nodes (fixed D, variable Q)

R-value	NCI1	NCI2	NCI3	NSI1	NSI2	NSI3	ADF_{avg}	NBI	I_n
AGD-dia	0.53	0.44	0.48	0.55	0.44	0.48	-0.05	0.21	0.76
GD	-0.60	-0.40	-0.21	-0.60	-0.40	-0.17	0.54	0.43	0.08
Deg_{avg}	-0.60	-0.40	-0.21	-0.60	-0.40	-0.17	0.54	0.43	0.08
BC_{avg}	0.53	0.46	0.50	0.55	0.46	0.50	-0.03	0.23	0.76
CC_{avg}	-0.54	-0.47	-0.51	-0.55	-0.47	-0.51	0.11	-0.18	-0.68
Cp_{avg}	-0.46	-0.34	-0.10	-0.45	-0.34	-0.06	0.40	0.33	-0.05
P-value	NCI1	NCI2	NCI3	NSI1	NSI2	NSI3	ADF_{avg}	NBI	I_n
AGD-dia	0.040	0.100	0.072	0.033	0.100	0.069	0.862	0.461	0.001
GD	0.017	0.143	0.459	0.018	0.143	0.537	0.037	0.108	0.776
Deg_{avg}	0.017	0.143	0.459	0.018	0.143	0.537	0.037	0.108	0.776
BC_{avg}	0.041	0.087	0.060	0.035	0.087	0.058	0.919	0.420	0.001
CC_{avg}	0.040	0.077	0.054	0.035	0.077	0.052	0.704	0.516	0.006
Cp_{avg}	0.087	0.215	0.723	0.092	0.215	0.840	0.136	0.226	0.869
Correlation	NCI1	NCI2	NCI3	NSI1	NSI2	NSI3	ADF_{avg}	NBI	I_n
AGD-dia	95%	R	R	95%	R	R	R	R	99%
GD	95%	R	R	95%	R	R	95%	R	R
Deg_{avg}	95%	R	R	95%	R	R	95%	R	R
BC_{avg}	95%	R	R	95%	R	R	R	R	99.9%
CC_{avg}	95%	R	R	95%	R	R	R	R	99%
Cp_{avg}	R	R	R	R	R	R	R	R	R

Table 6.29 Statistical analysis, 15 NGT networks of 151 nodes (fixed D, variable Q)

R-value	NCI1	NCI2	NCI3	NSI1	NSI2	NSI3	ADF_{avg}	NBI	I_n
AGD-dia	-0.13	0.03	-0.43	-0.13	0.03	-0.43	-0.45	-0.14	0.23
GD	0.19	0.14	0.72	0.19	0.14	0.72	0.49	0.33	-0.02
Deg_{avg}	0.19	0.14	0.72	0.19	0.14	0.72	0.49	0.33	-0.02
BC_{avg}	-0.13	0.03	-0.43	-0.13	0.03	-0.43	-0.45	-0.14	0.23
CC_{avg}	0.06	-0.12	0.33	0.06	-0.12	0.33	0.48	0.17	-0.17
Cp_{avg}	-0.12	-0.23	0.27	-0.12	-0.23	0.27	0.50	0.28	0.09
P-value	NCI1	NCI2	NCI3	NSI1	NSI2	NSI3	ADF_{avg}	NBI	I_n
AGD-dia	0.648	0.919	0.111	0.648	0.919	0.111	0.092	0.612	0.413
GD	0.508	0.621	0.003	0.508	0.621	0.003	0.066	0.232	0.940
Deg_{avg}	0.508	0.621	0.003	0.508	0.621	0.003	0.066	0.232	0.940
BC_{avg}	0.648	0.919	0.111	0.648	0.919	0.111	0.092	0.612	0.413
CC_{avg}	0.830	0.676	0.226	0.830	0.676	0.226	0.069	0.550	0.550
Cp_{avg}	0.659	0.400	0.336	0.659	0.400	0.336	0.060	0.310	0.759
Correlation	NCI1	NCI2	NCI3	NSI1	NSI2	NSI3	ADF_{avg}	NBI	I_n
AGD-dia	R	R	R	R	R	R	R	R	R
GD	R	R	99%	R	R	99%	R	R	R
Deg_{avg}	R	R	99%	R	R	99%	R	R	R
BC_{avg}	R	R	R	R	R	R	R	R	R
CC_{avg}	R	R	R	R	R	R	R	R	R
Cp_{avg}	R	R	R	R	R	R	R	R	R

Table 6.30 Statistical analysis, 15 NGT networks of 200 nodes (fixed D, variable Q)

R-value	NCI1	NCI2	NCI3	NSI1	NSI2	NSI3	ADF_{avg}	NBI	I_n
AGD-dia	-0.18	-0.34	-0.35	-0.16	-0.32	-0.32	-0.31	-0.12	-0.02
GD	0.58	0.71	0.82	0.60	0.73	0.84	0.42	0.18	0.28
Deg_{avg}	0.58	0.71	0.82	0.60	0.73	0.84	0.42	0.18	0.28
BC_{avg}	-0.18	-0.34	-0.35	-0.16	-0.32	-0.32	-0.31	-0.12	-0.02
CC_{avg}	0.13	0.28	0.28	0.12	0.27	0.27	0.29	0.09	0.07
Cp_{avg}	0.18	0.50	0.37	0.21	0.53	0.40	0.34	0.13	-0.06
P-value	NCI1	NCI2	NCI3	NSI1	NSI2	NSI3	ADF_{avg}	NBI	I_n
AGD-dia	0.522	0.217	0.207	0.574	0.250	0.239	0.261	0.664	0.947
GD	0.024	0.003	0.000	0.019	0.002	0.000	0.115	0.514	0.313
Deg_{avg}	0.024	0.003	0.000	0.019	0.002	0.000	0.115	0.514	0.313
BC_{avg}	0.522	0.217	0.207	0.574	0.250	0.239	0.261	0.664	0.947
CC_{avg}	0.641	0.307	0.307	0.669	0.326	0.326	0.288	0.754	0.793
Cp_{avg}	0.524	0.057	0.171	0.459	0.042	0.138	0.208	0.648	0.825
Correlation	NCI1	NCI2	NCI3	NSI1	NSI2	NSI3	ADF_{avg}	NBI	I_n
AGD-dia	R	R	R	R	R	R	R	R	R
GD	95%	99%	99.9%	95%	99%	99.99%	R	R	R
Deg_{avg}	95%	99%	99.9%	95%	99%	99.99%	R	R	R
BC_{avg}	R	R	R	R	R	R	R	R	R
CC_{avg}	R	R	R	R	R	R	R	R	R
Cp_{avg}	R	R	R	R	95%	R	R	R	R

Finally, it is remarkable to see that the results in Tables 6.25 to 6.27 show better correlations than those in Tables 6.28 to 6.30. Especially poor correlation is registered in relation to ADF_{avg} and NBI in case of the NGT networks, which could be a consequence of the impact of hydraulic parameters, influenced by the selected pipe diameters i.e. increased resistance.

6.10 CONCLUSIONS

The research in this chapter aims to analyse impacts of node connectivity on network reliability. Two approaches have been compared to evaluate this connectivity: one which does it in relative terms and in simplified way, by taking into consideration the connectivity of the two most dominant node categories (in three different ways), and the other which considers connectivity measures based on graph theory. The first approach has been coded within so called *network diagnostics tool* (NDT), while the NodeXL spreadsheet template in public domain has been used for the second approach. The mutual correlation has been analysed, and for each approach separately with the reliability measures discussed in Chapter 5, namely the ADF_{avg}, NBI and I_n. The test networks have included synthetic sets of various degrees of complexity and fixed/variable nodal demands and pipe diameters. The conclusions from the research can be formulated in the following bullets:

1. The results show variable correlations, which are mostly weak but indicate potential for further research. While it is obvious that the connectivity measures alone can hardly be correlated to the reliability measures that are also based on pipe geometry and network hydraulic performance, some of these could possibly be considered in combination with additional network input parameters influencing network energy balance. Although drawn

on statistically small sample of networks and possibly insufficiently different in the level of connectivity, it is believed that this conclusion would not change in case of larger and more diverse sets of networks.

2. The approach based on graph theory has shown more consistent results. It is however that more simple measures such as *GD* and *Deg~avg~*, correlate better than the more complicated ones. It is yet to be investigated if the elements of graph theory can sufficiently be integrated into the hydraulic aspects of water distribution networks, for instance by using weighted graphs to reflect pipe flows or resistances. The difference between undirected and directed graphs, the latter coinciding with flow directions in the pipes, analysed in this research has been insignificant. This is the reason why all the NodeXL results have been presented for undirected graphs.

3. The NDT connectivity measures show less consistency than the NodeXL results. They appear to be rather coarse and reflect the difference in connectivity only in case of visibly different network layouts. It is here as well that simpler *NGI* and *NCF~avg~* correlate often better than more complex *NCIx* and *NSIx*. Also, having the most of these values in the range between 0 and 1, regardless the number of nodes/links, gives impression that these factors/indices would work better as parameters of more complex formula including the network resistance and supplying heads. Having lower value of connectivity index for much larger network can be confusing at a time.

4. The preference shown for the Spearman test compared to the Pearson test would need to be verified on a larger sample of data sets i.e. evaluated networks. Nevertheless, the preliminary results obtained by the Pearson test show that the degree of correlation would not lead to different conclusions in case this test was used for more complete statistical analysis instead of the Spearman test.

5. Although used mostly for social networks, NodeXL has proven to be pretty convenient although lacking direct link with EPANET software. The range of features enabling good visual presentation of graphs gives impression that this MS Excel template can be used for more water distribution analysis, for instance for network zoning, district metering areas, etc.

REFERENCES

1. *Evolving Objects (EO)*, distributed under the GNU Lesser General Public License by SourceForge (http://eodev.sourceforge.net)
2. Jain, S. P., and Gopal, K. (1988). *An efficient algorithm for computing global reliability of a network*. IEEE Transactions on Reliability, 37(5), p 488-492.
3. Jamaković, A., and Uhlig, S. (2007). *Influence of the network structure on robustness*. Networks, 2007. ICON 2007. 15th IEEE International Conference, p 278-283
4. Jamaković, A., and Uhlig, S. (2008). *On the relationships between toplogical measures in real-world network*. Networks and heterogeneous media, 3(3), p 345-359.
5. Kansal, M. L., Kumar, A., and Sharma, P. B. (1995). *Reliability analysis of water distribution systems under uncertainty*. Reliability Engineering & System Safety, 50(1), p 51-59.
6. Kansal, M. L., and Devi, S. (2007). *An improved algorithm for connectivity analysis of distribution networks*. Reliability Engineering & System Safety, 92(10), p 1295-1302.
7. Prasad, T. D., and Park, N.S. (2004). *Multiobjective genetic algorithms for design of water distribution networks*. Journal of Water Resources Planning and Management. ASCE, 130(1), p 73-82.
8. Smith, M., Hansen, D., and Shneiderman, B. (2009). *Analysing social media networks: learning by doing with NodeXL*. e-print: www.codeplex.com/nodexl
9. Yazdani, A., and Jeffrey, P. (2010). *A complex network approach to robustness and vulnerability of spatially organized water distribution networks*. Cornell University Library, e-print: http://arxiv.org/abs/1008.1770v2/

CHAPTER 7

Diagnostics of Regular Performance of Water Distribution Networks and its Relation to the Network Reliability[1]

The research presented in this chapter aims to assess the potential of water distribution networks to sustain a certain level of failure based on analysis of their operational parameters under regular supply conditions, namely the pressures, flows and head loss distribution. For this purpose, the PDD network diagnostics tool (NDT) developed initially for the connectivity analysis has been expanded by adding three indices proposed to describe the network operation, namely the *network power index* (NPI), the *pressure buffer index* (PBI), and the *network residence time* (NRT). Correlation analyses have been done on the sets of networks studied in Chapter 6, using the same reliability measures, namely the ADF_{avg}, NBI and I_n. Furthermore, several statistical analyses have been conducted to check the correlations between various pipe properties and the loss of demand after the pipe failure. The conclusions based on the simulation results point fairly good matching between the reliability measures. The pipe correlations of geometric and hydraulic parameters show less consistent patterns. Different network categories have been showing different level of correlations of particular parameters but no consistent conclusions could have been made; the patterns will largely be related to the geometrical properties of pipes and their connectivity with nodes.

[1] Paper submitted by Trifunović, N. and Vairavamoorthy, K., (2012) under the title *Diagnostics of Resilience During Regular Operation of Water Distribution Networks*, to the Journal of Urban Water; under review.

7.1   INTRODUCTION

That the early diagnostics of diseases is essential for successful cure is all too obvious in medicine but is generally true for treating any other problem. Water distribution networks (WDNs) are complex systems that can be compared with live organisms and their failure to provide expected level of service can result from combination of factors related to mistakes made in any single step of the system development, such as improper planning, design, construction, operation and maintenance.

Hydraulic models can point problems caused by poor planning, design and operation. In early days of hydraulic computer modelling this would be done for limited number of scenarios or after some damage has already been inflicted. With the development of faster algorithms and tools, more substantial reliability analyses could be conducted for preventive purposes, usually by assuming a number of possible irregular scenarios, and then prepare an adequate response in advance; yet, bearing in mind that such a 'library of calamities' is never complete. This is to large extent still the practice and there are several reasons for that. Most importantly, every WDN is different, more or less, in the way it looks and is operated. Consequently, the attempts to categorise WDNs according to their reliability resulting from the way they have been designed and operated under regular conditions are rare in literature. Next to the uniqueness of WDNs, the latest optimisation techniques are still short of computational power to process large network models (say, over 1000 pipes) within reasonable period of time. Furthermore and also important, the libraries of real case networks for reliability analyses are rarely large and consistently variable, as far the properties, to lead to credible conclusions out of hydraulic simulation results. It is mostly the synthetic networks used for benchmarking, or isolated case studies used to illustrate particular concept or method, which are available in the literature. The need therefore exists to develop a large sample of networks where the impact of alterations of network simulation parameters can be monitored continuously, which has been discussed in Chapter 4. Last but not the least, quite some failures result from poor workmanship in the construction and maintenance phase, which is practically impossible to build accurately into a hydraulic model input. In addition, it is often that the failure records are maintained with little care or are insufficiently long to draw firm conclusions about vulnerability of particular network components.

The concept of network resilience developed by Todini (2000) and improved by Prasad and Park (2004) deals with the capacity of network to sustain certain level of calamity based on the network configuration and provision of energy at the sources or booster stations, where applicable. This concept, discussed in Chapter 5, is initially promising but relies too much on the supplying heads and not enough on the network configuration i.e. the geometry. It appears therefore to be less successful in the description of network buffer than the proposed network buffer index (NBI). The derivation of *NBI* however asks for pressure-demand driven hydraulic simulations under the failure conditions. The attempt in this chapter was to explore other measures of network buffer, which can be derived based on statistical analyses of the common parameters under regular operation. These measures have been mostly based on the hydraulic properties of the network, taking also into consideration the energy balance between the source(s) and discharge point(s).

7.2   HYDRAULIC PROPERTIES AS INDICATORS OF NETWORK RELIABILITY

7.2.1 Network Power Balance

As mentioned in Chapter 6, the power balance in any water distribution network consists of the power input P_i made at location of supply points (sources and service tanks discharging their volume) and pumping stations:

$$P_i = \sum_{s=1}^{l} Q_s H_s + \sum_{p=1}^{k} Q_p h_p \qquad\qquad 7.1$$

the power loss P_l inflicted by the friction- and minor losses:

$$P_l = \sum_{j=1}^{m} Q_j h_{f,j} + \sum_{v=1}^{t} Q_v h_{m.v} \qquad\qquad 7.2$$

and residual power P_r at discharge points:

$$P_r = \sum_{i=1}^{n} Q_i H_i \qquad\qquad 7.3$$

The parameters in Equations 7.1 to 7.3 have the following meaning: H_x ($x = s, i$) indicate the piezometric heads at l sources, and n nodes, which includes all the junctions with discharge and tanks that are supplied from the network, respectively. The values of h_x ($x = p, f, m$) include the heads of k pumps, the pipe friction losses of m pipes and the minor losses of t major valves, respectively.   Finally, Q_x ($x = s, p, i, j, v$) stands for corresponding flow supplying the network (s), being conveyed through it (p, j and v), or withdrawn from it (i), respectively. Hence, $P_i = P_r + P_l$. The network potential to supply the nodal demands Q_i and heads H_i as per design can be assessed against the ratio of the power loss and power input. The *network power index* (NPI) would therefore be:

$$NPI = 1 - \frac{P_l}{P_i} \qquad\qquad 7.4$$

The hypothesis tested in this research is that there is a correlation between *NPI* and network reliability. For theoretical range of values between 0 and 1, the higher value of *NPI* will indicate the more reliable system. This index is simplified version of the resilience indices of Todini (2000) and Prasad and Park (2004), which does not take the pressure threshold i.e. the surplus head/power but the total input head/power into consideration. In this case, the impact of the pressure threshold is observed while running the pressure demand driven (PDD) simulations that is not the situation with the two resilience indices from the literature.

7.2.2 Network Pressure Buffer

The length of pipes is indirectly included into the power balance in Equation 7.2 (and consequently in Equation 7.4) through the amount of friction loss h_{fj} that grows proportionally to the pipe length. Analyses of the network pressure buffer expose more significant impact of pipe lengths and nodal elevations i.e. the topography on the network reliability.

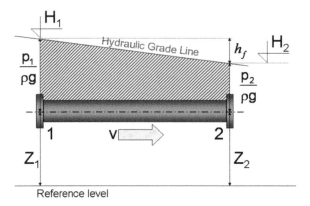

Figure 7.1 Pipe pressure buffer

The dashed area shown in Figure 7.1 represents this buffer. For a pipe section 1 – 2 having the length of L_{1-2}, the pressure buffer is calculated from the dashed polygon:

$$pb_{1-2} = \frac{p_1 + p_2}{2\rho g}\sqrt{L_{1-2}^2 - (z_1 - z_2)^2} \qquad 7.5$$

Similarly, the area of the triangle defined by the friction loss is calculated as:

$$hl_{1-2} = \frac{h_{f,1-2}}{2}\sqrt{L_{1-2}^2 - (z_1 - z_2)^2} \qquad 7.6$$

An index that describes the ratio of the values from Equations 7.5 and 7.6 will eventually be indirectly dependent on the pipe length hidden in the friction loss value, as is the case in Equations 7.2 and 7.4. The *pressure buffer index* of pipe j, PBI_j, is calculated as:

$$PBI_j = 1 - \rho g \frac{h_{f,j}}{p_{1,j} + p_{2,j}} \qquad 7.7$$

and for the entire network:

$$PBI = \frac{1}{m}\sum_{j=1}^{m} PBI_j \qquad 7.8$$

Alternatively, the value of *PBI* can be determined by summing the pressure buffer and head loss in the entire network, which is done in Equations 7.9 and 7.10.

$$pb_n = \frac{1}{2\rho g} \sum_{j=1}^{m} \left(p_{1,j} + p_{2,j}\right)\sqrt{L_j^2 - \Delta z_j^2}$$ 7.9

Similarly, the area of the triangle defined by the friction loss for the entire network is calculated as:

$$hl_n = \frac{1}{2} \sum_{j=1}^{m} h_{f,j}\sqrt{L_j^2 - \Delta z_j^2}$$ 7.10

Possible negative values of pressure (in either or both nodes) in Equations 7.5, 7.7 and 7.9, which can occur both in DD- and PDD mode of hydraulic simulation, mean that the buffer will be reduced, and can theoretically be negative for the entire network; this is a signal of serious calamity when calculated in DD-mode. Furthermore, Equation 7.7 can in theory result in division by zero, if sum of the two pressures equals zero. Alternatively, the *PBI* can be calculated as:

$$PBI_n = 1 - \frac{hl_n}{pb_n}$$ 7.11

Mathematically, Equations 7.7 and 7.11 do not yield the same result but will show the same trend; index *n* has been added to the *PBI* in Equation 7.11 to emphasize this difference. As is the case with *NPI*, the higher value of *PBI* will indicate more reliable system.

7.2.3 Network Residence Time

The parameter that could also serve as potential indicator of network reliability assessed under regular operational conditions is the *network residence time* (NRT). The *NRT* value can be determined as the ratio of total volume and total demand of the network, which yields the result expressed in units of hrs. By adding the weighting proportional to the pipe flow rates Q_j under regular conditions, to the individual residence time for *m* pipes, the network value becomes:

$$NRT = \sum_{j=1}^{m} \frac{V_j}{Q_j}\frac{Q_j}{Q_{tot}} = \frac{V_{tot}}{Q_{tot}}$$ 7.12

As such, the residence time combines in a simplified way the network geometry with its hydraulic performance. It is mostly used as an indicator of water quality in distribution networks but may have potential link with network reliability, as well. For the same level of demand, the networks with larger volume will have larger conveying capacity assuming that the pipe flow rates i.e. the head loss will be reducing after adding bigger- or additional pipes. Eventually, the higher residence time would mean the more reliable network.

7.2.4 Network Diagnostics Tool

To be able to check the correlations of the proposed measures, the network diagnostics tool (NDT) used in the analyses done in Chapter 6 has been expanded with additional results, namely:

1. network residence time (NRT),
2. network power input (P_i),
3. network power loss (P_l),
4. total residual power in discharge points (P_r),
5. network power index (NPI),
6. network pressure buffer (pb_n),
7. network friction loss (hl_n),
8. pressure buffer index (PBI),

All the additional measures can be calculated both in the demand-driven (DD) and pressure-driven demand (PDD) mode. The PDD calculations done for single network include the values of residence time, power loss, pressure buffer, and pressure loss for each pipe.

7.3   SIMULATION RUNS

The simulation runs have been executed on a sample of ten networks of different properties, selected from the batches used in the analyses in Chapter 6. The properties of the networks are given in Table 7.1 and the layouts shown in Figure 7.2.

Table 7.1 Synthetic networks used for comparison of reliability measures

No.	Name	Nodes	Links	Loops	L_{tot}	V_{tot}	Q_{pdd}	D range	p/ρg range	H_s / z range
		(-)	(-)	(-)	(km)	(m³)	(l/s)	(mm)	(mwc)	(msl)
1	sn73	77	79	3	10.4	2045	3179	500 fixed	20-81	160 / 50 fixed
2	sn80	77	109	33	16.0	3142	4210	500 fixed	20-37	160 / 50 fixed
3	O20sn73	77	79	3	10.4	740	2280	100-650	20-97	160 / 50 fixed
4	O20sn80	77	109	33	16.0	824	2280	100-750	20-101	160 / 50 fixed
5	sn0179	77	79	3	11.3	2214	2044	500 fixed	20-48	160 / 50 fixed
6	sn0199	77	99	23	13.4	2630	4127	500 fixed	20-45	160 / 50 fixed
7	sn0109	77	109	33	18.4	3618	2011	500 fixed	20-57	160 / 50 fixed
8	ngt01-50	50	66	17	77.4	608	15	100 fixed	20-49	60 / 0.2-18.8
9	ngt01-151	151	233	83	254.5	1999	15	100 fixed	20-41	60 / 0.2-20
10	ngt01-200	200	299	100	290.9	2285	19	100 fixed	20-47	60 / 0.2-19.8

These networks have been selected to provide a variety of connectivity, demands, pressures and pipe resistances. The *O20sn73* and *O20sn80* are the versions of *sn73* and *sn80* with GA-optimised pipe diameters for the minimum pressure of 20 mwc. In all other networks, the diameter has been kept fixed with the demands adjusted to satisfy the same pressure threshold. The last three networks have pipes with considerably longer lengths and smaller diameters. Also, the nodal elevations have been randomly generated in these networks. Finally, the pipe roughness value of 0.5 mm has been used in all the cases.

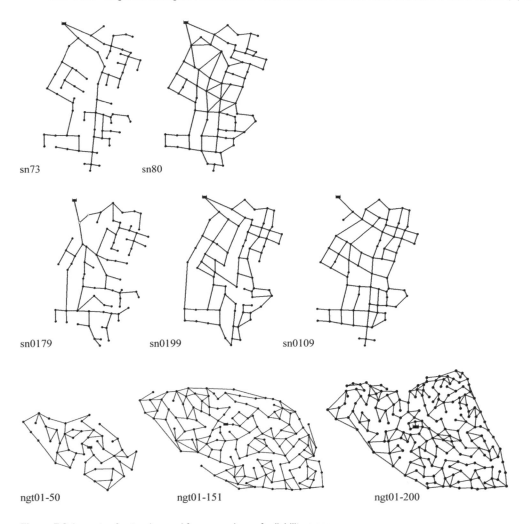

Figure 7.2 Layouts of networks used for comparison of reliability measures

7.4 CORRELATION OF RELIABILITY MEASURES WITH DEMAND GROWTH

In the first run of simulations, each network has been assessed for the demand growth of 32%. This has been achieved by adjusting the general demand multiplier exponentially at the increment of 2% (gradually, in 14 steps). The average value of available demand fraction (ADF_{avg}) has been calculated for the PDD threshold of 20 mwc. The results in Figures 7.3 to 7.6 show the correlation of *NPI*, *PBI* and *NRT* with ADF_{avg}; the trends of growth are obvious for all three measures.

The results in Figure 7.3 show the drop of ADF_{avg} from 0.99 to approximately 0.79, caused by the demand increase. Furthermore, the better connected network *sn80* is more reliable although with the lower values of *NPI*, *PBI* and *NRT* compared to *sn73*; this is the result of larger demand in *sn80*, which can be observed in Table 7.1.

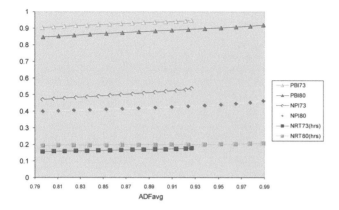

Figure 7.3 Reliability measures for network sn73 and sn80 at uniform demand growth of 32%

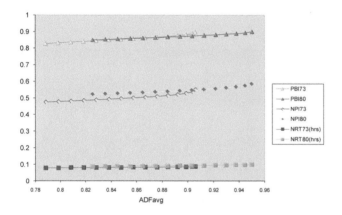

Figure 7.4 Reliability measures for network O20sn73 and O20sn80 for demand growth of 32%

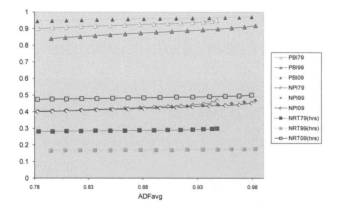

Figure 7.5 Reliability measures for networks sn 0179, 0199 and 0109 for demand growth of 32%

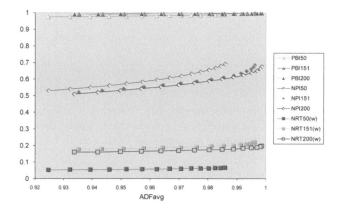

Figure 7.6 Reliability measures for networks ngt 01-50, 01-151 and 01-200 for demand growth of 32%

Figure 7.4 shows two networks with GA optimised pipe diameters supplying the same total demand. Again, the network *O20sn80* is more reliable but in this case has rather similar buffer compared to *O20sn73*. Although optimised for lower total demand, the network *O20sn73* is less reliable than *sn73*, as well as is the case with *O20sn80* and *sn80*, which can be seen by comparing the results in Figures 7.3 and 7.4.

Figure 7.5 shows also that the better connected networks, *sn0199* and *sn0109*, are more reliable than *sn0179*. These two perform within similar ADF_{avg} range despite the fact that *sn0199* supplies more than twice the demand of *sn0109*. This is possible as this network is the only of the three that is connected with two pipes to the source. Nonetheless, the higher demand results in the lower *PBI* and *NRT* values, also compared to *sn0179*.

Although kept with fixed diameters of 100 mm only, the geometry of the networks *ngt 01-50*, *01-151* and *01-200*, shows more realistic values of *NRT*, in the range of 9 to 37 hours. In Figure 7.6, these values have been expressed in weeks, to maintain the same scale of the Y-axis as in the previous three diagrams. Like in the case of three *sn01* networks, the better connected networks, *ngt01-151* and *ngt01-200*, are more reliable than *ngt01-50*. Equally, these two networks have the similar ADF_{avg} range because the bigger one, *ngt01-200*, also supplies the higher demand than *ngt01-151*. Consequently, the smaller network has a bit more buffer expressed in the difference between the values for *NPI* and *NRT*. The higher *NPI* value of *ngt01-50* compared to the other two is due to lower nodal elevations.

The similar trend can be observed analysing the *PBI* values. The two bigger networks show similar values and more favourable range of ADG_{avg} than the smallest one. The difference in the *PBI* values is however much smaller than in case of *NPI* and *NRT*, likely resulting from central positions of the source i.e. the lower level of friction losses. Last but not the least, the generally higher values of ADG_{avg} observed in case of the three *ngt* networks are also caused by good connectivity of the source. Hence, the preliminary analysis of all the figures confirms the correlations that can be summarised in the following bullets:
- Demand increase of a network causes higher friction losses (i.e. the lower nodal pressures/demands) leading to lower values of ADF_{avg}, *PBI*, *NPI* and *NRT*.
- At the same level of demand, better connected networks will have more favourable range of ADF_{avg}.

- The values of *PBI*, *NPI* and *NRT* appear to differ less in optimised networks with fixed number of nodes and variable number of pipes.
- *NPI* values are more susceptible to nodal elevations (i.e. the pressures) than are the *PBI* values.
- *PBI* values are more susceptible to the demand growth (i.e. the friction losses) than are the *NPI* values.

The observations from the figures confirm logical responsiveness given the nature of the analysed measures. The statistical analysis of the correlation between the reliability measures has been done by applying the Pearson's test; the results are shown in Table 7.2.

Table 7.2 Pearson correlation of reliability measures for demand growth of 32 %

	sn73Q2			sn80Q2			O20sn73Q2			O20sn80Q2		
	PBI	NPI	NRT	PBI	NPI	NRT	PBI	NPI	NRT	PBI	NPI	NRT
ADF_{avg}	0.99	0.98	0.89	1.00	0.98	0.98	0.99	0.95	0.97	0.99	0.97	0.99
NBI	0.99	0.98	0.89	1.00	0.97	0.97	0.98	0.94	0.96	0.98	0.96	0.98
I_n	0.97	1.00	0.96	1.00	0.97	0.97	0.99	0.96	0.98	0.99	0.96	0.98

	sn0179Q2			sn0199Q2			sn0109Q2		
	PBI	NPI	NRT	PBI	NPI	NRT	PBI	NPI	NRT
ADF_{avg}	0.99	0.96	0.95	1.00	0.98	0.98	1.00	0.99	0.99
NBI	0.99	0.94	0.94	0.99	0.96	0.96	1.00	0.98	0.98
I_n	0.99	0.96	0.96	1.00	0.97	0.97	1.00	0.98	0.98

	ngt01-50Q2			ngt01-151Q2			ngt01-200Q2		
	PBI	NPI	NRT	PBI	NPI	NRT	PBI	NPI	NRT
ADF_{avg}	0.99	0.98	0.96	0.98	0.95	0.93	0.99	0.97	0.95
NBI	0.99	0.98	0.96	0.98	0.95	0.93	0.98	0.97	0.94
I_n	1.00	1.00	0.99	1.00	1.00	0.99	1.00	1.00	0.99

The results in the table, which includes the network buffer index (*NBI*) and the network resilience (I_n) discussed in Chapters 5 and 6, show generally good correlation: specifically good in case of the PBI values and less good in case of the NRT values. As such, the table verifies the impressions obtained from Figures 7.3 to 7.6.

7.5   CORRELATION OF RELIABILITY MEASURES WITH DIAMETER INCREASE

The above networks have been analysed for the scenario of stress gradually increased by adapting the demand multiplier. Very soon after the first few increments applied, the PDD simulation will cause the loss of demand, even in no failure condition. In this case, the actual demand increase is lower than 32% which has been calculated against the initial i.e. baseline demand. Still, the 'reduced' demand increase will lead to negative values of some reliability measures, in the first place the *NBI* and I_n.

The following set of simulations has been conducted to analyse the sensitivity of the reliability measures when buffer is added to the networks. This has been achieved by increasing the diameters of each network for 32%, in the same way as with the demand: 2% in 14 consecutive runs. A general diameter multiplier has been built in the NDT code in the similar way as the demand multiplier exists in EPANET. This multiplier has been applied uniformly for all the pipes and the change of the reliability measures was monitored against the increase of the total network volume. The selection of results is shown in Figures 7.7 to 7.9.

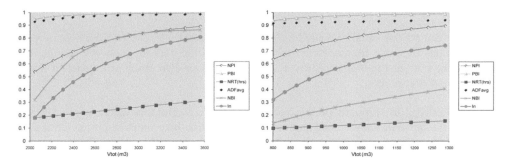

Figure 7.7 Reliability measures for networks sn73 (left) and O20sn73 (right) for D increase of 32%

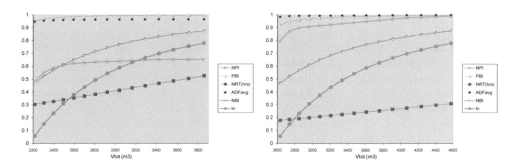

Figure 7.8 Reliability measures for networks sn0179 (left) and sn0199 (right) for D increase of 32%

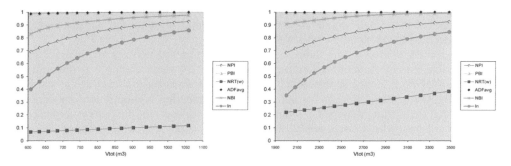

Figure 7.9 Reliability measures for networks ngt01-50 (left) and ngt01-151 (right) for D increase of 32%

The figures show the values of the reliability measures for enlarged network capacity that is described by the total pipe volume. Good correlations are visible in almost all the cases and they look specifically good between the values of *NPI* and I_n.

More in details, the optimised layout of the network *sn73* in Figure 7.7 (right) shows clearly the lower values of ADF_{avg} and *NBI* compared to the overdesigned network on the left, despite the fact that the optimised network supplies the lower demand. This fact will also be the biggest contributor to the comparable values of *NPI*, *PBI* and I_n for the two networks, while the lower NRT value of the optimised layout results from the lower total volume.

Figure 7.8 depicts the situation where the network of bigger capacity (on the right) supplies also the bigger demand. Some discrepancy exists in the fact that this network has the lower NRT values and yet the higher *NBI* and ADF_{avg}, the latter also being an impact of the two connections to the source.

The most difficult to compare are the results in Figure 7.9. Both networks supply the same demand with the same source head and the one the right is three times bigger in terms of the number of pipes. This is visible in the difference between the values of *NRT*, but one would expect substantial difference in the other measures, too. Still, the values of *NPI* and I_n look slightly lower on the right diagram although the value of *NBI* is expectedly higher; the reason for this has to be found in lower elevations of the smaller network.

The entire networks look fairly to very reliable from the perspective of ADF_{avg}, which gives almost horizontal curve in all the diagrams. The initial impression is that the substantial increase of the pipe diameters and subsequent increase of all other reliability measures does little to improve this index, as it already has quite high value. To get more insight into the correlation between ADF_{avg} and other reliability measures, the similar diagrams as for the growth of demand have been produced in Figures 7.10 to 7.12. The conclusions while observing these figures can be summarised in the following bullets:

-   Regardless the level of demand, all three diagrams show logical response as far the values of ADF_{avg}: the network *sn73* is more reliable than its optimised version (in Figure 7.10), as well as are the networks with larger number of pipes, shown in Figures 7.11 and 7.12.
-   The higher values of *PBI* and *NPI* for optimised layout in Figure 7.10 are the consequence of lower demands i.e. the higher nodal pressures.
-   The different form of the curves for the same index results from the presentation of that index in the different range of ADF_{avg}. After continuous increase of pipe diameters, each network enters the state where this action has little implication for the increase of ADF_{avg}. It is specifically visible in the results of Figure 7.11.
-   Further extrapolation of the index curves of the optimised network in Figure 7.10, by increasing the pipe diameters, would eventually yield the shape ending with the saturation zone of ADF_{avg}. How big this zone will be also depends on the head at the source(s). Equally, the extrapolation of the index curves representing non-optimised networks, by reducing their pipe diameters, would bring them closer to the curves representing optimised networks. Consequently, the anticipated shape of a curve covering the entire range, from optimised diameters to the diameters resulting in no further increase of ADF_{avg}, would be a kind of S-shape. The results shown in Figure 7.8 represent the very end of that curve, zoomed-in.
-   As discussed in relation with Figure 7.9, the higher values of *NPI* shown in Figure 7.12 for *ngt01-50* results from lower nodal elevations of this network.

As was the case with the demand growth analyses, the statistical analysis of the correlation between the reliability measures has been done for the diameter increase in all ten networks, by applying the Pearson's test. The results are shown in Table 7.3.

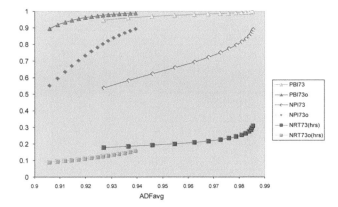

Figure 7.10 Correlation between reliability measures for sn73 and O20sn73 for D increase of 32%

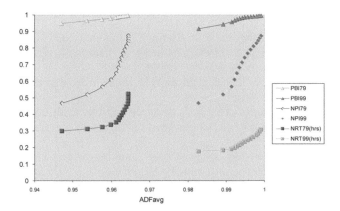

Figure 7.11 Correlation between reliability measures for sn0179 and sn0199 for D increase of 32%

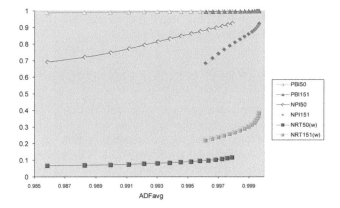

Figure 7.12 Correlation between reliability measures for ngt01-50 and ngt01-151 for D increase of 32%

Table 7.3 Pearson correlation of reliability measures for diameter increase of 32 %

	sn73D2			sn80D2			O20sn73D2			O20sn80D2		
	PBI	NPI	NRT	PBI	NPI	NRT	PBI	NPI	NRT	PBI	NPI	NRT
ADF_{avg}	0.99	0.98	0.89	0.93	0.99	0.96	0.94	0.99	0.99	0.98	1.00	0.97
NBI	0.99	0.98	0.89	0.93	0.99	0.97	0.94	0.99	0.99	0.98	1.00	0.97
I_n	0.97	1.00	0.96	0.93	1.00	0.96	0.98	1.00	0.96	0.98	1.00	0.96

	sn0179D2			sn0199D2			sn0109D2		
	PBI	NPI	NRT	PBI	NPI	NRT	PBI	NPI	NRT
ADF_{avg}	1.00	0.91	0.79	0.98	0.96	0.90	0.77	0.53	0.40
NBI	1.00	0.91	0.78	0.98	0.96	0.90	0.77	0.53	0.39
I_n	0.94	1.00	0.96	0.94	1.00	0.96	0.93	1.00	0.96

	ngt01-50D2			ngt01-151D2			ngt01-200D2		
	PBI	NPI	NRT	PBI	NPI	NRT	PBI	NPI	NRT
ADF_{avg}	1.00	1.00	0.95	0.99	1.00	0.97	0.92	0.88	0.76
NBI	1.00	1.00	0.95	0.99	1.00	0.97	0.92	0.88	0.76
I_n	1.00	1.00	0.97	1.00	1.00	0.97	0.99	1.00	0.97

The results in the table are less consistent than those in Table 7.2, but this comes as no surprise knowing the initial configurations. The correlations seem to start weakening the more overdesigned the network becomes. The table shows:

- the stronger correlations in initially optimised *O20snXXD2* than in corresponding *snXXD2* designed initially with uniform 500 mm diameter;
- the stronger correlations in optimised *ngt01-XX(X)D2*, designed initially with uniform 100 mm diameter, than in corresponding *sn01XXD2*, designed initially with uniform 500 mm diameter;
- hence, the weaker correlations in the networks designed with more/bigger pipes and/or lower demand; the examples are *sn0109D2* against *sn0199D2*, and *ngt01-200D2* against *ngt01-151D2*.

The table also shows generally better correlations of the *PBI* and *NPI* compared to the *NRT*. Excellent correlation in all the cases appears to be between the I_n index and *NPI*, even in the networks where other correlations are much weaker, and despite the fact that the I_n index is the product of DD simulations while the NPI is calculated from the PDD simulations. Such result verifies the impression obtained from Figures 7.7 to 7.9 and is not a surprise in view of the similarities in the approach to determine these two indices.

7.6 NETWORK PROPERIES AND THEIR RELATION TO DEMAND LOSS

Flows in pipes under regular operation are taken as the parameters for calculation of *NBI*, as demonstrated in Chapter 5. It has been shown there that the correlation between the loss of demand, $1 - ADF_j$, caused by the failure of pipe j, and the corresponding flow Q_j under regular supply has a linear tendency in GA-optimised looped networks. The example in Figure 7.13 shows this correlation in the optimised network *O20sn73* (on the left), together with the correlation of the pipe volumes (on the right).

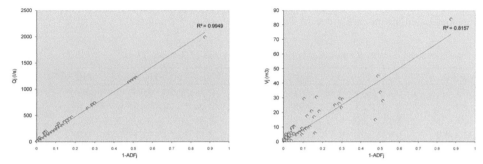

Figure 7.13 Correlation of pipe flows (left) and pipe volumes (right) with the loss of demand - O20sn73

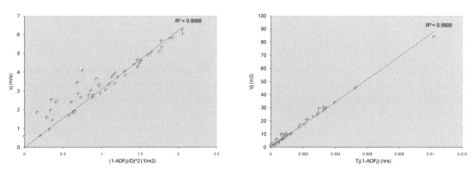

Figure 7.14 Linear correlations of pipe velocities (left) and pipe volumes (right) - O20sn73

The figure shows that the pipe volumes expectedly correlate visibly worse than the flows. Nevertheless, showing them against the loss of demand multiplied by the pipe residence time, T_j, will improve the correlation, as can be seen in Figure 7.14 on the right. Equally, the relation between the pipe velocity and the loss of demand divided by the squared diameter will show linear correlation (Figure 7.14, left). Thus, any pipe parameter when linked to the flow can be expressed against the loss of demand. In networks having more buffer this correlation will however be weaker.

Statistical analyses have been conducted to explore how typical pipe (hydraulic) properties or their derivatives correlate to the loss of demand, and are there specific patterns existing for various types of networks. The following pipe parameters were analysed:
1. volume - V_j,
2. flow - Q_j,
3. velocity - v_j,
4. friction loss - $h_{f,j}$,
5. hydraulic gradient - $S_j = h_{f,j} / L_j$,
6. power loss - $P_j = \rho g Q_j h_{f,j}$,
7. residence time - $T_j = V_j / Q_j$,
8. pressure buffer - pb_j calculated as in Equation 7.5,
9. unit pressure buffer - pb_j / L_j,
10. hydraulic loss area - hl_j, calculated as in Equation 7.6,
11. pressure buffer index - PBI_j calculated as in Equation 7.7.

The results for simulation runs of the ten networks, keeping the same initial parameters as in the previous calculations, are given in Table 7.4.

Table 7.4 Pearson correlation of network pipe properties to the loss of demand, $1 - ADF_j$

No. Name	V_j	Q_j	v_j	$h_{f,j}$	S_j	P_j	T_j	pb_j	pb_j/L_j	hl_j	PBI_j
1 sn73	0.45	0.96	0.96	0.87	0.96	0.77	-0.58	0.50	0.30	0.77	-0.80
2 sn80	0.39	0.83	0.83	0.99	0.98	1.00	-0.08	0.19	-0.17	0.98	-1.00
3 O20sn73	0.91	1.00	0.72	0.18	0.08	0.85	-0.36	0.66	0.67	0.16	-0.01
4 O20sn80	0.93	0.98	0.53	0.00	-0.02	0.87	-0.16	0.29	0.34	0.04	0.07
5 sn0179	0.26	0.85	0.85	0.97	0.97	0.95	-0.30	0.29	0.14	0.96	-0.96
6 sn0199	0.39	0.84	0.84	0.99	0.98	0.98	-0.10	0.31	0.05	0.93	-0.98
7 sn0109	0.02	0.90	0.90	0.95	0.99	0.96	-0.09	0.10	0.39	0.88	-0.90
8 ngt01-50	-0.20	0.80	0.80	0.93	0.96	0.99	-0.07	-0.25	-0.32	0.87	-0.98
9 ngt01-151	-0.10	0.80	0.80	0.95	0.97	0.98	-0.02	-0.16	-0.28	0.89	-0.94
10 ngt01-200	-0.11	0.66	0.66	0.83	0.87	0.95	-0.03	-0.12	-0.08	0.73	-0.91

Although the table shows inconsistent results, a few interesting conclusions can be made:
- There is a clear distinction in the correlation between the GA-optimised networks, O20snXX, and other oversized networks. In general, the correlation between the loss of demand and pipe parameters which involve friction loss is fairly good in oversized nets, whilst it is very bad in the two optimised nets. Oppositely, the optimised nets show reasonable correlation of pipe volumes with the demand loss, whilst this correlation is weak in oversized networks.
- The weaker correlation of flows in oversized nets is not a surprise; those have been used to assess the buffer through *NBI*. Hypothetically, the weaker the correlation between the flows and loss of demand could mean the more reliable network. The lowest value of the Person's index of 0.66 for the biggest network, *ngt01-200*, is signalling this.
- The most consistent correlations across the entire range of networks are for the power loss, P_j, which are pretty good for oversized networks and moderately good for optimised ones. Furthermore, the correlations of the pipe residence time, T_j, are consistently bad, hinting this parameter as inappropriate indicator of potential loss of demand in case of the pipe failure.

To check the consistency of results for optimised networks, all 30 *SN*-nets analysed in Chapter 6 have been optimised and the correlation of the same parameters done; the results are shown in Table 7.5. Eventually, the same trend has been observed for all the optimised networks as concluded based on the analysis of the results in Table 7.4. The best correlation is made with pipe flows, the reasonable one with the volumes and the moderate one with power loss. All other parameters show poor correlation.

To get some insight into the pattern of correlation in case of the demand increase, the two optimised networks from Table 7.4 have been tested for the total demand increase of 32% (*O20sn73*) and 51% (*O20sn80*). The results of correlations are given in Tables 7.6 and 7.7, respectively. No matter how good or bed the correlations are, the trend of improvement following the growth of demand is visible in both tables. This of course comes with the fact that both networks are becoming increasingly unreliable with the growth of demand.

Table 7.5 Correlation of GA-optimised network pipe properties to the loss of demand, $1 - ADF_j$

No.	Name	V_j	Q_j	v_j	$h_{f,j}$	S_j	P_j	T_j	pb_j	pb_j/L_j	hl_j	PBI_j
1	O20sn2	0.90	0.98	0.54	-0.05	-0.06	0.90	-0.20	0.24	0.29	-0.03	0.09
2	O20sn6	0.93	0.98	0.64	0.07	-0.02	0.85	-0.09	0.29	0.28	0.14	0.03
3	O20sn8	0.95	0.99	0.62	-0.06	-0.06	0.87	-0.08	0.30	0.46	-0.06	0.12
4	O20sn12	0.93	0.99	0.46	0.00	-0.01	0.79	-0.07	0.37	0.49	0.00	0.09
5	O20sn13	0.94	0.97	0.53	-0.05	-0.12	0.85	-0.11	0.49	0.37	0.03	0.13
6	O20sn14	0.94	0.98	0.52	-0.05	-0.07	0.94	-0.05	0.29	0.34	-0.02	0.11
7	O20sn16	0.88	0.92	0.55	-0.04	-0.07	0.86	-0.21	0.34	0.36	-0.01	0.10
8	O20sn17	0.89	0.99	0.61	-0.08	-0.08	0.84	-0.25	0.34	0.57	-0.07	0.14
9	O20sn19	0.94	0.99	0.61	-0.01	-0.04	0.92	-0.16	0.26	0.40	-0.04	0.10
10	O20sn21	0.88	0.97	0.59	-0.03	-0.03	0.75	-0.21	0.24	0.39	-0.05	0.13
11	O20sn23	0.94	0.98	0.60	-0.01	-0.10	0.85	-0.29	0.50	0.44	0.07	0.10
12	O20sn26	0.95	0.95	0.50	-0.03	-0.08	0.90	-0.14	0.37	0.34	-0.02	0.12
13	O20sn31	0.95	0.99	0.55	0.03	0.01	0.89	-0.18	0.43	0.43	0.05	0.09
14	O20sn33	0.96	0.98	0.55	0.03	-0.03	0.92	-0.18	0.43	0.32	0.03	0.07
15	O20sn34	0.93	0.99	0.49	0.02	-0.08	0.86	-0.10	0.49	0.44	0.08	0.09
16	O20sn37	0.74	0.86	0.58	0.03	0.03	0.71	-0.08	0.36	0.52	0.01	0.10
17	O20sn40	0.95	0.98	0.55	0.06	0.03	0.92	-0.07	0.36	0.30	0.03	0.04
18	O20sn46	0.96	0.99	0.55	0.05	0.01	0.87	-0.08	0.36	0.34	0.01	0.04
19	O20sn53	0.85	0.98	0.59	-0.05	-0.04	0.81	-0.06	0.27	0.44	-0.04	0.13
20	O20sn59	0.91	0.98	0.53	-0.04	-0.06	0.86	-0.13	0.28	0.44	-0.04	0.14
21	O20sn69	0.90	0.99	0.62	-0.02	-0.08	0.81	-0.23	0.46	0.44	0.07	0.11
22	O20sn70	0.91	1.00	0.66	0.04	-0.07	0.85	-0.25	0.57	0.54	0.13	0.09
23	O20sn72	0.95	1.00	0.62	0.02	-0.05	0.89	-0.24	0.44	0.51	0.08	0.09
24	O20sn73	0.91	1.00	0.72	0.18	0.08	0.85	-0.36	0.66	0.67	0.16	-0.01
25	O20sn77	0.95	0.95	0.47	-0.01	-0.11	0.90	-0.10	0.39	0.23	0.10	0.06
26	O20sn79	0.84	0.95	0.53	0.06	-0.02	0.72	-0.07	0.34	0.38	0.14	0.04
27	O20sn80	0.93	0.98	0.53	0.00	-0.02	0.87	-0.16	0.29	0.34	0.04	0.07
28	O20sn82	0.94	0.99	0.62	0.04	0.02	0.89	-0.25	0.39	0.40	0.03	0.05
29	O20sn85	0.91	0.98	0.48	-0.10	-0.11	0.83	-0.16	0.33	0.43	-0.07	0.17
30	O20sn88	0.84	0.98	0.58	-0.04	-0.07	0.83	-0.06	0.32	0.39	-0.01	0.13

Table 7.6 Correlation patterns of GA-optimised network at the demand increase of 32% - O20sn73

Name	V_j	Q_j	v_j	$h_{f,j}$	S_j	P_j	T_j	pb_j	pb_j/L_j	hl_j	PBI_j
o20sn73	0.91	1.00	0.72	0.18	0.08	0.85	-0.36	0.66	0.67	0.16	-0.01
o20sn7302	0.91	1.00	0.72	0.18	0.09	0.85	-0.37	0.68	0.68	0.16	0.00
o20sn7303	0.91	1.00	0.72	0.19	0.09	0.85	-0.37	0.69	0.69	0.16	0.00
o20sn7304	0.91	1.00	0.72	0.19	0.09	0.85	-0.37	0.70	0.70	0.17	0.00
o20sn7305	0.92	1.00	0.72	0.20	0.10	0.85	-0.37	0.71	0.70	0.17	0.00
o20sn7306	0.91	1.00	0.72	0.20	0.11	0.86	-0.37	0.71	0.71	0.17	0.01
o20sn7307	0.91	1.00	0.72	0.21	0.11	0.86	-0.37	0.72	0.71	0.17	0.01
o20sn7308	0.91	1.00	0.72	0.21	0.12	0.86	-0.38	0.73	0.72	0.18	0.01
o20sn7309	0.91	1.00	0.73	0.22	0.12	0.86	-0.38	0.73	0.72	0.18	0.01
o20sn7310	0.91	1.00	0.73	0.22	0.13	0.86	-0.38	0.74	0.73	0.18	0.01
o20sn7311	0.91	1.00	0.73	0.23	0.14	0.86	-0.38	0.74	0.73	0.18	0.01
o20sn7312	0.91	1.00	0.73	0.23	0.14	0.86	-0.38	0.75	0.74	0.19	0.02
o20sn7313	0.91	1.00	0.73	0.24	0.15	0.86	-0.38	0.75	0.74	0.19	0.02
o20sn7314	0.91	1.00	0.73	0.24	0.15	0.86	-0.38	0.75	0.74	0.19	0.02
o20sn7315	0.91	1.00	0.73	0.24	0.15	0.86	-0.38	0.76	0.75	0.19	0.02

Table 7.7 Correlation patterns of GA-optimised network at the demand increase of 51% - O20sn80

Name	V_j	Q_j	v_j	$h_{f,j}$	S_j	P_j	T_j	pb_j	pb_j/L_j	hl_j	PBI_j
o20sn80	0.93	0.98	0.53	0.00	-0.02	0.87	-0.16	0.29	0.34	0.04	0.07
o20sn8002	0.93	0.98	0.53	0.01	-0.02	0.87	-0.16	0.30	0.36	0.04	0.08
o20sn8003	0.93	0.98	0.53	0.01	-0.02	0.88	-0.15	0.31	0.37	0.04	0.09
o20sn8004	0.93	0.98	0.54	0.01	-0.01	0.88	-0.15	0.32	0.38	0.04	0.09
o20sn8005	0.93	0.98	0.54	0.02	-0.01	0.88	-0.11	0.33	0.39	0.05	0.10
o20sn8006	0.93	0.98	0.55	0.02	0.00	0.88	-0.07	0.34	0.40	0.05	0.10
o20sn8007	0.93	0.98	0.55	0.03	0.00	0.89	-0.04	0.35	0.41	0.05	0.10
o20sn8008	0.93	0.99	0.55	0.03	0.01	0.89	-0.08	0.35	0.42	0.06	0.11
o20sn8009	0.93	0.99	0.56	0.03	0.01	0.89	-0.11	0.36	0.43	0.06	0.11
o20sn8010	0.93	0.99	0.56	0.04	0.01	0.89	-0.14	0.36	0.43	0.06	0.11
o20sn8011	0.93	0.99	0.57	0.04	0.02	0.90	-0.16	0.37	0.44	0.07	0.12
o20sn8012	0.93	0.99	0.57	0.04	0.02	0.90	-0.18	0.37	0.45	0.07	0.12
o20sn8013	0.93	0.99	0.57	0.05	0.03	0.90	-0.19	0.38	0.45	0.07	0.13
o20sn8014	0.92	0.99	0.58	0.05	0.03	0.90	-0.20	0.39	0.46	0.08	0.13
o20sn8015	0.92	0.99	0.58	0.06	0.03	0.90	-0.21	0.39	0.47	0.08	0.14

The similar test has been done with the two oversized networks: *sn0199* and *ngt01-151*, but in two steps. First, the simulations were run for the initial networks and the total demand increase of 51%, at increments of 3%; the results of these calculations are shown in Tables 7.8 and 7.10 respectively. The repeated runs were done at the demand increment of 5%, i.e. total 98% but starting from the lower level of demand than the original demand. The corresponding results in Tables 7.9 and 7.11 therefore show the situation where the added buffer is slowly being lost until the networks have reached the same initial state. In Table 7.9, it is the simulation *sn019914* done with the same demand as of the network *sn0199* in Table 7.8. In Table 7.11, it is the simulation *ngt01-15113* done with the same demand as of the network *nbt01-151* in Table 7.10. Consequently, Tables 7.8 and 7.10 show the stress situations where the minimum pressure drops below the threshold, yet in non-optimised network, while Tables 7.9 and 7.11 show low demand/high buffer situations.

Table 7.8 Correlation patterns of GA-optimised network at the demand increase of 51% - sn0199

Name	V_j	Q_j	v_j	$h_{f,j}$	S_j	P_j	T_j	pb_j	pb_j/L_j	hl_j	PBI_j
sn0199	0.39	0.86	0.86	0.99	0.99	0.98	-0.11	0.35	0.12	0.93	-0.98
sn0199x02	0.39	0.86	0.86	0.99	0.99	0.98	-0.11	0.35	0.12	0.93	-0.98
sn0199x03	0.39	0.86	0.86	0.99	0.99	0.98	-0.10	0.38	0.17	0.93	-0.97
sn0199x04	0.40	0.87	0.87	0.99	0.99	0.98	-0.09	0.40	0.20	0.93	-0.97
sn0199x05	0.40	0.87	0.87	0.99	0.99	0.98	-0.09	0.41	0.22	0.93	-0.97
sn0199x06	0.40	0.87	0.87	0.99	0.99	0.98	-0.09	0.42	0.24	0.93	-0.97
sn0199x07	0.40	0.87	0.87	0.99	0.99	0.98	-0.08	0.44	0.26	0.93	-0.97
sn0199x08	0.40	0.87	0.87	0.99	0.99	0.98	-0.08	0.45	0.27	0.93	-0.97
sn0199x09	0.40	0.87	0.87	0.99	0.99	0.98	-0.07	0.46	0.29	0.93	-0.97
sn0199x10	0.40	0.87	0.87	0.99	0.99	0.98	-0.06	0.48	0.30	0.93	-0.97
sn0199x11	0.40	0.87	0.87	0.99	0.99	0.98	-0.05	0.49	0.31	0.93	-0.97
sn0199x12	0.40	0.87	0.87	0.99	0.99	0.98	-0.04	0.50	0.33	0.93	-0.97
sn0199x13	0.40	0.87	0.87	0.99	0.99	0.98	-0.03	0.51	0.34	0.93	-0.97
sn0199x14	0.40	0.87	0.87	0.99	0.99	0.98	-0.03	0.52	0.35	0.93	-0.97
sn0199x15	0.40	0.87	0.87	0.99	0.99	0.98	-0.04	0.53	0.36	0.93	-0.97

Table 7.9 Correlation patterns of GA-optimised network at the demand increase of 98% - sn0199

Name	V_j	Q_j	v_j	$h_{f,j}$	S_j	P_j	T_j	pb_j	pb_j/L_j	hl_j	PBI_j
sn019901	0.15	0.48	0.48	0.64	0.70	0.71	-0.03	-0.06	-0.68	0.50	-0.65
sn019902	0.22	0.57	0.57	0.77	0.80	0.83	-0.04	-0.03	-0.79	0.65	-0.78
sn019903	0.28	0.63	0.63	0.85	0.86	0.91	-0.05	-0.01	-0.86	0.76	-0.87
sn019904	0.31	0.66	0.66	0.90	0.89	0.94	-0.05	0.01	-0.89	0.82	-0.91
sn019905	0.33	0.67	0.67	0.92	0.90	0.96	-0.05	0.02	-0.89	0.86	-0.94
sn019906	0.34	0.68	0.68	0.94	0.91	0.97	-0.05	0.04	-0.87	0.88	-0.95
sn019907	0.35	0.69	0.69	0.95	0.91	0.98	-0.05	0.05	-0.84	0.90	-0.96
sn019908	0.36	0.69	0.69	0.95	0.91	0.98	-0.06	0.06	-0.79	0.91	-0.97
sn019909	0.36	0.69	0.69	0.96	0.91	0.99	-0.06	0.08	-0.72	0.92	-0.97
sn019910	0.37	0.69	0.69	0.96	0.91	0.99	-0.06	0.10	-0.62	0.92	-0.98
sn019911	0.37	0.70	0.70	0.97	0.92	0.99	-0.06	0.13	-0.49	0.93	-0.98
sn019912	0.38	0.74	0.74	0.98	0.94	1.00	-0.07	0.19	-0.27	0.94	-0.99
sn019913	0.39	0.82	0.82	0.99	0.98	0.99	-0.09	0.29	0.00	0.94	-0.98
sn019914	0.39	0.86	0.86	0.99	0.99	0.98	-0.11	0.36	0.14	0.93	-0.98
sn019915	0.40	0.87	0.87	0.99	0.99	0.98	-0.09	0.40	0.20	0.93	-0.97

Table 7.10 Correlation patterns of GA-optimised network at the demand increase of 51% - ngt01-151

Name	V_j	Q_j	v_j	$h_{f,j}$	S_j	P_j	T_j	pb_j	pb_j/L_j	hl_j	PBI_j
ngt01-151	-0.10	0.80	0.80	0.95	0.97	0.98	-0.02	-0.16	-0.28	0.89	-0.94
ngt01-151x02	-0.10	0.81	0.81	0.95	0.98	0.98	-0.02	-0.16	-0.26	0.89	-0.94
ngt01-151x03	-0.11	0.82	0.82	0.95	0.98	0.98	-0.02	-0.16	-0.25	0.89	-0.93
ngt01-151x04	-0.11	0.83	0.83	0.96	0.98	0.97	-0.03	-0.15	-0.23	0.90	-0.93
ngt01-151x05	-0.11	0.83	0.83	0.96	0.98	0.97	-0.03	-0.15	-0.21	0.90	-0.93
ngt01-151x06	-0.11	0.84	0.84	0.96	0.99	0.97	-0.04	-0.15	-0.19	0.91	-0.92
ngt01-151x07	-0.11	0.84	0.84	0.96	0.99	0.97	-0.04	-0.14	-0.17	0.91	-0.92
ngt01-151x08	-0.11	0.85	0.85	0.96	0.99	0.97	-0.04	-0.14	-0.15	0.91	-0.92
ngt01-151x09	-0.11	0.85	0.85	0.97	0.99	0.97	-0.04	-0.14	-0.13	0.91	-0.91
ngt01-151x10	-0.11	0.85	0.85	0.97	0.99	0.97	-0.05	-0.13	-0.12	0.91	-0.91
ngt01-151x11	-0.11	0.86	0.86	0.97	0.99	0.97	-0.04	-0.13	-0.10	0.92	-0.90
ngt01-151x12	-0.11	0.86	0.86	0.97	0.99	0.96	-0.01	-0.12	-0.08	0.92	-0.90
ngt01-151x13	-0.11	0.86	0.86	0.97	0.99	0.96	-0.04	-0.12	-0.07	0.92	-0.90
ngt01-151x14	-0.12	0.86	0.86	0.97	0.99	0.96	-0.04	-0.11	-0.06	0.92	-0.89
ngt01-151x15	-0.12	0.87	0.87	0.97	0.99	0.96	-0.03	-0.11	-0.04	0.92	-0.89

All four tables show fairly consistent patterns where in most cases the correlations improve with the growth of demand. This improvement is faster with the reduction of buffer (Tables 7.9 and 7.11), and shows certain saturation with the increase of stress (Tables 7.8 and 7.10). These conclusions are not entirely true for the pipe residence time, T_j, where the trend of increase seems to be reversing once the network $sn0199$ enters the stress conditions, and does not show clear pattern of correlation in $ngt01-151$; also in this network, the correlation patterns for (unit) pressure buffer show reverse trend once the minimum pressure starts dropping below the threshold. This altogether is not quite disturbing because all three parameters show weak correlations with the loss of demand, in general. The only parameter that follows the reversed pattern amongst those who correlate better is the pressure buffer index, PBI_j. It is also to be observed that all reverse trends are more visible in the network $ngt01-151$ that has much smaller diameters/higher resistance than $sn0199$.

Table 7.11 Correlation patterns of GA-optimised network at the demand increase of 98% - ngt01-151

Name	V_j	Q_j	v_j	$h_{f,j}$	S_j	P_j	T_j	pb_j	pb_j/L_j	hl_j	PBI_j
ngt01-15101	-0.02	0.45	0.45	0.66	0.64	0.76	-0.02	-0.08	-0.32	0.63	-0.75
ngt01-15102	-0.03	0.50	0.50	0.71	0.70	0.81	-0.02	-0.09	-0.34	0.69	-0.81
ngt01-15103	-0.04	0.56	0.56	0.77	0.76	0.87	-0.02	-0.11	-0.36	0.74	-0.86
ngt01-15104	-0.05	0.61	0.61	0.82	0.82	0.91	-0.02	-0.13	-0.38	0.78	-0.90
ngt01-15105	-0.07	0.66	0.66	0.86	0.87	0.95	-0.03	-0.14	-0.39	0.82	-0.93
ngt01-15106	-0.07	0.70	0.70	0.89	0.90	0.97	-0.03	-0.15	-0.40	0.85	-0.95
ngt01-15107	-0.08	0.73	0.73	0.91	0.93	0.98	-0.02	-0.16	-0.39	0.86	-0.96
ngt01-15108	-0.09	0.75	0.75	0.93	0.94	0.99	-0.02	-0.16	-0.39	0.87	-0.97
ngt01-15109	-0.09	0.77	0.77	0.94	0.96	0.99	-0.02	-0.16	-0.37	0.88	-0.97
ngt01-15110	-0.10	0.78	0.78	0.94	0.96	0.99	-0.03	-0.16	-0.36	0.88	-0.96
ngt01-15111	-0.10	0.79	0.79	0.94	0.97	0.98	-0.02	-0.16	-0.34	0.88	-0.95
ngt01-15112	-0.10	0.79	0.79	0.95	0.97	0.98	-0.03	-0.16	-0.31	0.88	-0.95
ngt01-15113	-0.10	0.80	0.80	0.95	0.97	0.98	-0.03	-0.16	-0.29	0.88	-0.94
ngt01-15114	-0.10	0.81	0.81	0.95	0.98	0.98	-0.02	-0.16	-0.26	0.89	-0.93
ngt01-15115	-0.11	0.83	0.83	0.96	0.98	0.97	-0.03	-0.15	-0.23	0.90	-0.93

Finally, the similar test has been repeated on the real life network of Amsterdam North, presented in Chapter 5. The correlations calculated for the sample of 5044 pipes in this case at four different demands are shown in Table 7.12 for the PDD threshold of 15 mwc.

Table 7.12 Correlation patterns of Amsterdam North network at various demand increase

Name	V_j	Q_j	v_j	$h_{f,j}$	S_j	P_j	T_j	pb_j	pb_j/L_j	hl_j	PBI_j
initial Q	0.15	0.33	0.20	0.06	0.08	0.12	0.00	0.01	-0.05	0.02	-0.05
increase 32%	0.29	0.66	0.43	0.15	0.17	0.33	0.00	0.03	-0.02	0.06	-0.13
increase 73%	0.39	0.83	0.52	0.19	0.20	0.44	0.00	0.06	0.04	0.09	-0.15
increase 98%	0.42	0.88	0.55	0.21	0.21	0.47	0.00	0.07	0.08	0.10	-0.15

In a way, the results comply with the conclusions from the results in previous tables. The network of Amsterdam North is fairly overdesigned and with lots of buffer, which is reflected by rather weak correlations. Moreover, in such a big network the implications of a single pipe failure on the total loss of demand are mostly insignificant except for a few bigger/connecting pipes, as was illustrated in Chapter 5. Expecting therefore strong correlations in this case may be unrealistic. Yet, Table 7.12 shows the same trend of improvement in correlations with the growth of demand, leaving room to speculation that a unique transition pattern of correlation exists and in optimised layout that pattern would look similar as shown in the results for the networks *O20snXX*.

7.7   PIPE FLOWS AND VOLUMES AS INDICATOR OF NETWORK RELIABILITY

If the hypothesis of some pipe parameters correlating better or worse with the loss of demand in networks of particular level of buffer is true, this would mean that the mutual correlation of those parameters can potentially be taken as an indicator of network reliability. That the values of $h_{f,j}$, S_j and P_j in Table 7.4 show similar correlation in the oversized networks is not a big surprise because the friction loss is directly used to calculate the hydraulic gradient and the power loss. Hence, correlating these parameters with each other makes little sense. On the other hand, a good correlation of volumes and flows in optimised networks observed in the results in Tables 7.4 and 7.5 could be more meaningful, because the link between these two parameters is not direct. Thus, in case each of these two parameters shows consistently good correlation with the loss of demand, their mutual good or bad correlation could suggest the level of demand loss even without running pipe failure analyses.

The results in Table 7.5 do not entirely come as a surprise knowing that GA-optimised looped networks transform into 'quasi-branched' layouts, which was elaborated in Chapter 5. In perfectly designed branched network, the relation between the pipe flow and volume becomes clearer. This can be seen in the simple example shown in Figures 7.15 and 7.16, and Table 7.13.

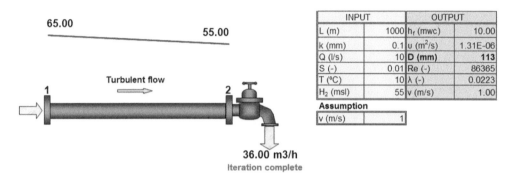

INPUT		OUTPUT	
L (m)	1000	h_f (mwc)	10.00
k (mm)	0.1	u (m^2/s)	1.31E-06
Q (l/s)	10	D (mm)	113
S (-)	0.01	Re (-)	86365
T (°C)	10	λ (-)	0.0223
H_2 (msl)	55	v (m/s)	1.00
Assumption			
v (m/s)	1		

Figure 7.15 Calculation of pipe diameter at fixed value of flow and hydraulic gradient (Trifunović, 2006)

A single pipe of fixed length and roughness value can be optimised for fixed value of hydraulic gradient and the range of flows. It is a simple iterative process that can be easily conducted in a spreadsheet like the one in Figure 7.15 (Trifunović, 2006). The results for the pipe shown in the figure, and the range of flows between 10 and 150 l/s at fixed value of hydraulic gradient of $S = 0.01$, give the range of diameters between 113 and 315 mm, respectively, as can shown in Table 7.13. The corresponding volumes are calculated for the pipe length of 1000 m and the roughness value $k = 0.1$ mm. The correlation between the flows and diameters/volumes is obvious. The Pearson's correlation between the values of Q and V yields the coefficient 0.997 leading to a perfect exponential curve shown in Figure 7.16. For selected units, $V = 0.0018Q^{0.7562}$.

Assuming that this pipe belongs to a serial or branched network, its failure would mean the loss of demand equal to its flow under regular conditions. Consequently, for selected distribution of demands, and selected source head and minimum pressure in the network, the diameter of each pipe can also be calculated for the known flow and the friction loss value; thus, each demand scenario will have its optimal network. Eventually, this could mean that ideal correlation between the pipe flow and volume actually hints optimised branched network. In all other cases, this correlation should worsen: in case of GA-optimised looped network to a lesser extent, depending on the choice of diameters used for optimisation, while the extent would grow by adding buffer to the network. Strictly speaking, this hypothesis is false because it neglects the fact that one volume can be a product of different combinations of pipe diameters and lengths. For instance, the pipe flow of 100 l/s in Table 7.13, transported along the 1000 m pipe with $D = 270$ mm and at the friction loss of 10 mwc leading to $S =$

0.01 m/km, will require different diameter if the pipe length changes; this can be seen in Table 7.14 and Figure 7.17.

Table 7.13/Figure 7.16 Correlation between optimal diameter and pipe volume, L = 1000 m, S = 0.01

Q (l/s)	D (mm)	V (m³)
10	113	10.029
20	147	16.972
30	171	22.966
40	191	28.652
50	207	33.654
60	222	38.708
70	236	43.744
80	248	48.305
90	259	52.685
100	270	57.256
110	280	61.575
120	289	65.597
130	298	69.746
140	306	73.542
150	315	77.931

Table 7.14/Figure 7.17 Correlation between optimal diameter and pipe volume, Q = 100 l/s, h_f = 10 mwc

L(m)	S(-)	D (mm)	V (m³)	v (m/s)
100	0.100	172	2.324	4.31
200	0.050	197	6.096	3.29
300	0.033	213	10.690	2.81
400	0.025	225	15.904	2.51
500	0.020	235	21.687	2.30
600	0.017	244	28.056	2.14
700	0.014	252	34.913	2.01
800	0.013	258	41.823	1.91
900	0.011	264	49.265	1.82
1000	0.010	270	57.256	1.75
1100	0.009	275	65.335	1.69
1200	0.008	279	73.363	1.63
1300	0.008	284	82.351	1.58
1400	0.007	288	91.202	1.53
1500	0.007	292	100.449	1.49

Hence, the pipe lengths also influence the correlation between the flows and volumes, next to the diameters. Nevertheless, common engineering practice deals with some design principles in which not every combination of parameters would be applied. For instance, the velocities much above 1 m/s would be causing too high energy losses/consumption. Equally, the hydraulic gradients much above 0.01 mean increased operational costs. Furthermore, neither very long pipes of small diameter nor short pipes of large diameter would be common in water distribution networks. The issue is therefore how good is the correlation between the parameters of commonly designed pipes.

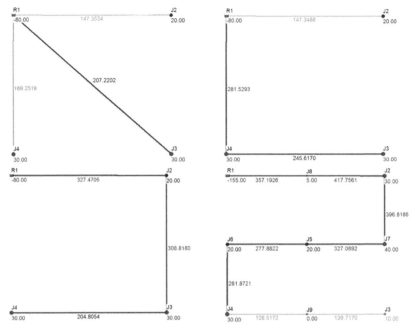

Figure 7.18a GA optimised nets 1 to 4 (EO-optimiser) – pipes: D (mm), nodes: Q (l/s)

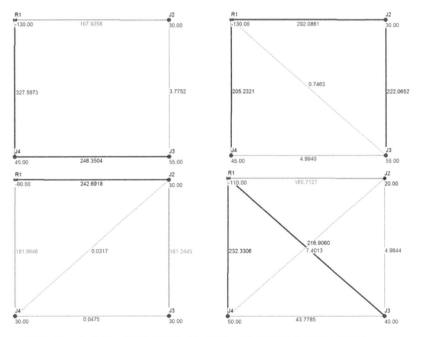

Figure 7.18b GA optimised nets 5 to 8 (EO-optimiser) – pipes: D (mm), nodes: Q (l/s)

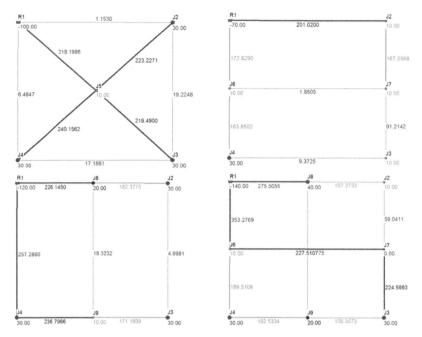

Figure 7.18c GA optimised nets 9 to 12 (EO-optimiser) – pipes: D (mm), nodes: Q (l/s)

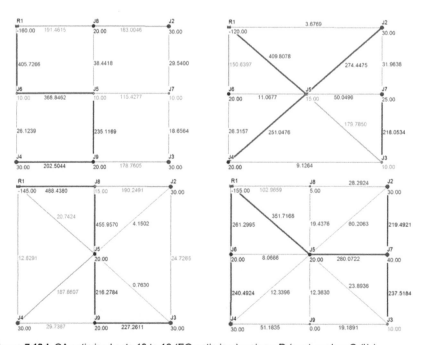

Figure 7.18d GA optimised nets 13 to 16 (EO-optimiser) – pipes: D (mm), nodes: Q (l/s)

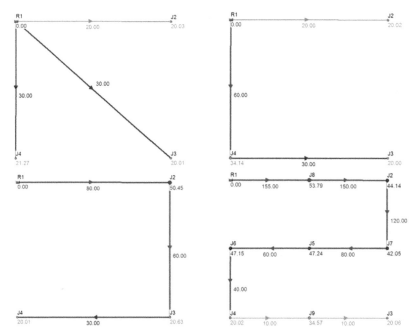

Figure 7.19a No-failure condition, nets 1 to 4 – pipes: Q (l/s), nodes: p/ρg (mwc)

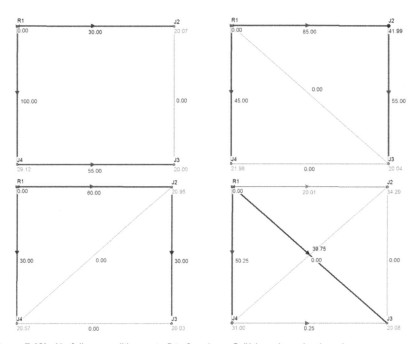

Figure 7.19b No-failure condition, nets 5 to 8 – pipes: Q (l/s), nodes: p/ρg (mwc)

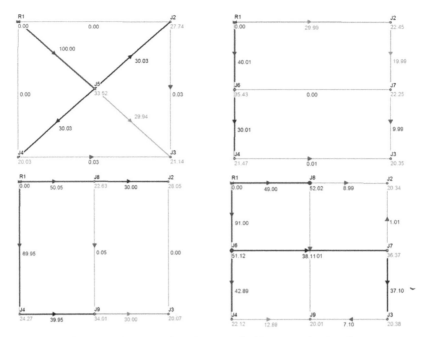

Figure 7.19c No-failure condition, nets 9 to 12 – pipes: Q (l/s), nodes: p/ρg (mwc)

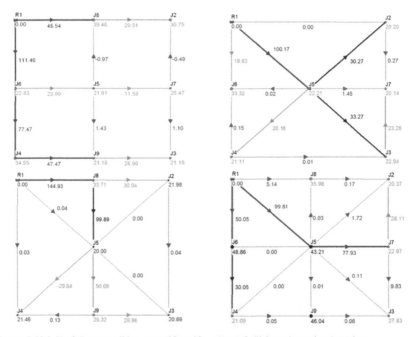

Figure 7.19d No-failure condition, nets 13 to 16 – pipes: Q (l/s), nodes: p/ρg (mwc)

First of all, the correlations between the pipe volumes and flows have been tested on the set of 16 simple networks used in the analyses discussed in Chapters 5 and 6. Here, the networks have been more stringently optimised, on continuous range of diameters and without specified limits. This has been done to check to which extent the results would comply with those in Table 5. The GA optimisation done by EO-optimiser shows the layouts and hydraulic performance of the nets as displayed in Figures 7.18 and 7.19.

Obviously, the figures show more refined selection of pipe diameters done to fit the minimum pressure of 20 mwc more accurately, which eventually has been achieved, whilst the loped layout has been brought even more close to a branched configuration. Table 7.15 shows the results of correlation between the pipe volumes and flows with the loss of demand and mutually (left side), which is eventually compared to the values of the six indices used as reliability measures (right side).

Table 7.15 Correlation between pipe flows and volumes and its relation to loss of demand

No. Name	V_j	Q_j	V_jQ_j	NPI	PBI	NRT	ADF_{avg}	NBI	I_n
1 CO20Net01	0.45	1.00	0.45	0.53	-0.41	0.80	0.67	0.00	0.02
2 CO20Net02	1.00	1.00	1.00	0.61	0.37	1.37	0.54	0.00	0.17
3 CO20Net03	0.83	1.00	0.83	0.65	0.83	2.15	0.29	0.00	0.26
4 CO20Net04	0.91	1.00	0.91	0.71	0.92	1.37	0.50	0.00	0.42
5 CO20Net05	0.98	1.00	0.98	0.64	0.54	0.65	0.64	0.00	0.12
6 CO20Net06	0.96	1.00	0.96	0.63	0.49	0.52	0.72	0.00	0.17
7 CO20Net07	0.92	1.00	0.92	0.48	0.32	0.77	0.73	0.00	0.01
8 CO20Net08	0.96	1.00	0.96	0.68	0.71	0.65	0.83	0.01	0.11
9 CO20Net09	0.98	1.00	0.98	0.67	0.79	1.03	0.76	0.00	0.10
10 CO20Net10	0.68	1.00	0.68	0.49	0.63	0.80	0.73	0.00	0.05
11 CO20Net11	0.95	1.00	0.95	0.54	0.60	0.84	0.74	0.00	0.08
12 CO20Net12	0.85	1.00	0.87	0.66	0.76	0.96	0.81	0.08	0.20
13 CO20Net13	0.96	1.00	0.97	0.53	0.65	0.76	0.77	0.01	0.07
14 CO20Net14	0.97	1.00	0.97	0.78	0.72	1.18	0.84	0.02	0.09
15 CO20Net15	0.98	1.00	0.97	0.75	0.82	0.94	0.78	0.00	0.15
16 CO20Net16	0.97	1.00	0.97	0.78	0.91	0.89	0.88	0.02	0.21

Expectedly, the results for volumes and flows show stronger correlation than those in Table 5. Still, this correlation is much weaker in case of *CO20Net 01, 03, 10* and *12* than for the rest of the networks. The relation between the pipe volume and flow can be easily described using the Darcy-Weisbach equation:

$$V = \sqrt{\frac{\lambda L^3}{2gDh_f}}Q = \frac{L}{v}Q \qquad\qquad 7.13$$

Hence, the correlation between the pipe volumes and flows will be perfect in theory in case the ratio of the pipe length and velocity is equal for all network pipes. It is then easy to find out why the correlation will be weak. For example, network *CO20Net01* (Figure 7.18, upper left) could have been optimised manually so that the entire available pipe head loss is utilised, leading to the minimum pressure of exactly 20 mwc, in all three discharge nodes. Table 7.16 shows the results obtained by EO-optimiser, by manual calculation of pipe diameters at maximum available pipe head loss, and the calculation of pipe lengths and diameters that correlate the pipe volumes and flows based on the optimised volume of P01.

Table 7.16 Pipe flows and volumes in optimised network - CO20Net01

Pipe	h_f(mwc)	L(m)	D(mm)	V(m^3)	Q(l/s)	v(m/s)	L/v(s)
P01-GA	39.97	3000	147.35	51.160	20	1.17	2558
P02-GA	19.99	4000	207.22	134.900	30	0.89	4497
P03-GA	28.73	2000	169.25	44.996	30	1.33	1500
P01-maD	40.00	3000	148	51.610	20	1.16	2581
P02-maD	20.00	4000	207	134.614	30	0.89	4487
P03-maD	30.00	2000	168	44.334	30	1.35	1478
P01-maD	40.00	3000	148	51.610	20	1.16	2581
P02-maV	20.00	2680	192	77.594	30	1.04	2586
P03-maV	30.00	3000	181	77.191	30	1.17	2573

The results in the third segment of the table show that both the pipe length and diameter have to be modified if the full correlation between the volumes and flows is to be achieved. As a result, *P02* will become shorter and of smaller diameter, while *P03* becomes longer and with bigger *D*. Satisfying at the same time the condition of the maximum available head loss in each pipe means that, next to the nodal demand, this combination will also be influenced by the nodal elevations and the source head(s). A 'snap-shot' conclusion is that the correlation between pipe volumes and flows in optimized networks will be stronger the more even the distribution of geometric and hydraulic parameters is. This point can also be illustrated on a simple serial network in Figure 7.20.

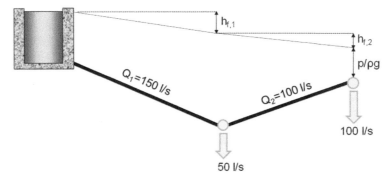

Figure 7.20 Serial network

Table 7.17 Correlation between pipe flows and volumes and its relation to loss of demand

$h_{f,1}$(mwc)	$h_{f,2}$(mwc)	L_1(m)	L_2(m)	D_1(mm)	D_2(mm)	V_1(m^3)	V_2(m^3)	Q_1(l/s)	Q_2(l/s)	v_1(m/s)	v_2(m/s)	L_1/v_1(s)	L_2/v_2(s)	V_{tot}(m^3)
10	10	1000	1000	315	270	77.931	57.256	150	100	1.92	1.75	519.5	572.6	135.187
9	11	1000	1000	321	264	80.928	54.739	150	100	1.85	1.83	539.5	547.4	135.667
8.8	11.2	1000	1000	323	264	81.940	54.739	150	100	1.83	1.83	546.3	547.4	136.679
10	10	1500	500	341	235	136.990	21.687	150	100	1.64	2.31	913.3	216.9	158.677
15	5	1500	500	315	270	116.897	28.628	150	100	1.92	1.75	779.3	286.3	145.524
19.5	0.5	1500	500	299	426	105.323	71.265	150	100	2.14	0.70	702.2	712.7	176.589
10	10	500	1500	275	292	29.698	100.449	150	100	2.53	1.49	198.0	1004.5	130.147
5	15	500	1500	315	270	38.966	85.883	150	100	1.92	1.75	259.8	858.8	124.849
0.3	19.7	500	1500	551	256	119.224	77.208	150	100	0.63	1.94	794.8	772.1	196.432

The results in Table 7.17 show several combinations of the pipe lengths and diameters for the total available head loss of 20 mwc and the demand of 250 l/s. In all cases, the pipe diameters have been minimised to deliver this service, and the correlation of the volumes and flows achieved in three iterations. The most realistic situation is described in the first segment of the table where both pipes are of 1000 m and with similar diameters and friction losses. As soon the length of the pipes is significantly different, the corresponding diameters will also change and achieving the correlation between the pipe volumes and flows will become more difficult. Thus, the application of the concept becomes limited to the networks of specific configuration and topography.

To test further the hypothesis, the correlations between the pipe flows and volumes have been assessed in ten initial networks against their average loss of demand. The results are shown in Table 7.18. The diagrams related to the results in Tables 7.15 and 7.18 are shown in Figures 7.21 and 7.22, respectively.The correlation between the volumes and flows, and other reliability measures is not strong. The best pattern has been with the *PBI* values of the ten networks that are consistently lower if the correlation between the pipe volumes and flows improves. This is to lesser extent the case with the values of *NBI* and *ADF*$_{avg}$, and certainly weak in case of *NRT*, I_r and *NPI*. The results from the 16 optimised nets show even less consistency although there as well the best correlation has been achieved with the *PBI* values.

Table 7.18 Correlation between pipe flows and volumes and its relation to loss of demand

No.	Name	V_j	Q_j	V_jQ_j	NPI	PBI	NRT	ADF$_{avg}$	NBI	I_n
1	sn73	0.45	0.96	0.45	0.54	0.95	0.18	0.93	0.32	0.18
2	sn80	0.39	0.83	0.39	0.46	0.92	0.21	0.99	0.84	0.04
3	O20sn73	0.91	1.00	0.92	0.55	0.90	0.09	0.91	0.08	0.18
4	O20sn80	0.93	0.98	0.92	0.59	0.90	0.10	0.95	0.33	0.20
5	sn0179	0.26	0.85	0.30	0.47	0.95	0.30	0.95	0.48	0.06
6	sn0199	0.39	0.84	0.31	0.47	0.92	0.18	0.98	0.79	0.05
7	sn0109	0.02	0.90	0.02	0.45	0.97	0.50	0.98	0.76	0.03
8	ngt01-50	-0.20	0.80	-0.27	0.69	0.99	11	0.99	0.83	0.40
9	ngt01-151	-0.10	0.80	-0.19	0.69	1.00	37	1.00	0.91	0.35
10	ngt01-200	-0.11	0.66	-0.10	0.67	1.00	33	1.00	0.96	0.36

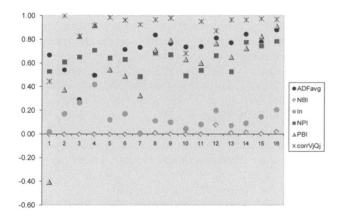

Figure 7.21 Correlation between pipe flows and volumes and its relation to loss of demand - Table 7.15

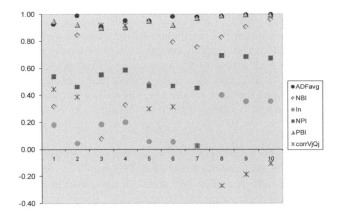

Figure 7.22 Correlation between pipe flows and volumes and its relation to loss of demand - Table 7.18

7.8   CONCLUSIONS

The research presented in this chapter aims to assess the potential of water distribution networks to sustain a certain level of failure based on analysis of their operational parameters under regular supply conditions, namely the pressures, flows and head loss distribution. Three different reliability measures have been proposed to describe the network operation, namely the network power index (NPI), the pressure buffer index (PBI), and the network residence time (NRT). Comparisons have been done with the reliability measures discussed in Chapter 6, namely the available demand fraction (ADF_{avg}), the network buffer index (NBI) and the network reliability index, I_n. Furthermore, several statistical analyses have been conducted to check the correlations between the pipe properties, specifically the volumes and flows, as possible indicators of the loss of demand after the pipe failure. The conclusions based on the simulation results can be summarised in the following bullets:

1. The results point fairly good matching between the reliability measures. The logical response of the indices occurs after the demand and diameter increase, or having various degree of network connectivity applied in the model.
2. NPI values are generally more susceptible to nodal elevations (i.e. the pressures) than are the PBI values. Oppositely, the PBI values are more susceptible to the demand growth (i.e. the friction losses) than are the NPI values.
3. Additional buffer achieved by the increase of pipe diameters shows different effect on the improvement of network resilience, which can be observed on the diagrams showing the reliability measures. The different form of the curves for the same reliability index results from the presentation of that index in different range of ADF_{avg}. After sufficient increase of pipe diameters, each network enters the state where furthering this action has little implication for the increase of ADF_{avg}. A curve covering the entire range, starting from optimised diameters, would be of S-shape ending with the saturation zone of ADF_{avg}. How big this zone will be, also depends on the available head in the network. Nevertheless, more research should be done to solidify this conclusion.
4. The pipe correlations of geometric and hydraulic parameters show less consistent patterns. Different network categories have been showing different level of correlations of particular parameters. GA-optimised networks show generally better correlations between

the volumes and flows, while the networks with more buffer have better correlations of the hydraulic parameters involving friction losses. In most of the cases, the results have also shown the trend of improved correlations after gradually increasing the level of demand, but no firm conclusions could have been drawn. At the moment the networks are put under significant stress, these correlations are still rather weak despite positive trends.

5. Generally good correlations of pipe volumes and flows emerging from the most of optimised networks appear to be more of a consequence of particular network configuration and the level of demand. It is indeed that these correlations are better in configurations that are more logical in practice but it would be far too premature to present this as a pattern. More analysis is necessary regarding the boundary conditions of such configurations. The same can be concluded for the correlation of the pipe volumes and flows, or any other pipe property, as an indicator of network reliability.

The consistency of some of the results makes further research in this direction sensible. Nevertheless, as it has been also suggested in the conclusions of Chapter 6, this should be done with more substantial integration of hydraulic and geometrical parameters that influence network resilience. The statistical analyses done hint some patterns but mostly visible in extreme cases of optimised diameters or increased demand. To continue with this approach, many more case networks of different characteristics will have to be analysed. In parallel, an attempt should be made to build more solid mathematical foundation into the concept, than was the case in this research.

REFERENCES

1. *Evolving Objects (EO)*, distributed under the GNU Lesser General Public License by SourceForge (http://eodev.sourceforge.net)
2. Prasad, T. D., and Park, N.S. (2004). *Multiobjective genetic algorithms for design of water distribution networks.* Journal of Water Resources Planning and Management. ASCE, 130(1), p 73-82.
3. Trifunović, N. (2006). *Introduction to Urban Water Distribution*, Taylor & Francis Group, London, UK, 509 p.

CHAPTER 8

Economic Aspects of Decision Making in Reliability Assessment of Water Distribution Networks[1]

The research presented in this chapter compares implications of increased investment and operational costs for the reliability of networks with direct pumping and those combined with balancing storage. Economic aspects have been taken into consideration by applying the Present Worth method. Using this approach, the reliability of two simple network layouts has been analysed manipulated through five different topographical patterns, three altitude ranges and four economic scenarios. Each of the 120 scenarios was further tested on 20 additional designs comprising gradual increase of pipe capacity combined with reduction in pumping capacity, while targeting the similar minimum pressure. The network reliability has been assessed with the resilience indices introduced in Chapter 5, namely the *NBI* and I_n. To process efficiently total 2520 scenarios, the network diagnostics tool (NDT) used in the analyses in Chapters 6 and 7 has been upgraded with cost calculations. The results show that cheapest design scenarios are not necessarily the least reliable ones. Furthermore, better reliability for the same level of investment is achieved if done into additional pipe capacity rather than into pump capacity/operation. Moreover, both strategies confirm that the point after which the further investment is useless indeed depends on the choice of economic parameters.

---

[1] Extended abstract submitted by Trifunović, N., Kanowa, L. and Vairavamoorthy, K., under the title *Impact of Balance between Investment and Operational Costs on Resilience of Water Distribution Networks*, for the 14th Water Distribution Systems Analysis Conference in Adelaide, Australia, 24-27 September 2012.

8.1   INTRODUCTION

Water distribution networks will be designed to provide satisfactory level of service under regular conditions, and also in number of irregular supply scenarios. This additional safety factor comes at additional costs that should be paid for investments and/or operational measures with the highest effect for the service levels. The least cost design should in theory be the least reliable, but this conclusion cannot be analysed from the perspective of asset value alone, because the operational costs and repayment of loans can participate significantly in the cost of water supply. Thus, how much is worthwhile to invest into additional safety and what would be the impact on the service levels is a complex question.

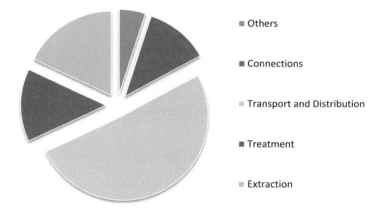

■ Others

■ Connections

▨ Transport and Distribution

■ Treatment

▨ Extraction

Figure 8.1 Proportion of value of Dutch water supply assets (VEWIN, 1990)

Distribution network is the most expensive part of any water supply system asset. Figure 8.1 shows the value of assets of Dutch water supply assessed in 1988 at approximately five billion US$, in which it can be seen that transportation and distribution component account for over 50% of the total value (VEWIN, 1990). Similar proportion will be found in any other country or particular drinking water system.

On the other hand, energy input in operation of distribution networks can also be very high. This will depend on water extraction but in extreme topographic conditions of zone supply, even more on water distribution. Belgrade Waterworks and Sewerage (BWS) estimated the average consumption of electric energy of 0.86 kWh per m^3 of produced water, which in 2007 was approximately 211 million m^3 for the city of Belgrade in Serbia (Cvjetković, 2008). This includes the costs of pumping from deep wells and also the drainage pumping stations, in the city of 1.5 million inhabitants and distribution network located over five pressure zones maintained by 28 drinking water pumping stations. Hence, the annual electricity bill of the company is in the order of millions of US$.

This participates around 5% in the total annual costs of the company, as shown in Figure 8.2. The total operational costs are estimated at 13% while the maintenance and materials occupy

12% of the annual budget. The significant component is the depreciation of assets that contributes 39% to the total cost.

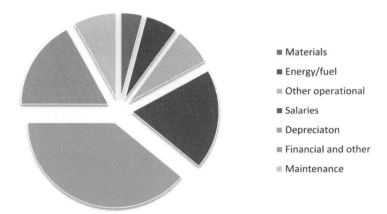

Materials

Energy/fuel

Other operational

Salaries

Depreciaton

Financial and other

Maintenance

Figure 8.2 Total cost structure of BWS (Cvjetković, 2008)

The proportion of the costs as in Figure 8.2 may look different in different systems. Water supply in developing countries is normally run by relatively large number of staff and cheap labour, while developed countries will strive to higher level of automation due to much higher costs of labour. Furthermore, next to topographical conditions, the energy input may also be influenced by strategic choices driven by the costs and reliability of electricity supply. For instance, one implication of erratic electricity supply could be to build additional emergency storage volume in the network. Finally, the depreciation is based on the proper estimates of the technical and economic lifetime for the selected level of maintenance, but also depends on the conditions agreed for repayment of loans used for construction; here again, the distribution network will utilise the most of annuities. The technical lifetime of the network components can vary significantly as shown in Table 8.1 (Trifunović, 2006), and in well maintained networks it can be extended even further. Hence, the choice amongst particular technical solutions that will improve network reliability has to take into consideration all these aspects before being made.

Table 8.1 Technical lifetime of distribution system components (Trifunović, 2006)

Component	Period (years)
Transmission mains	30 - 60
Distribution mains	30 - 80
Reservoirs	20 - 80
Pumping station - facilities	20 - 80
Pumping station - equipment	15 - 40

8.2 PLANNING OF COSTS IN WATER DISTRIBUTION

Investment costs of water distribution networks are estimated during the design process by applying the following simplified approach (Trifunović, 2006):

1. Pipes: the first cost (FC) is proportional to the diameter. Linear or exponential approach can be applied where $FC = aD$, or $FC = b + cD^n$, with D representing the pipe diameter, say, in millimetres.
2. Storage: the first cost of network storage will be proportional to its volume. Hence, $FC = aV^n$, where V is the storage volume, say, in m^3.
3. Pumping stations: Similar as for the storage, $FC = aQ^n$, where Q is the maximum installed pumping capacity, say, in m^3/h.

Factors a, b, c, and n in the above equations will depend on the type of materials and local manufacturing conditions, and units of measure used for D, V and Q.

Annual operation and maintenance costs are normally planned as a certain percentage of the investment costs. From the practice of developed countries, these costs can be roughly budgeted in the order 0.5% for pipes, 0.8% for storage and 2.0% for pumping stations, per annum. This does not include the energy used for pumping, which is calculated based on the actual water quantity pumped in the system.

Various design alternatives may consider phased investment. A preliminary cost comparison can be carried out using the *Present Worth* method (also known as the *Annual Worth* method). Applying this method, all actual and future investments are calculated back to a reference year, which in general is the year of the first investment. The basic parameter used in the calculation is the *single present worth factor*, $p_{n/r}$:

$$p_{n/r} = \frac{1}{s_{n,r}} = \frac{1}{(1+r)^n} \qquad\qquad 8.1$$

where $s_{n/r}$ is the *single compound amount factor*, which represents the growth of the present worth (PW) after n years with a *compounded interest rate* of r. If the first cost, FC, has been made in the future, after n years, the present worth, PW, of the future sum becomes $PW = p_{n/r}FC$.

A loan taken for the investment has to be repaid after n years. The annual sum, A, which is to be allocated for that purpose will be calculated as:

$$A = \frac{r(1+r)^n}{(1+r)^n - 1} PW = a_{n/r}PW \qquad\qquad 8.2$$

where, $a_{n/r}$ represents the *capital recovery factor (annuity)*.

The *ideal interest rate*, i, can replace the true interest rate, r, in Equations 8.1 and 8.2, if the annual inflation rate, f, is to be taken into consideration. This is done in the following way:

$$i = \frac{r - f}{1 + f} \qquad\qquad 8.3$$

The above equations are commonly applied while comparing various design alternatives. The field of Engineering Economy offers more detailed cost evaluations that can be further studied in appropriate literature, e.g. in De Garmo et al. (1993).

8.3   COMPONENTS OF MOST ECONOMIC DESIGN

The most economic alternative is usually the best compromise between the investment and operational costs. For example, if a transportation pipe convening the flow, Q, (in m^3/s) is to be designed along the length, L, the head loss that will be generated during the operation is ΔH (in mwc). The cost of energy needed to deliver the flow at required pressure will be calculated as:

$$EC = \frac{\rho g Q H_p}{1000 \times \eta} T \times e \qquad\qquad 8.4$$

where H_p is the pumping head that includes the static head H_{st} composed of the minimum pressure $p_{end}/\rho g$ at the pipe end, and elevation difference of the upstream and downstream end ΔZ i.e.:

$$H_p = H_{st} + \Delta H \quad ; \quad H_{st} = \frac{p_{end}}{\rho g} \pm \Delta Z \qquad\qquad 8.5$$

Furthermore, e is the unit price (per kWh) of the energy needed to compensate the pipe head loss, T is the time component, and η the pumping efficiency. The static head can be neglected in the calculation of optimal pipe diameter. Hence, the annual costs of the energy wasted per metre length of the pipe become:

$$EC = \frac{9.81 \times 24 \times 365 \times Q \Delta H}{3600 \times L} \frac{e}{\eta} \approx 24 \times Q \frac{e}{\eta} \frac{\Delta H}{L} \qquad\qquad 8.6$$

By using the Darcy-Weisbach equation and assuming the friction factor λ of 0.02:

$$EC = 24 \times Q \frac{e}{\eta} \frac{0.02 \times Q^2}{12.1 \times D^5 \times 3600^2} \approx 3 \times 10^{-9} \frac{e}{\eta} \frac{Q^3}{D^5} \qquad\qquad 8.7$$

Applying liner relation between the pipe cost and diameter, the total annual costs including investment and operation of the pipe become:

$$A = a \times D \times a_{n/r} + 3 \times 10^{-9} \frac{e}{\eta} \frac{Q^3}{D^5} \qquad\qquad 8.8$$

The most economic diameter will finally be calculated from $\delta A/\delta D = 0$ for Equation 8.8:

$$a \times a_{n/r} = 5 \times 3 \times 10^{-9} \frac{e}{\eta} \frac{Q^3}{D^6} \quad \Rightarrow \quad D = 0.05\sqrt{Q} \sqrt[6]{\frac{e}{\eta a_{n/r} a}} \qquad\qquad 8.9$$

The above simplified derivation considers fixed energy costs and fixed flow rate over the entire design period. The growth of these parameters should obviously be taken into account, if any.

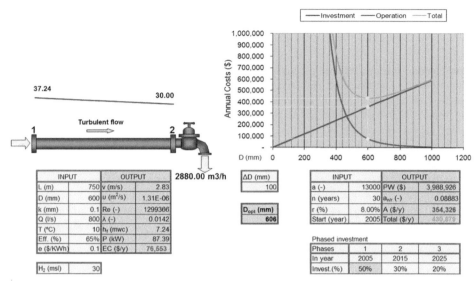

INPUT		OUTPUT	
L (m)	750	v (m/s)	2.83
D (mm)	600	u (m²/s)	1.31E-06
k (mm)	0.1	Re (-)	1299366
Q (l/s)	800	λ (-)	0.0142
T (°C)	10	h_f (mwc)	7.24
Eff. (%)	65%	P (kW)	87.39
e ($/KWh)	0.1	EC ($/y)	76,553

H₂ (msl)	30

2880.00 m3/h

ΔD (mm)
100

D_opt (mm)
606

INPUT		OUTPUT	
a (-)	13000	PW ($)	3,988,926
n (years)	30	a_nfr (-)	0.08883
r (%)	8.00%	A ($/y)	354,326
Start (year)	2005	Total ($/y)	430,879

Phased investment

Phases	1	2	3
In year	2005	2015	2025
Invest.(%)	50%	30%	20%

Figure 8.3 The most economic diameter (Trifunović, 2006)

A spreadsheet application, like the one shown in Figure 8.3 (Trifunović, 2006), can be used to compare various possibilities with more or less favourable energy costs and conditions of the loan. The diagram in the figure shows that increased investment costs i.e. the bigger pipe diameter (the blue line), have as a consequence the reduced operational costs (the red curve). Some kind of optimal diameter will correspond to the minimum sum of the two costs, which is represented by the lowest point on the green curve. Favourable energy costs and unfavourable loan conditions could suggest further reduction of the diameter, whilst the opposite situation will cause its increase.

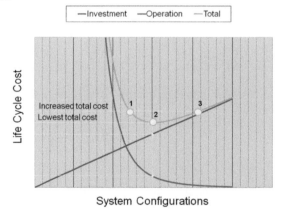

Figure 8.4 The most economic option and reliability considerations

Similar considerations will be made in case of a design of whole network or its portion; striving towards the best solution means thorough comparisons between investment- and operational costs. The diagram like the one in Figure 8.3 will therefore also be prepared for various design alternatives and their corresponding costs, as is the case in Figure 8.4. When the most economic solution is finally selected from such a diagram (option 2), the question is how that solution can be perceived from the perspective of supply in irregular situations. Moreover, in case additional funds are to be invested to improve the network reliability, some kind of guidelines should suggest either investing into increasing of the pumping i.e. operational costs (option 1), or additional pipe capacity (option 3) would yield better results.

8.4   DESIGN ALTERNATIVES BASED ON TOPOGRAPHY

Good hydraulic design of water distribution network will aim to utilise the topography of the area covered by the network to the largest possible extent, in order to reduce the energy input. This can be taken as a principle, because the situations where this philosophy would lead to more expensive design alternative are mostly theoretical. Even if the energy supply would prove to be so much cheaper than the investment costs of the network, high energy input into the water distribution may not be justified from the environmental point of view, the least. Hence, were gravity supply is possible, it should also be provided, as Figure 8.5 shows.

Secondly, the topography plays important role as far the pressure in the network. Additional pumping or storage may be needed to supply water to the areas far away from the source(s). This eventually results in combined supply configurations where the storage will be mostly elevated at higher elevations and pumping at lower elevations.

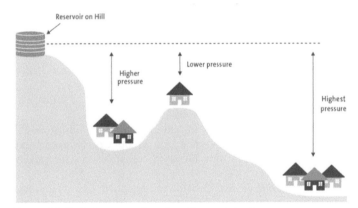

Figure 8.5 Typical cross section of gravity supply landscape

Reservoirs and pumping stations are always designed to operate in synchronised way. Pumps will either supply from the source reservoir to the network or fill a reservoir that has to supply (part of) the network by gravity. To provide sufficient balancing of the demand, the volume of the reservoir and size of the pump will be designed to satisfy the minimum pressure, accordingly. Location of the storage, pump and consumers can be correlated to five typical terrain configurations:

1. *Flat*. Both the source and consumers are located on similar altitudes. The supply will mostly take place by direct pumping and the storage plays role only on the suction side of the pump, to balance the demand with the average source production. Possible tanks within the consumers' area will be elevated (water towers) with primary purpose to stabilise pressures rather than to balance the demand.

2. *Slope up.* In this setup, the consumers are located higher than the source(s). The supply scheme would be the same as for flat terrain except that higher-lift pumps are needed.
3. *Slope down.* In this setup, the consumers are located lower than the source(s), which gives possibility of supply by gravity. If the available elevation difference is not sufficient, additional pumping may be necessary. Nevertheless, those pumps will be of lower lift compared to the situations under 1 and 2.
4. *Hill.* When the source and the consumers are separated by an area of high altitude, the hill can be used to position a reservoir that should be supplied by constant flow from the pump, and supply the variable demand by gravity.
5. *Valley.* If the distribution area is located in a valley, a possibility exists to locate the reservoir on the opposite side of the source. The benefit is twofold: (1) the pumps can supply average flow constantly throughout 24 hours while part of the area on the side of the reservoir are supplied by gravity, and (2) as a consequence of such supply scheme more balanced selection of pipe diameters makes future extensions of the network easier.

Each situation in reality can be covered by one of the above setups or, in more complex networks, by their combination.

8.5 HYDRAULIC RELIABILITY AND ITS COSTS

The above five cases have different implications on the cost of operation and network reliability, which can be summarised in the following bullets (assuming single source of supply):

-   Pumps in any of the first three terrain configurations will be designed on the maximum consumption hour of maximum consumption day. This capacity can be reduced if the reservoir is introduced within the service area; the more volume has been introduced to balance the demand, the smaller the pumps can be.
-   Pipe diameters will be gradually reduced, from the source towards the end of the system in case of single source supply in the terrain configurations 1, 2 and 3.
-   In case of gravity supply applied in the terrain configuration 3, larger pipe diameters will be needed to utilise the topography as much as possible, on account of reduced head-losses. The more of pumping capacity has been introduced, in combination with gravity, the smaller the pipes can be.
-   The pumps in the terrain configurations 4 and 5 will be designed on the average consumption hour of maximum consumption day. This capacity can be increased if the volume of the balancing reservoir has been reduced.
-   Pipes connecting pumps and reservoirs in the terrain configurations 4 and 5 will be more balanced in diameter than in case of the terrain configurations 1, 2 and 3.
-   The reliability of supply with balancing tank should be higher in general, compared to the direct pumping supply scheme.

Nevertheless, the cost of storage comes on top of the cost of pipes once the investment costs are to be considered with the operational costs. More reservoir/pipe volume means:

-   increased investment costs,
-   reduced operational costs,
-   more buffer in the network,
-   higher maintenance costs (for storage).

The latter aspect is important from the perspective of water quality in distribution networks, which may deteriorate due to stagnation of water. In case of reliable and affordable electricity supply, the practice of water distribution in some West European countries suggest that the (large) reservoirs are not economic due to expensive cleaning and/or expensive land use where they are to be constructed. That logic maybe completely different in developing countries, where the supply of electricity is much less reliable and the problem of reservoir maintenance can be mitigated by cheap labour. Still, the (extreme) topographic conditions will be prevailing while making the final decisions.

As far the pipe capacity, much of spare capacity in urban water distribution networks is often resulting from (sometimes overdesigned) hydrant capacity. That in itself produces more reliable networks but also the situation where more frequent pipe cleaning is necessary. The network of Amsterdam North discussed in Chapter 5 is a typical case. Last but not least, network designers easily add comfortable safety factors; a slight (linear) increase in pipe costs will result in large (exponential) increase of conveying capacity.

8.6   HYDRAULIC SIMULATIONS AND RELIABILITY ASSESSMENT

To find out how the parameters described in the previous paragraphs interact with network reliability, a series of hydraulic and reliability simulations has been conducted on two simple networks: one with direct pumping supply (A), and the other where pumping is combined with balancing storage (B), which actually is an adaptation of the A-layout. Both networks are shown in Figure 8.6. Depending on the terrain configuration, the tank connected in the figure to node $N10$, can also be connected to nodes $N4$ or $N6$.

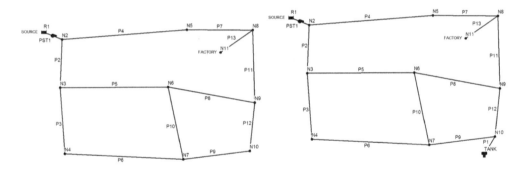

Figure 8.6 Network layouts A (left) and B (right)

The basic information about the network geometry and hydraulic performance has been given in Tables 8.2 and 8.3, respectively. The total network lengths are similar: 8.16 km (A) and 8.31 km (B), and all pipes have the uniform k-value of 0.5 mm. As the nodes are not densely connected, increased network resistance has been generated by selecting relatively large pipe lengths. Furthermore, the EPANET pump curve has been defined by single pair of duty flow and duty head points: 300 l/s at 85 mwc, in case of network A, and 255 l/s by 63 mwc, in case of network B. In both cases, the efficiency curve has been calculated for the maximum efficiency of 75% (achieved at the duty flow).

Table 8.2 Initial geometry and hydraulic performance of network A

Nodes ID	Demand (l/s)	Head (msl)	Pressure (mwc)	Links ID	Length (m)	Diameter (mm)
N2	44	64	53	P2	430	500
N3	27	62	46	P3	600	400
N4	40	61	38	P4	1100	500
N5	80	62	44	P5	950	400
N6	56	60	37	P6	1050	300
N7	64	57	28	P7	580	300
N8	28	61	36	P8	780	300
N9	44	59	27	P9	590	200
N10	25	57	22	P10	660	150
N11	25	59	35	P11	650	150
R1 (Pump)	-436	10	0	P12	420	150
				P13	350	200

Table 8.3 Initial geometry and hydraulic performance of network B

Nodes ID	Demand (l/s)	Head (msl)	Pressure (mwc)	Links ID	Length (m)	Diameter (mm)
N2	44	54	44	P2	430	500
N3	27	54	37	P3	600	400
N4	40	53	30	P4	1100	500
N5	80	53	35	P5	950	400
N6	56	53	30	P6	1050	300
N7	64	53	23	P7	580	300
N8	28	52	28	P8	780	300
N9	44	52	20	P9	590	200
N10	25	58	24	P10	660	150
N11	25	51	27	P11	650	150
R1 (Pump)	-388	10	0	P12	420	150
Tank 12	-48	65	5	P13	350	200
				P1	150	200

Both layouts have been optimised for various scenarios at the minimum pressure of 20 mwc using optiDesigner GA-software (http://www.optiwater.com), and the range of 14 pipe diameters varying between 25.4 and 609.6 mm (1 to 24 inch). The purpose of this optimisation has been to set a starting layout of the lowest investment costs, which has further been enlarged in 20 steps using diameter multipliers. In parallel with the increase of network conveying capacity, the pump speed has been reduced accordingly to keep the operation in the area of minimum pressure, as much as possible. The diagrams in Figures 8.7 and 8.8 show typical effect of combined change of network resistance and pumping capacity on the change of minimum pressure for one of the analysed scenarios and five different terrain types.

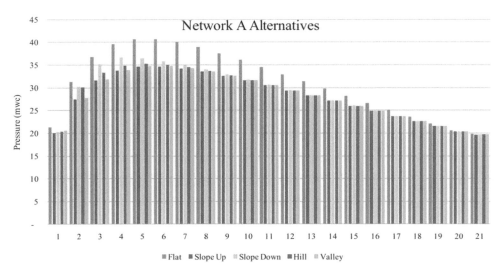

Figure 8.7 Minimum pressure in network A based on combined change of resistance pumping capacity

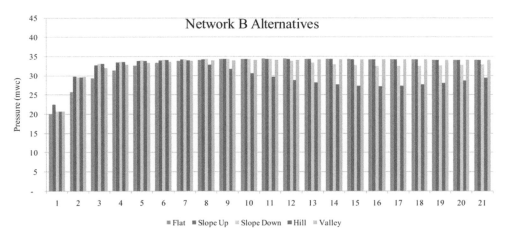

Figure 8.8 Minimum pressure in network B based on combined change of resistance pumping capacity

The first result in both figures is from GA-optimised layout whilst all other results originate from progressive increase of pipe diameters (up to a maximum factor of 2.65) and reduced duty flow and duty head of the pump (up to a minimum factor for the pump speed of 0.87). As expected, the lower minimum pressure variation in case of network options *B* results from the available balancing tank.

All hydraulic simulations have been run both in demand driven (DD) and in pressure demand driven mode (PDD threshold pressure = 20 mwc) for total 2520 different scenarios that are summarised in Table 8.4. The four investment/operation scenarios deal with extreme combinations of loan conditions and energy costs, which are shown in Table 8.5.

Table 8.4 Network scenarios for reliability analysis

VARIABLES	LABEL	OPTIONS	TOTAL
Supplying Scheme	A	Direct Pumping (Net A)	2 scenarios
	B	Pumping & balancing tank (Net B)	
Topography pattern	F	Flat	10 scenarios
	U	Slope up	
	D	Slope down	
	H	Hill	
	V	Valley	
Altitude range	L	Low altitude range *(10 - 20m)*	30 scenarios
	M	Medium altitude range *(10- 30m)*	
	H	High altitude range *(10 - 40m)*	
Investment/operation combinations	HH	High investment - high operation	120 scenarios
	HL	High investment - low operation	
	LH	Low investment - high operation	
	LL	Low investment - low operation	
Design scenarios		21 combination of network resistance and pumping capacity	2520 scenarios

Table 8.5 Network scenarios for reliability analysis

Label	Investment cost	Operation cost	Annual interest	Loan repayment	Annual inflation	Energy cost (US$/kWh)
HH	High	High	12%	10 years	10%	0.45
HL	High	Low	12%	25 years	10%	0.05
LH	Low	High	4%	10 years	4%	0.45
LL	Low	Low	4%	25years	4%	0.05

The first costs have been calculated applying uniform equation aX^n where X stands for pipe diameter (in mm), storage volume (in m³) and maximum installed capacity of pumping station (in l/s), respectively. Corresponding factors a and n have been applied as shown in Table 8.6.

Table 8.6 Factors used for the first cost

Factor	D (mm)	V (m³)	Q (l/s)
A	1	222	22,222
N	1	1	0.8

All the simulations have been run for single steady state, and at constant demand. Consequently, the volume of the balancing tank in B of 2124 m³ has been kept constant all the time (also in the cost calculations). Finally, the same percentage of the investment costs has been adopted for annual costs of operation and maintenance, as suggested above: 0.5% for pipes, 0.8% for storage and 2.0% for pumping station. To enable efficient work while processing large number of scenarios, the network diagnostics tool (NDT) introduced in Chapter 6, has been further upgraded with options that deal with calculations of investment, operation and maintenance costs on annual basis. The reliability i.e. the resilience of each network/demand scenario has therefore been assessed from the perspective of total costs. The reliability measures that have been used in this assessment are the network buffer index (*NBI*) and network resilience (I_n), introduced in Chapter 5.

8.7   PERFORMANCE OF OPTIMISED NETWORKS

As mentioned, the two initial networks have been manipulated with five topographic patterns and three altitude ranges. Figures 8.9 and 8.10 give impression about the distribution of pressures in GA-optimised i.e. initial layouts in each of the topographic patterns and the range of nodal elevations taken as an example between 10 and 30 msl. As it can be seen, the pressures in all cases are represented with contour lines of 20 to 50 mwc; the fraction of red area in the second pattern shown on Figure 8.9 reflects the minimum pressure of 19.6 mwc. Both figures show logical response to the particular topographic patterns: the areas of higher pressure around the source and in low elevated areas, and the areas of low pressure more far away of the source and around high elevated areas. Networks *B* have higher pressures, which is the consequence of elevated tank selected to ensure the minimum pressure. The maximum pressure amongst the *A* options is 53.6 mwc, while for the *B* options it is 71.2 mwc, both in node *N2* and in hilly terrain configuration.

Figures 8.11 to 8.15 show the relation between the reliability measures and total annual costs taken as an average of all four investment options from Table 8.6. The initial layouts are categorised per topography pattern and altitude range. The observations from the figures can be summarised in the following bullets:
-    In all the cases, the optimised networks of option *A* are more expensive than those of option *B*. This is a direct consequence of the difference in supplying scheme in which the pump and the maximum pipe diameter have to be designed on the maximum peak demand, while in case of the tank existing in option *B* the design flow for the pump and the maximum pipe diameter can be reduced.
-    The costs of the networks in the flat topography shown in Figure 8.11 are the lowest of all, as well as they are the least dependent on the selected altitude. In other four topographic patterns the costs would generally grow with (significantly) wider altitude range. This trend is more visible in option *A* than in *B* and has again the likely cause in need of a bigger pump and maximum pipe diameter.
-    There is a difference in the values of *NBI* that are consistently lower than those of I_n. This emerges from the origin of the two indices, which has been discussed in Chapter 5. The network resilience, I_n, is more of a head-driven index whilst the network buffer index, *NBI*, is more responsive to the network connectivity and its resistance. Nonetheless, in both cases, the values of indices depict rather low reliability, which is expected for GA-optimised networks.
-    The network resilience is lower in case of flat and slope-down configurations (Figures 8.11 and 8.13, respectively) which is the result of generally lower heads/pressures occurring in those configurations. The other three terrain configurations inflict higher pressures, be it from the lower nodal elevations, higher pumping heads and/or larger pipe diameters. This additional buffer therefore makes those networks slightly more reliable.
-    The NBI-values have erratic patterns, yet in much lower range. Similarities with I_n are visible in Figures 8.12, 8.14 and 8.15, but the correlations are generally weak.

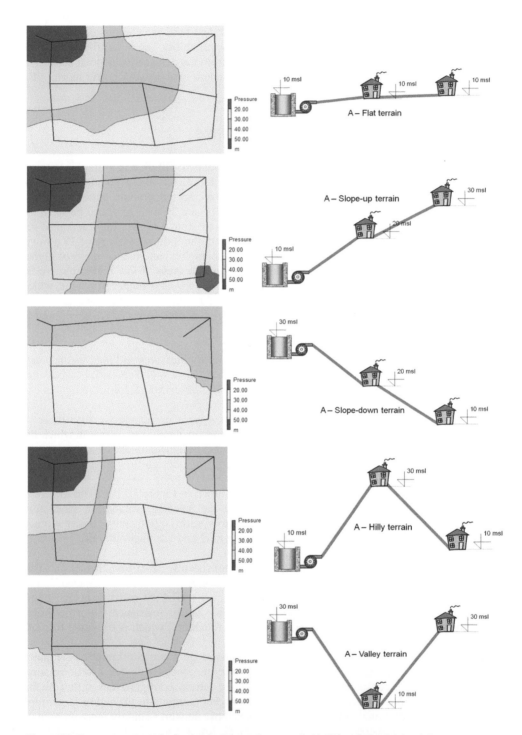

Figure 8.9 Pressure contours for five topography patterns applied in GA-optimised network A

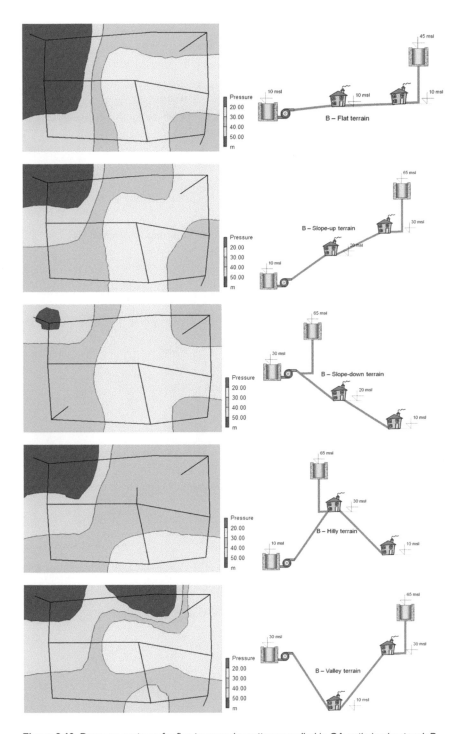

Figure 8.10 Pressure contours for five topography patterns applied in GA-optimised network B

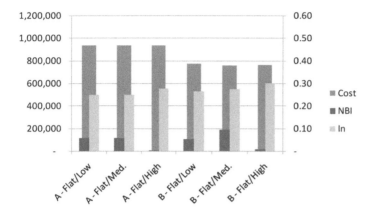

Figure 8.11 Average annual cost (US$) vs. reliability analysis of GA-optimised nets in flat terrain

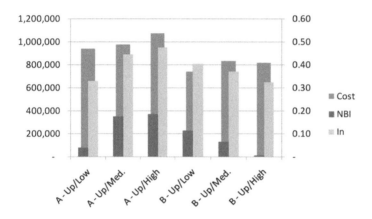

Figure 8.12 Average annual cost (US$) vs. reliability analysis of GA-optimised nets in slope-up terrain

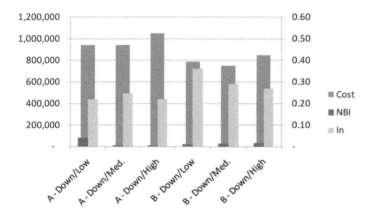

Figure 8.13 Average annual cost (US$) vs. reliability analysis of GA-optimised nets in slope-down terrain

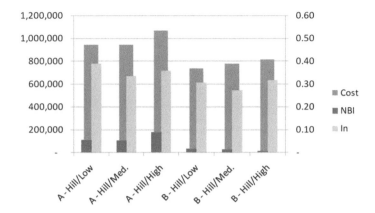

Figure 8.14 Average annual cost (US$) vs. reliability analysis of GA-optimised nets in hilly terrain

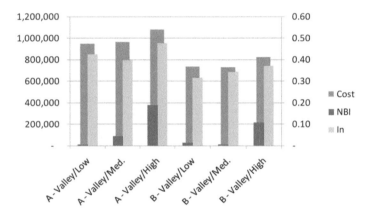

Figure 8.15 Average annual cost (US$) vs. reliability analysis of GA-optimised nets in valley terrain

8.8   RELIABILITY PATTERNS OF NETWORKS WITH REDUCED RESISTANCE

To get more insight about the development of reliability measures inflicted by gradual increase of the pipe diameters and reduction of the pump speed, the first series of 630 simulations was conducted without economic side taken into consideration. The results for *NBI* and I_n of network A and three altitude ranges are shown in Figures 8.16 to 8.18. Twenty one design scenarios that have been analysed for five topographic patterns have similar growth trend of pipe capacity, which has been for the sake of simplicity represented by the average network volume (V_{avg}, shown on the left Y-axis in m^3).

Following are the observations from the figures:
- The NBI-values grow more gradually with the increase of the network volume than is the case with I_n that approaches the maximum value already after 7 to 9 increments. The network resilience, I_n, gives therefore an impression that a certain reliability level can be satisfied at lower investment than is the case looking at the diagrams of *NBI*.

- The NBI-values appear to be much more sensitive to the selected terrain configuration and altitude range, than is the case with I_n. That difference however appears to be reduced with the increase of altitude range. Figure 8.18, on the left, shows three out of five curves ('Up', 'Hill' and 'Valley') eventually arriving at the shape of those curves on the right, showing I_n. On the other hand, the curves representing the flat- and slope-down terrain configuration arrive closer to the shape of the volume curve, suggesting a correlation between the network volume and its reliability, partly discussed in Chapter 7.
- Consequently, the NBI-values of Up-, Hill- and Valley-topography show rather marginal increase in the range where the network volume i.e. investment cost starts to grow significantly, while the trend for Flat- and Down-topography seems to be the opposite. This difference is not signalled by the values of I_n, which all show the same trend, regardless the terrain configuration.
- Almost all the results give impression that both networks A and B reach a certain maximum reliability measured by NBI and I_n, which cannot be further exceeded by additional increase of the volume i.e. at reasonable cost. The NBI-values indicate that this happens at much more different level of the network volume than is the case with I_n, the exception being the Up-curve in Figure 8.17.

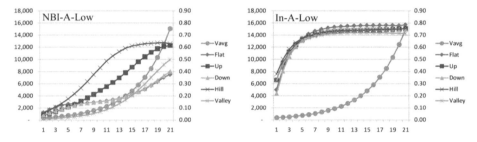

Figure 8.16 Average volume (m³) and reliability for 21 scenario of network A at low altitude range

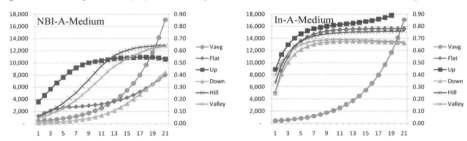

Figure 8.17 Average volume (m³) and reliability for 21 scenario of network A at medium altitude range

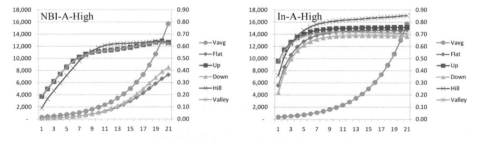

Figure 8.18 Average volume (m³) and reliability for 21 scenario of network A at high altitude range

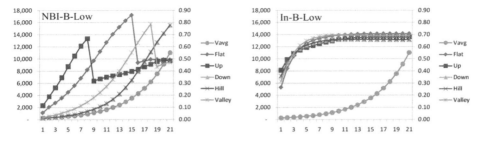

Figure 8.19 Average volume (m³) and reliability for 21 scenario of network B at low altitude range

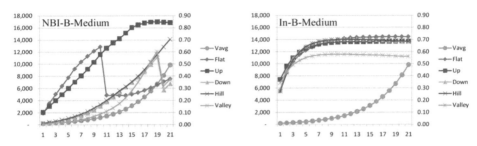

Figure 8.20 Average volume (m³) and reliability for 21 scenario of network B at medium altitude range

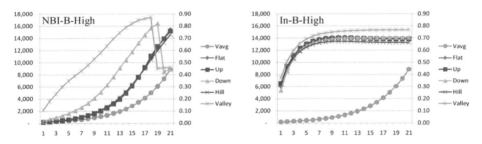

Figure 8.21 Average volume (m³) and reliability for 21 scenario of network B at high altitude range

The results of the same analysis for network *B* are shown in Figures 8.19 to 8.21 with following observations:

- The conclusions drawn in the analyses of network *A* are more or less similar in case of network *B*. There is a distinct difference in the results between *NBI* and I_n. Already after five to seven increments, a further increase of network volume does not seem to improve the network reliability significantly, according to more coherent values of I_n.
- On the other hand, in couple of situations expressed by *NBI*, an instant drop of the value has been registered before it continues to re-grow, however at lower pace. In the process of gradual increase of pipe diameters and reduction of pump speed, the supply from the tank will be covering gradually larger area in each new scenario, until the moment its elevation/head surpass the head of the pump eventually switching it off. The instant curve drop therefore indicates that the system continues to operate as a full gravity system and *NBI* will further grow only from pipe diameter increase.

- Whilst *NBI* is pretty responsive towards the change of supply conditions, the values of I_n do not reflect them; they look pretty similar to those of network A. Next to the different nature of *NBI* and I_n, the contributing factor to the difference in results is the demand-driven (DD) hydraulic calculation used to obtain I_n, whilst the *NBI* is calculated by pressure-driven demand (PDD) calculation. As a consequence, the curves created by the DD calculation are seemingly more 'neat' but actually less accurate representation of the network reliability.

8.9   ECONOMIC ASPECTS OF RELIABILITY ANALYSIS

Each of 630 scenarios (two supplying schemes in five different topographies, three altitude ranges and 21 combinations of pipe diameters and pump speed) was further evaluated on four different investment and operation cost options given in Table 8.5. The average costs of investment and operation and maintenance (O&M) per topographic pattern and altitude range are shown in Figures 8.22 and 8.23.

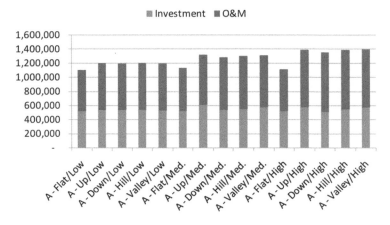

Figure 8.22   Average annual costs (US$) of investment and O&M of all A-scenarios

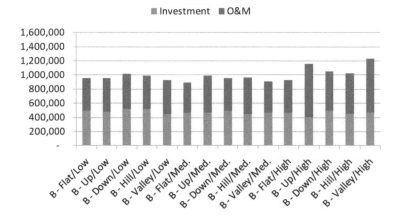

Figure 8.23   Average annual costs (US$) of investment and O&M of all B-scenarios

The figures show clearly that:
- The direct pumping options (A) are in general more expensive than the options that include the balancing tank (B).
- The contribution of O&M cost in the total annual cost is more significant in the case of options A than is the case with options B.
- The contribution of O&M cost in the total annual cost grows with the higher altitude range. This trend is steadier in case of options A than is the case with B.

Typical diagrams which show the relation between investment and O&M costs against reliability measures for particular topographic pattern, altitude range and investment scenario will look like in Figures 8.24 and 8.25. Figure 8.24 shows the annual costs against the reliability measures for 21 design scenarios of network A, located on the hilly topography (H) of low altitude range (L) using high investment/high operation cost scenario (HH) on the left, and high altitude range (H) using low investment/low operation cost scenario (LL) on the right. Figure 8.25 shows the same situation for the network in a valley (V).

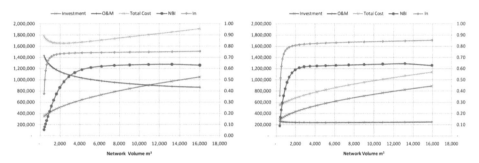

Figure 8.24 Annual costs (US$) and reliability for 21 scenario of network A/H: left - L/HH, right - H/LL

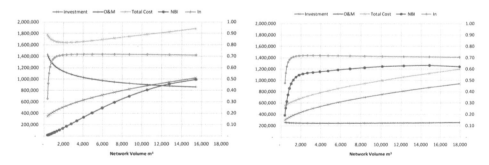

Figure 8.25 Annual costs (US$) and reliability for 21 scenario of network A/V: left - L/HH, right - H/LL

The figures reflect the following:
- The investment cost is much more determining factor, as far the reliability matters, than the O&M costs.
- The O&M costs become significant in the HH cost scenario in the area of low reliability and their increase is actually detrimental for the network reliability. It shows that the lack of conveyance capacity in pipes during failure conditions can hardly be compensated by the increased operation of the pump.

- The increase of pipe capacity as a means to improve the reliability has limited effect. From some point it makes no sense to apply this measure further apart from the fact that it inflicts possible water quality problems in the network, due to low velocities/water stagnation. The values of *NBI* evaluate this moment more rigorously that the values of I_n.
- Consequently, the most economic design is not necessarily the least reliable.

Similar diagrams for network options *B* are shown in Figures 8.26 and 8.27. Figure 8.26 shows the situation of the slope-up topography (U) of low altitude range (L) using high investment/high operation cost scenario (HH) on the left, and high altitude range (H) using low investment/low operation cost scenario (LL) on the right. Figure 8.27 shows the same situation if the network is located in a slope-down terrain (D).

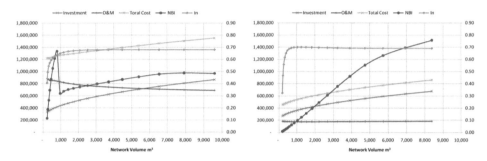

Figure 8.26 Annual costs (US$) and reliability for 21 scenario of network B/U: left - L/HH, right - H/LL

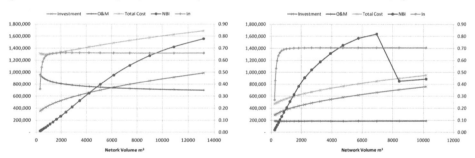

Figure 8.27 Annual costs (US$) and reliability for 21 scenario of network B/D: left - L/HH, right - H/LL

The results in Figures 8.26 and 8.27 support the conclusions drawn from the analysis of the results in Figures 8.24 and 8.25. Not surprisingly, the impact of investment costs becomes even more dominant compared to the O&M costs, both in the HH- and LL cost options. The reduced pumping in network options *B*, compared to *A*, is the reason for such an outcome.

There have been total 120 diagrams like in the above figures, coming out of the analysis, and they all support the above conclusions, more or less. Figures 8.28 to 8.30 show the correlations of all four cost options with the values of *NBI* calculated for 21 design scenarios analysed in different supply and terrain configurations, and classified for three altitude ranges. All diagrams show clearly that:
- the pipe capacity i.e. the pipe investment cost is the driving factor for network reliability,
- the tank in the *B*-networks has positive effect on the network reliability (when it operates properly in combination with pumping station), as well as it reduces the total annual cost.

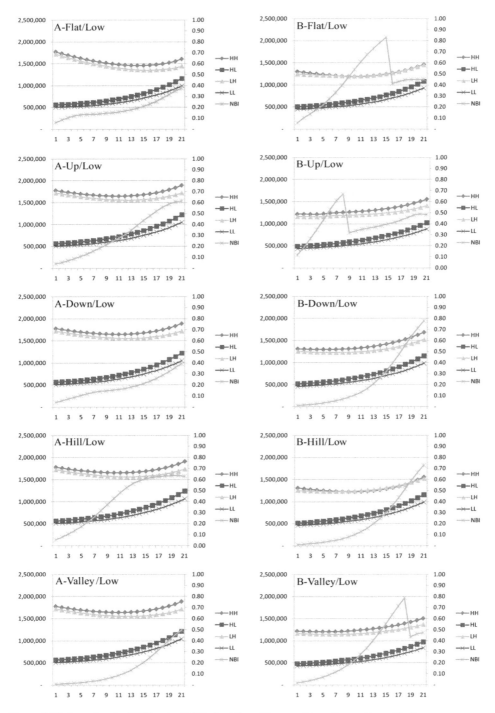

Figure 8.28 Annual costs (US$) and reliability for different topographic patterns at low altitude range

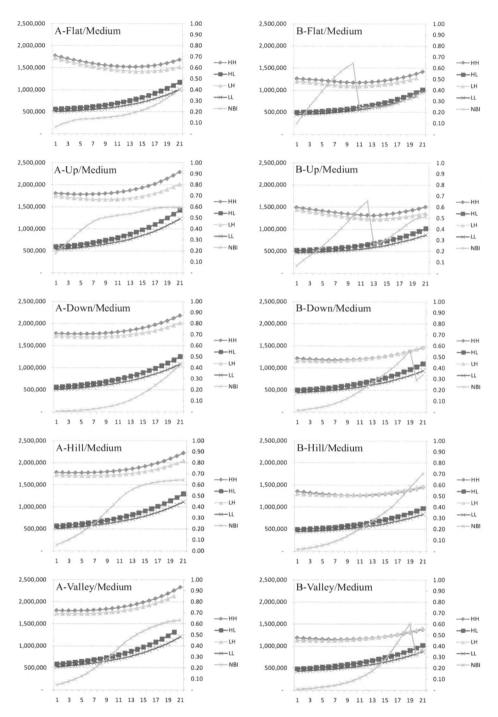

Figure 8.29 Annual costs (US$) and reliability for different topographic patterns at medium altitude range

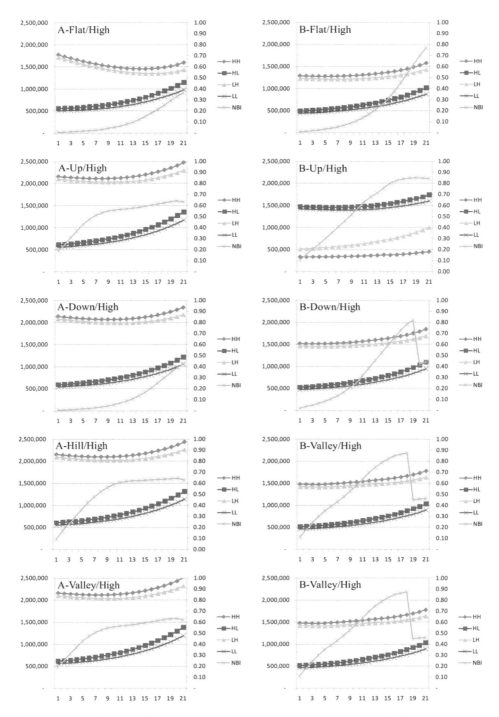

Figure 8.30 Annual costs (US$) and reliability for different topographic patterns at high altitude range

8.10 CONCLUSIONS

The research developed and discussed in this chapter has compared the implications of increased investment and/or operational costs for the reliability of two network configurations with different supplying scheme and with different topographic and design scenarios. The economical aspects have been taken into consideration by applying the Present Worth method that includes actual costs of operation for particular energy tariff and the costs of annual maintenance calculated in simplified way. With this, an attempt has been made to analyse the effect on the reliability if the same amount of money is invested to enlarge the network and/or to improve its operation. The network layouts have been manipulated through five different topographical patterns, three altitude ranges and four economic scenarios. Each of the 120 scenarios was further tested for additional 20 design scenarios comprising gradual increase of the pipe capacity combined with the reduction of pumping capacity, while targeting the similar minimum pressure. The network reliability has been assessed by the two resilience indices introduced in Chapter 5, namely the NBI and I_n.

The analyses of the results done on total 2520 network scenarios lead to the conclusions that can be summarised in the following bullets:
1. Reliability analysed on the level of whole network, or significant part of it, will be much more efficiently maintained by investing into additional pipe capacity rather than by installing additional pumping capacity. The solutions that favour additional pumping on account of reduced pipe diameters hide calamities with potentially serious implications for service levels.
2. In the economic scenario of high O&M costs and unfavourable loans leading to substantial annuities, the least cost design scenario is likely to be more reliable than the one with increased costs for O&M.
3. The design scenarios with balancing tank appear to be more reliable and less costly than those with direct pumping, provided that the tank is properly designed and operated in combination with pumping station.
4. The reliability of networks laid in topographies which allow substantial use of gravity for water conveyance, appears to be more dependent on network capacity i.e. pipe volume.

In the process of analyses using the NDT, the following observations have been made:
1. The tool allows quick execution of large number of design scenarios but has limitations in targeting the minimum pressure for different combinations of network conveying capacity and pumping capacity. This is (currently) not done by using GA optimisation, except for the initial scenario, because it would very likely be a time consuming process for any larger (sample of) networks.
2. The alternative method that is using diameter multiplier and pump speed multiplier is a very basic surrogate of optimiser that is pretty fast and reasonably effective. Nevertheless, the selection of increments asks for trial and error procedure to avoid results outside realistic operational range. Even with this, the fine-tuning of each and every design scenario that could result in the preferred (fixed) minimum pressure is virtually impossible if fixed multipliers are used across the entire range.

Finally, a few conclusions about the reliability measures used in the analyses:
1. The NBI-values have appeared to be much more responsive to the various terrain configurations that is the case with the values of I_n. Again, the conclusions made in

Chapter 5 come forward in the fact that the values of I_n (and I_r) are predominantly driven by the heads, while those of *NBI* are better responsive to the nodal connectivity. Consequently, the two (three) indices give the results that may suggest different economic choices, although they will follow the similar general trends.

2. The values of I_n (and I_r), look more consistent. They are mathematically 'neat' evaluating networks as reliable, as long sufficient head is available in the network, which is somewhat questionable. On the other hand, in case of supply under stress conditions, the formulas used to calculate these indices, based on demand driven simulations, allow negative values both in the numerator and/or denominator, leading to strange peaks or sudden negative values, as shown in Figure 8.31. Such results may have occurred by inappropriate combinations of diameter- and pump speed multipliers, but would be self-corrected in the next increment.

3. The NBI-values can reduce for the increased network volume, which may look weird but is again a result of a strange combination of diameter- and pump speed multipliers: the reducing pump speed causing more loss of head than it has been gained by the increase of the pipe diameters. These are the situations mostly occurring under stress conditions, as it can also be seen on Figure 8.31.

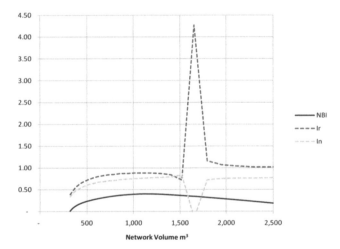

Figure 8.31 Discrepancy in the values of I_n and I_r for supply under stress conditions

4. Finally, the sudden drop of NBI-values, experienced during the analyses in several situations with the balancing tanks, can also be attributed to the situation that is to be avoided in reality. A substantial reduction of pump speed/capacity leading to supply controlled mostly or exclusively by the balancing tank is a signal of serious stress; the reduction in pressures around the pump may cause the loss of demand even without any pipe failure. Neither of the two resilience indices, I_n and I_r, clearly alarm this kind of situations, which is seen as serious deficiency, illustrated in Figure 8.32.

5. The NBI-values will therefore describe the situations of supply under stress better than the values of I_n and I_r, which is mostly the consequence of pressure driven demand nature of *NBI*. This is then also what makes the shape of NBI curves shown in the diagrams occasionally less 'neat' but more true representation of the network reliability than is the case with I_n and I_r.

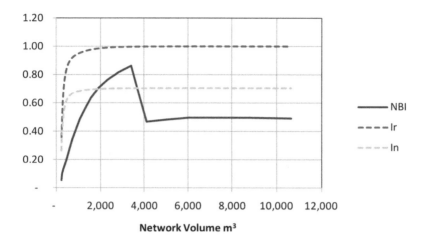

Figure 8.32 Discrepancy in the values of *NBI*, I_n and I_r for supply under stress conditions

Needless to say, all the conclusions have been based on just two simple networks, despite the fact that numerous simulations have been run. This underlines the fact that this kind of analyses can be highly laborious in case any larger (number of) networks. Nevertheless, the research illustrates that the real challenge to optimise between the investment and O&M costs from the perspective of network reliability exists only on a smaller network scale i.e. in particular situations of a few (larger) pipes. Any water distribution network will usually become more reliable in general, if additional funds have been invested into larger pipes rather than the larger pumps.

REFERENCES

1. Cvjetković, M. (2008). *Belgrade Waterworks and Severage - Development and Reforms from 2000 until 2008*. Belgrade Waterworks and Severage, Serbia.
2. De Garmo, P., Sullivan,W.G., and Botandelli, J.A. (1993). *Engineering Economy*. Macmillan Publishing Co., New York.
3. *optiDesigner software*, developed by OptiWater (http://www.optiwater.com)
4. Trifunović, N. 2006. *Introduction to Urban Water Distribution*, Taylor & Francis Group, London, UK, 509 p.
5. VEWIN. *Waterleidingstatistiek 1960-2001.* (in Dutch language).

CHAPTER 9

Decision Support Tool
for Design and Reliability Assessment
of Water Distribution Networks

This chapter describes the decision support tool for **ne**twork **d**esign and **r**eliability **a**ssessment (NEDRA), compiled from the software applications specifically developed for the research presented in Chapters 5 to 8. The reliability analyses that have been conducted on geometric-, hydraulic- and economic grounds all reflect the complexity of network resilience that is still difficult to generalise or classify, as well as it is impossible to asses without full integration of all three aspects. It has been therefore appropriate to conclude the research on pattern recognition for network reliability assessment by a piece of software that should be able to generate sufficient number of design scenarios, assess them in various aspects and let the user eventually draw his/her own conclusions about the best alternative, also from the reliability perspective. NEDRA is composed of the modules for network generation, filtering, initialisation, optimisation on the least-cost diameter, diagnostics of network geometric-, hydraulic-, and economic parameters and finally the assessment of network resilience. Computations can be done for single network in one or multiple scenarios, or a sequence of networks of virtually unlimited size and number, mutually similar or different in properties. The programme has been illustrated on a design example of 50-node network.

9.1   INTRODUCTION

Design of water distribution networks aims at solutions that will guarantee acceptable service levels. The common understanding of the word 'acceptable' will depend on the availability of water at the source, needs and habits of consumers and their ability and willingness to pay for expected level of service. Once all boundary conditions have been properly assessed in the planning phase, the design would initially be approached from hydraulic perspective, the main objective being sufficient water quantity delivered at guaranteed minimum pressure. Until 15-20 years ago, the issues of water quality in distribution networks and network hydraulic reliability were less considered due to absence of powerful methods and tools for such analyses. As far the network reliability, the story would be kept short by including comfortable safety factors in design solutions. Those were driven mostly by designers' engineering experience combined with reasonable economic logic applied during assessment of several scenarios. That however, done with little worries about potentially detrimental impact of large pipe diameters on the water quality.

IT-revolution has significantly changed such a practice, whilst the awareness of consumers about their service levels has grown in parallel, practically all over the world. Consequently, water companies are increasingly pressurised to improve their efficiency at affordable costs. Next to the consumers, the second main driver of this transformation has been the need to protect ever limited drinking water sources in some countries and prevent their uncontrolled depletion. The result of this new paradigm, possible due to better computer tools available, has been a more stringent design process with significantly sharpened economic and reliability considerations. This assumes design parameters taking into considerations reduced pressures (and leakages), reduced fire demands, the effects of water demand management programmes, quick isolation of the network areas affected by calamity or regular maintenance, etc. The quality of the related analyses has improved mainly due to application of pressure-driven demand (PDD) algorithm, which eventually enabled more insight into irregular supply scenarios caused by various calamities, pipe bursts, electricity failures, water rationing, etc. Consequently, the safety factors applied in the design could have received more or less justification once the accompanying costs have been compared with the network performance in stress situations, while assessing many more alternatives.

9.2   DESIGN PARAMETERS AND RELIABILITY CONSIDERATIONS

The aim of any network design is to provide conveyance of sufficient water quantities and then to preserve the water quality obtained at the source i.e. after the treatment process has been completed. From the perspective of public health, this is somewhat twisted logic driven by assumption that sufficient pressure safeguards the quality of water preventing intrusion of pollutants. Furthermore, the problem of overdesigned pipes leading eventually to water stagnation can be mitigated by additional disinfection and pipe cleaning activities, which is mostly true but costs money. More threatening water quality problems actually arrive from inadequate construction and operation of the network, such as poor laying and jointing of pipes, poor selection of materials and appurtenances, intermittent operation or poor protection of pipes against internal and external corrosion, etc. This all is difficult to predict and/or prevent in the hydraulic design phase.

Trifunović (2006) states two postulates of water distribution network hydraulic design (quote):
1. Water flows to any discharge point choosing the easiest path: either the shortest one or the one with the lowest resistance.
2. Optimal design from the hydraulic perspective results in a system that demands the least energy input for water conveyance.

In practice, this means:
-   maximum utilisation of the existing topography (gravity),
-   use of pipe diameters that generate low friction losses,
-   as little pumping as necessary to guarantee the design pressures, and
-   valve operation reduced to a minimum. (end quote)

Hydraulic design parameters mostly concern pressures, hydraulic gradients and pipe velocities, the latter having the largest impact on water quality in distribution networks under regular supply conditions. Acceptable pressures must be maintained throughout the system while satisfying the demand. This is needed to protect the water from pollution entering through damaged sections of the pipes. The pressures in water distribution networks may vary from case to case; they are largely dependent on the availability of water and conditions of the network. Therefore, there is no universally acceptable pressure range. For instance, The Office of Water Services (OFWAT, 1996) in England specifies that pressures of seven meters water column (mwc) above street level can be considered as the minimum acceptable standard, below which the consumers may be entitled to compensation for unsatisfactory service. Typically, the water companies in developed countries set the minimum pressures between 15 and 20 mwc, measured in the most critical part of the system (usually at the boundary of the network), which applies during a peak daily demand. The practice of many water companies in The Netherlands is to maintain the minimum pressures around 20 mwc allowing temporary drops in irregular situations to around 10 to 15 mwc. Tanyimboh et al. (1999) indicate the required minimum service pressure to be as high as 25 mwc, to allow for possible increases in the demand. This is more or less sufficient to supply a 'standard-high' building without internal boosting installation, which in most urban areas is three to five floors high. The maximum pressure in the network should normally be around 60 to 70 mwc. According to Chase (2000), the pressures during normal operations should be kept above 20 mwc and below 70 mwc. Typical range of pressures observed in some distribution networks worldwide is shown in Table 9.1 (Source: Kujundžić, 1996).

Table 9.1 Pressures in world cities (Source: Kujundžić, 1996)

City/Country	Min. - Max. (mwc)
Amsterdam / NL	±25
Wien / Austria	40 - 120
Belgrade / Serbia	20 - 160
Brussels / Belgium	30 - 70
Chicago / USA	±30
Madrid / Spain	30 - 70
Moscow / Russia	30 - 75
Philadelphia / USA	20 - 80
Rio de Janeiro / Brasil	±25
Rome / Italy	±60
Sophia / Bulgaria	35 – 80

There is a clear hydraulic dependency between the demand and pressures in any water distribution system. Delivering a satisfactory level of service will mean that the maximum demand expected in the system under regular supply conditions will be satisfied maintaining the pressures above accepted threshold. Further increase of the demand for any reason, for instance a fire fighting, will cause a corresponding drop of the pressures. The service will also be violated in case of a failure where parts of the system become supplied though alternative routes, causing increased energy losses. As a consequence, the drop of pressure below the threshold will start to affect the demands causing their reduction. As the pressure in many systems in the developing countries is below the above-listed limits, for various reasons, its impact on the demand is even more profound than in case of well-operated systems.

As for the hydraulic gradients, the design criteria depend on adopted minimum and maximum pressures, the distance over which the water needs to be transported, local topographic circumstances and the size of the network, including possible future extensions. Trifunović (2006) suggests the following values as a rule of thumb:
- 5-10 m/km, for small diameter pipes,
- 2-5 m/km, for mid-range diameter pipes,
- 1-2 m/km, for large transportation pipes.

Hydraulic gradients tell something about the network conveyance i.e. the balance between the energy input and energy loss and as such, the balance between the investment and operational costs. They eventually reflect whether the minimum pressure in the network has been created through increased pumping or enlarged pipes, which also has the implications on network resilience against the whole economy of water distribution, as discussed in Chapter 8.

Finally, the velocity range is also to be assessed in the network design process. Too low velocities have potential implications for water quality (sediment accumulation, low chlorine residuals, increased corrosion), while too high velocities are mostly corresponding to high hydraulic gradients, indicating exceptional head-losses. The recommended design values are:
- ± 1m/s, in distribution systems,
- ± 1.5 m/s, in transportation pipes,
- 1-2 m/s, in pumping stations.

These values are however not easily achievable where particular (high) reliability is deemed necessary, be it in the form of fire demand, provision of capacity in case of repair and maintenance, or even as a consequence of overwhelming demand forecasts.

Figure 9.1 shows the pipe flushing diagram in one location of the city of Rotterdam in The Netherlands, which is a result of sediment removal accumulated by very low velocities (Source: Water Supply Company 'Europort', currently 'Evides', Rotterdam). The diagram shows normal operating pressure of approximately 2.7 bar (27 mwc) while the velocity is barely 0.1 m/s. By opening and closing the hydrants, a sufficient velocity is generated (close to 1.5 m/s) to remove the sediment from the pipe. During those moments, the pressure drops down to one bar. The pipe flushing will be conducted in normal working hours, which does not affect the consumers in surrounding areas significantly because of sufficient extra capacity in the rest of the system. Evidently overdesigned network in this case originates from two facts: (1) high fire demand requirement and (2) specific demand prognosis done a couple of decades ago that did not take into account the water demand management programme successfully implemented in the last two decades. Consequently, the reliability guidelines in the Netherlands are mostly focused to the repair processes during which are the consumers effectively disconnected. The moment of failure will hardly be felt, except a major one, due to extra storage in the pipes.

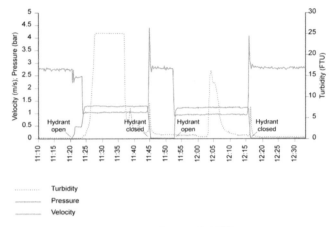

- - - - - - - Turbidity
————— Pressure
————— Velocity

Figure 9.1 Pipe flushing diagram (Trifunović, 2006)

Ideal matching of recommended design values will mostly depend on the topographic conditions that are the crucial factor for determination of supply scheme. The design parameters will further be influenced by the pipes' condition. The form of the Darcy-Weisbach equation shown for pipe j in Equation 9.1 proves this point. Figure 9.2 shows the relation between the pipe velocities, v, and hydraulic gradients, S, for the range of diameters between 50 and 1500 mm and four k-values between 0.1 and 5 mm.

$$v_j = \sqrt{\frac{2gD_jS_j}{\lambda_j}}$$

9.1

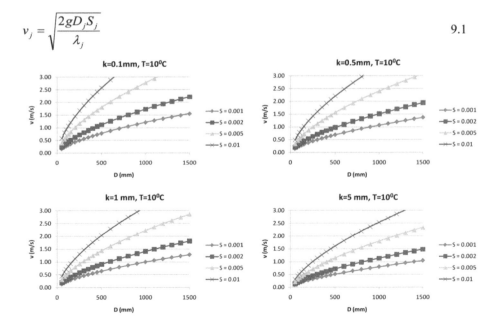

Figure 9.2 Relation between pipe velocities and hydraulic gradients

It is visible on the graphs that the matching of design velocities and hydraulic gradients can be achieved over the wider range of diameters, for lower S-values and/or higher k-values. The first condition could be considered favourably from the perspective of reliability, while the second one has its limitations. It is normal design practice to include the effect of minor losses though somewhat increased k-values, but if those are much higher than a few millimetres, that would signal corroded- rather than properly designed pipes.

9.3   NETWORK DESIGN AND RELIABILITY ASSESSMENT TOOL

The **n**etwork **d**esign and **r**eliability **a**nalysis tool (NEDRA) has been developed to facilitate design process of water distribution networks by adding more of reliability assessment perspective to it. The tool has been compiled from the software applications specifically developed for the research presented in Chapters 5 to 8. By adding the filtering and initialisation feature to the network generation tool elaborated in Chapter 4, and using the hydraulic solver for pressure-driven demand calculations, a coherent set of routines has been created which can analyse numerous design scenarios considering different connectivity, pipe diameters, demand distribution (both spatial and temporal) and assess these against several reliability measures and economic scenarios. Although not necessarily relevant for the design process, the features related to random generation of some input data, namely the connectivity and nodal elevations, have been preserved keeping NEDRA still applicable for research purposes.

The programme has been coded in Microsoft Visual C++ 2010 Express, using EPANET toolkit library to communicate with EPANET software of US EPA (Rossman, 2000). The GA-optimiser that has been integrated into NEDRA is the *Evolving Objects (EO)*, used for network optimisations in Chapters 4 to 8.

NEDRA consists of five main modules dealing with network (1) generation, (2) filtering, (3) initialisation, (4) optimisation and (5) diagnostics. A brief description of each module is given in the following sections.

9.3.1 Network Generation Module

The network generation module has been developed applying the concepts of graph theory, which has been elaborated in Chapter 4. The module, presented there as NGT (network generation tool) generates different network layouts, based on selected location of nodes and sources, which is initially prepared in EPANET in INP-file format. The programme reads the nodal information prepared in EPANET, which includes the coordinates, the elevations and baseline demands. The latter two can be modified, also by specifying a range of values that will be randomly assigned in both cases.

The network generation takes place in two ways: (1) non-randomly and (2) randomly. The graph theory principles applied in the non-random generation mode result in an algorithm that explores all possible elements of a matrix formed on the limitation that each node in the process of generation can be connected to maximum three additional nodes. Next to this condition, the algorithm is developed to connect first the closest available nodes, avoiding the crossings with the existing pipes, and duplications that could occur as a result of the reversed order of the pipe nodes .The degree of complexity of generated networks can be influenced by picking a combination out of total seven available matrix columns, some of which can be skipped in the process of generation. That eventually leads to seven different levels of complexity (1 - lowest, 7 - highest). Furthermore, the number of connections to the nodes can be maximised, as well as the range in number of links to be generated in the process can be specified. The obvious correlation between these settings asks for some caution while choosing their combination, because some of these can lead to impossible assignment: for instance, a high complexity with low maximum number of nodal connections and/or pipes, or the other way round.

The random network generation is more straight forward process but also with more unpredictable result in terms of configuration i.e. the complexity of generated networks. This is logical, because the process is less controlled although the settings to influence the complexity are also available in this mode, through the preset maximum and minimum number of links and the maximum number of connections to the nodes, however with the same concern as in case of non-random network generation. More detailed information about the settings and the pros and cons of the non-random and random network generation has been given in Chapter 4.

The generated links further receive the properties in the following way:
1. Lengths, L, which are calculated from the available nodal coordinates, or automatically assigned as uniform- or randomised values (in specified range).
2. Diameters, D, automatically assigned as uniform- or randomised values (in specified range). If the layouts are going to be GA-optimised, additional files will be created to control the GA-simulation. Hence, the GA-optimisation does not take place within the network generation module but is to be run, if preferred, after the modules of network filtering and initialisation have been run (or skipped).
3. Pipe roughness, k, automatically assigned as uniform- or randomised values (in specified range).

Hence, all generated links will be initially treated as ordinary pipes, meaning that the pumps and valves have to be added separately, which can be done during the initialisation phase. The output of the network generation process is a number of INP-files that can be directly loaded and further analysed in EPANET. Their names (without the extensions) will be stored in a separate file with unique name *EPAinpfiles.txt* which can be further used for handling of multiple files in other modules in one go.

9.3.2 Network Filtering Module

The network filtering module is in fact an add-on to the network generation module, which has been specifically developed for the design purposes. It is a short programme that compares all the networks generated in the previous module with the network template prepared by connecting the nodes of the initial INP-file with links that could suggest the preferred or the only possible routes i.e. the streets. The degree of similarity with the template file will be summarised in the specified output text file. This file lists the number of links in each of the generated networks and the number of those that match the template. The IDs of the mismatch links will be shown in the networks that have less than ten of these. Finally, the list of all names of the files that match the template will be printed at the end, for possible adaptation of the contents of the file *EPAinpfiles.txt*. For that purpose, the generated networks that contain all the pipes corresponding to the links in the template shall also have their name saved in the *EPAinpfilter.txt* file. The user can also decide to add the networks below certain number of mismatched links to this list, qualifying them as 'passed'. The rest of the files will be eliminated from further analyses but they are not deleted from the folder, for possible re-use at later stage.

9.3.3 Network Initialisation Module

The major network parameters will be assigned either directly in EPANET and/or during the process of network generation. The network initialisation module leaves a possibility to further modify the model input by comparing the content of the files with the name on the list of *EPAinpfiles.txt* file with the template used for the filtering of generated files. The template is a network whose basic properties can be assigned to the generated networks if this was not done during the process of generation, or their data have been modified in the course of various analyses, for instance during the optimisation. The data conversion will also include pumps and valves if existing in the template. This module can therefore be run before or after the optimisation.

Next to the practical reasons, this module also serves to correct what seems to be a bug in the EPANET toolkit function *ENsaveinpfile*. This function omits the nodal coordinates from the saved INP-file, restricting visual representation in EPANET of such-saved file. The coordinates will be reinstated in the process of data initialisation, as well as can be the case with nodal demands, elevations and emitter coefficients. The module further restores the pipe lengths, diameters, k-values and in addition, the status of deliberately closed pipes, if existing in the template. Currently, the contents of INP-files will be over-written, which has to be taken into consideration for later use (no back-up file is made).

9.3.4 Network Optimisation Module

The generated- and possibly filtered/initialised networks can be further optimised by using the Evolving Objects (EO) optimiser, which is a single objective oriented GA-optimization tool developed by Keijzer et al. (2002), and distributed under the GNU Lesser General Public License by SourceForge (http://eodev.sourceforge.net).

The goal of the optimization module is to get optimal solutions for the network pipe diameters. The main reason to use the EO has been the fact that it is an open source package for evolutionary computations, which can be adapted for specific research purposes and used for relatively large number of nodes and links in a network. Hence, the EO has proven to be convenient choice for the specific purpose of this research, too. Furthermore, an investigation on the best performing GA-optimiser to be built into NEDRA, or on the list of optimal default GA-parameters, was not seen as an objective of this research.

The networks listed in the *EPAinpfiles.txt* file are optimised to keep the selected threshold pressure at each demand node; this is done with help of information available in SNxxx files prepared in the network generation process, and their list stored in the *Inp.txt* file, which can be also adapted during the network filtration process. In case the pressure at node is below the threshold pressure, the penalty will be added. After a number of iterations, the GA-process will provide optimal diameters for the least cost design. The default boundary conditions of the objective function, constraint and the penalty cost are shown in Table 9.2.

Table 9.2 Default boundary conditions for GA-optimisation of generated/filtered networks

Description	EO-Optimizer
Objective function	Least-cost diameter
Fitness	Threshold pressure (20mwc)
Penalty	For pressure < 20mwc, +200,000

The process starts by importing three types of files:
1. The generated/filtered INP-files (*name.inp*) whose names are specified in the accompanying *EPAinpfiles.txt* file.
2. The TXT-files stored in the *Inp.txt* file that have been generated in case the GA optimisation setting has been selected in the network generation module. Each TXT-file labelled as *SNxxx.txt*, xxx being the serial number, contains the EPANET input filename, with INP and RPT extension, followed by the total number of links, and their IDs, and finally the total number of nodes, followed by the node ID.
3. The file *GApara_sample.txt* which contains the definition of GA-parameters: population size, selection type, total generation number and GA-operators (crossover and mutation rates). Based on the information in this file, each network will automatically receive the file *SNxxx_GApara.txt* with its own settings based on its own size i.e. number of links.

The default GA-parameters applied for optimization in this module are shown in Table 9.3.

Table 9.3 Default GA-optimisation parameters

Description	GA parameters
Vector size	20~200
Population size	100
Maximum no. of generations	5,000
No. of generations with no improvement	3,500
Selection	Tournament
Crossover probability	0.55
Mutation probability	0.4 ~ 0.45

The list of diameters available for optimisation can be created in three ways: (1) by specifying continuous range, (2) by specifying a range and the diameter increment, and (3) by specifying the number of specific diameters manually. The latter is the option closest to the reality, while the first one is the fastest and the most accurate one in view of targeting the threshold pressure. The final output from the simulation is (a sequence of) optimised INP-file(s) coded as *Y_Optimized_name.inp*. Based on the above selection of diameters, Y can assign *C* (for continuous *D*), *I* (for incremental *D*) or *M* (for manual *D*).

9.3.5 Network Diagnostics Module

The network diagnostics module, introduced initially in Chapter 6 as NDT (network diagnostics tool) and further upgraded for the analyses presented in Chapters 7 and 8 analyses geometric and hydraulic properties, both in DD- and PDD mode, of a network or sequence of networks. It further assesses the network resilience using a number of measures and puts it in the prospective of the total costs of investment and operation and maintenance. The calculations for single network are twofold. A selection can be made between the calculations where detailed results will be given for one specific scenario by tabulating all the pipes, or alternatively a particular network parameter can be modified in number of incremental steps, defining the multipliers in similar way as it is done in EPANET with the general multiplier. In addition to the uniform multipliers, a feature has been added to specify a range of randomised multipliers for number of parameters. The list of options in the module is shown in Figure 9.3.

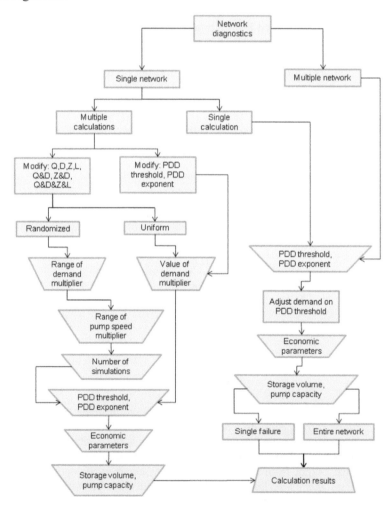

Figure 9.3 Menu structure in the network diagnostics module

In the calculations of multiple networks, the results for the networks in the list of *EPAinpfiles.txt* file include the following parameters, discussed in Chapters 6:
1. number of nodes and links, (n and m, respectively),
2. number of basic loops and complex loops (l and L_n, respectively),
3. the network grid index (*NGI*),
4. the network connectivity factor and index (*NCF, NCI*), in all three variants,
5. the network shape index (*NSI*), in all three variants,
6. total network length and volume (L_{tot} and V_{tot}, respectively),
7. total demand, both in the demand-driven mode (DD) and pressure-driven mode (PDD),
8. minimum pressure and corresponding node ID (in DD- and PDD mode),
9. maximum pressure and corresponding node ID (in DD- and PDD mode),
10. the resilience index of Todini (I_r),
11. the network resilience of Prasad and Park (I_n),
12. the network buffer index (*NBI*) in PDD mode,
13. the average available demand fraction (ADF_{avg}) calculated by failing individual links consecutively,
14. the minimum available demand fraction (ADF_{min}) and corresponding link causing the highest loss of demand.

Some other interim results will be printed in the output file as well, such as the hydraulic indicators of the PDD simulations used for determination of ADF-values, the number of intentionally closed pipes, the values of some input parameters (multipliers), etc.

In the calculations of single network, the results include the following parameters discussed in Chapter 7, which is shown in the output file for each pipe:
1. volume - V_j,
2. flow - Q_j,
3. velocity - v_j,
4. friction loss - $h_{f,j}$,
5. hydraulic gradient - $S_j = h_{f,j} / L_j$,
6. power loss - $P_j = \rho g Q_j h_{f,j}$,
7. residence time - $T_j = V_j / Q_j$,
8. pressure buffer - pb_j,
9. unit pressure buffer - pb_j / L_j,
10. hydraulic loss area - hl_j,
11. pressure buffer index - PBI_j.

In cases the GA-optimisation has been bypassed, a possibility exists to adjust the demand multiplier, both in single and multiple network simulations. A simple iterative process has been built in the programme that attempts to eliminate the difference between the minimum pressure in the most critical node and specified PDD threshold pressure within ten interpolation steps. In this way, the general multiplier in EPANET will be changed rather than the pipe diameters. This form of 'synchronised' demand optimisation has been introduced as a result of network generation process in which the geometric properties of the nets are not necessarily matched with reasonable level of demands causing the pressure range that is actually useless for any viable hydraulic analysis.

All calculation results of the network diagnostics module are tabulated in a text file with extension 'xls', which makes it easily transferable in MS Excel for further processing and graphical representation of the results.

9.3.6 Available Demand Fraction and Coverage

The network diagnostics module establishes the values of available demand fraction based on the results of PDD simulations for consecutive failure of links, one at a time. This includes pumps and valves, except those in initially closed status. The drop of demand will be compared with the full demand i.e. the one calculated in DD mode. This means that the demand drop from the low pressures in PDD mode, caused by low supplying heads despite all links being functional, will also be considered as an irregular situation.

In the single network menu option, the *ADF* can be displayed for all the pipes, while for the multiple network simulations two values will be shown: the average one and the one for the most critical pipe failure, causing the lowest *ADF*. Next to that, the nodal *ADF* will be divided into ten categories, at the increments of 10%, displaying the number of nodes within each category, as well as the percentage of total original demand affected with the failure. This analysis, done for the worst case pipe failure, will depict the spatial distribution of demand loss in the first case, and its severity, in the second case. Based on these results, it becomes more clear whether the calamity affects large or low number of nodes, and if it has happened to moderate or severe extent.

9.3.7 Economic Considerations

Economic considerations in the network diagnostics module are founded on the engineering economy concepts discussed in Chapter 8. Based on the selected loan conditions and inflation, the investment costs will include the annuities, while the operational costs will include the actual energy consumption in case of pumping used for water conveyance; this is done from the results of PDD simulations. The pipe costs are calculated for actual length and diameter, while the investment costs into the pumps and storage are based on the storage volume and maximum installed capacity of pumping stations. For the time being, there is no automatic link between the pipe costs used in the network diagnostics module and the GA optimiser (which is also the case with PDD threshold pressure). These values therefore have to be synchronised manually during the analyses. Logically, all the figures will be shown for the entire network, and all the networks listed in the *EPAinpfiles.txt* file.

9.4 CASE STUDY

NEDRA has been illustrated on a synthetic case of a reservoir and 50 demand nodes. The location of the source and the nodes prepared in EPANET is shown in Figure 9.4a showing the nodal elevations (in msl), and 9.4b showing the baseline demands (in l/s).

The case reflects a gravity supply situation on a terrain descending from the reservoir for some 45 m. The total demand of 305 l/s is distributed over relatively wide area within the distance to the most faraway node of roughly 16 km, calculated based on the nodal coordinates. The head of the reservoir is 50 msl.

The network generation has taken place in various configurations settings, both in random and non-random mode. Total 13,000 networks have been generated.

Figure 9.4a Case network area processed by NEDRA package - elevations (msl)

Figure 9.4b Case network area processed by NEDRA package - baseline demands (l/s)

9.4.1 Preliminary Network Generation

The first 3000 networks have been generated in 30 batches of 100 configurations applying different settings. This was done to explore the type of configurations and generation time, and overall robustness of the network generation module. Table 9.4 gives the overview of all the settings used, and the corresponding time to generate the first 100 configurations (in seconds), using an ordinary i.e. average-speed laptop.

Table 9.4 Network batches and corresponding generation time (in seconds for 100 configurations)

Batch code	Description of generation settings	Max.4-5 con.	Free con.
R55-65	Random, between 55 and 65 links	5983 (4 con)	389
R65-75	Random, between 65 and 75 links	475 (4 con)	92
R55-75	Random, between 55 and 75 links	385 (4 con)	96
Rfull	Random, no restriction on number of links	380 (4 con)	88
NfullC2	Non-random, no link restriction, complexity 2	135 (4 con)	130
NfullC3	Non-random, no link restriction, complexity 3	150 (4 con)	149
NfullC4	Non-random, no link restriction, complexity 4	58 (5 con)	57
NfullC5	Non-random, no link restriction, complexity 5	71 (5 con)	72
NfullC6	Non-random, no link restriction, complexity 6	64 (5 con)	66
N55-65C2	Non-random, 55 to 65 links, complexity 2	133 (4 con)	142
N55-65C3	Non-random, 55 to 65 links, complexity 3	196 (4 con)	165
N55-75C2	Non-random, 55 to 75 links, complexity 2	142 (4 con)	144
N55-75C3	Non-random, 55 to 75 links, complexity 3	196 (4 con)	162
N55-75C4	Non-random, 55 to 75 links, complexity 4	58 (5 con)	66
N55-75C5	Non-random, 55 to 75 links, complexity 5	69 (5 con)	70

To illustrate the type of configurations, each batch is represented by one network, mostly the first one or the last one in the sequence of 100, which is shown in Figures 9.5a to 9.5o. The figures give visual impression that complies with the selected complexity level in the non-random generation. Moreover, this mode generates pretty similar layouts within the same complexity level, which suggests that the non-random generation needs much larger number of simulations to arrive at significantly different layouts if all matrix columns are to be visited; just a few dozen layouts may differ in one or two connections, only. The random generation produces more variety within the batch, and is less dependent on the combinations of selected number of pipes and/or the maximum number of connections. Specific settings of both of these parameters, which have clear relation to the number of nodes, can quite easily impact the generation time and the number of available configurations. This is particularly visible in the significantly higher generation time of the first two random batches (R55-65). On average, the random generation has been longer than the non-random one.

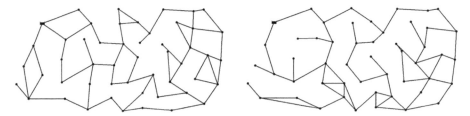

Figure 9.5a Sample of generated networks R55-65: 4 connections - left, free connectivity - right

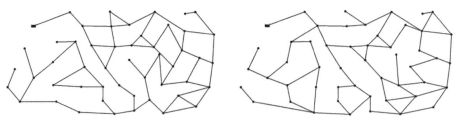

Figure 9.5b Sample of generated networks R65-75: 4 connections - left, free connectivity - right

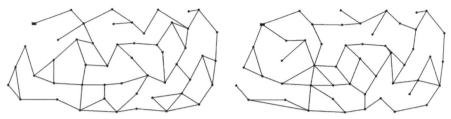

Figure 9.5c Sample of generated networks R55-75: 4 connections - left, free connectivity - right

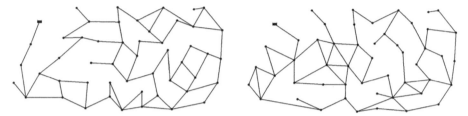

Figure 9.5d Sample of generated networks Rfull: 4 connections - left, free connectivity - right

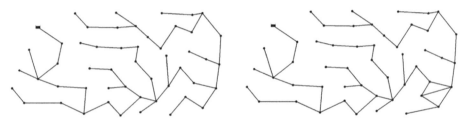

Figure 9.5e Sample of generated networks NfullC2: 4 connections - left, free connectivity - right

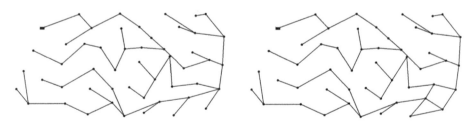

Figure 9.5f Sample of generated networks NfullC3: 4 connections - left, free connectivity - right

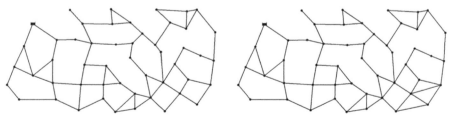

Figure 9.5g Sample of generated networks NfullC4: 5 connections - left, free connectivity - right

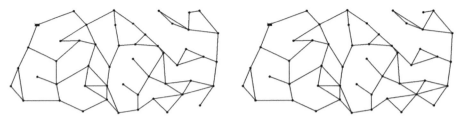

Figure 9.5h Sample of generated networks NfullC5: 5 connections - left, free connectivity - right

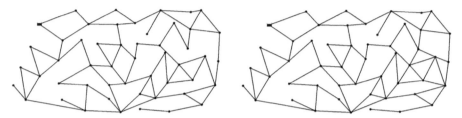

Figure 9.5i Sample of generated networks NfullC6: 5 connections - left, free connectivity - right

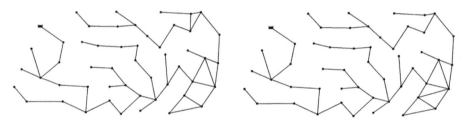

Figure 9.5j Sample of generated networks N55-65C2: 4 connections - left, free connectivity - right

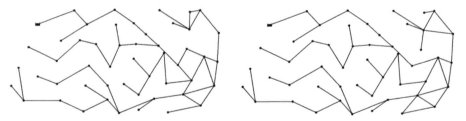

Figure 9.5k Sample of generated networks N55-65C3: 4 connections - left, free connectivity - right

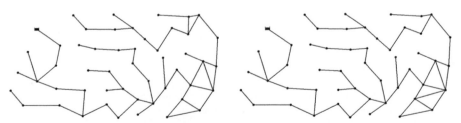

Figure 9.5l Sample of generated networks N55-75C2: 4 connections - left, free connectivity - right

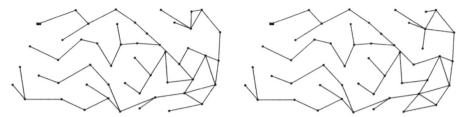

Figure 9.5m Sample of generated networks N55-75C3: 4 connections - left, free connectivity - right

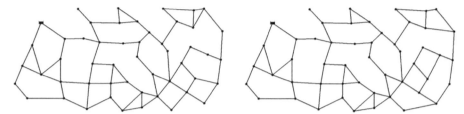

Figure 9.5n Sample of generated networks N55-75C4: 5 connections - left, free connectivity - right

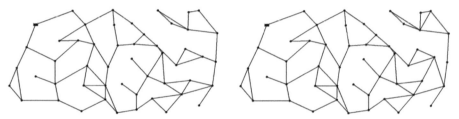

Figure 9.5o Sample of generated networks N55-75C5: 5 connections - left, free connectivity - right

9.4.2 Network Filtering and Additional Generation

To proceed further in the design process, all the generated layouts have been filtered using the template network shown in Figure 9.6. This layout, consisting of 130 links, suggests the routes that are permitted for the pipes of generated layouts.

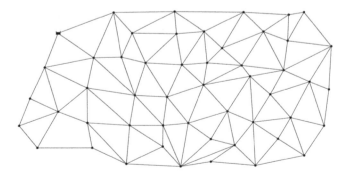

Figure 9.6 Network template for the case area in Figure 9.4

Table 9.5 Results of network filtering for 30 batches of 100 nets

Batch code	Limited nodal connectivity (4 or 5)					Unrestricted nodal connectivity				
	Pipe range	Avg.	Net match	MisM range	Avg. miss	Pipe range	Avg.	Net match	MisM range	Avg. miss
R55-65	55-65	63.65	0	1-9	5.04	55-65	63.63	0	2-9	4.94
R65-75	65-75	68.82	0	2-8	4.97	65-75	71.04	0	2-9	5.75
R55-75	55-75	68.10	0	1-8	5.11	65-75	71.33	0	3-9	5.85
Rfull	62-74	67.99	0	1-9	4.99	63-81	71.78	0	1-9	5.72
NfullC2	50-54	51.46	54	0-1	0.46	50-54	51.40	36	0-1	0.64
NfullC3	50-53	51.76	0	7-8	7.66	50-54	51.67	0	7-8	7.81
NfullC4	74-77	75.13	0	1-2	1.73	74-77	75.13	0	1-2	1.73
NfullC5	74-77	74.94	0	9	9.00	74-78	75.21	0	9	9.00
NfullC6	74-80	77.16	0	9	9.00	77.16	74-80	0	9	9.00
N55-65C2	55-56	55.10	40	0-1	0.60	55-57	55.16	14	0-1	0.86
N55-65C3	55-56	55.04	0	7-8	7.59	55-57	55.16	0	7-8	7.87
N55-75C2	55-56	55.10	40	0-1	0.60	55-57	55.16	14	0-1	0.86
N55-75C3	55-56	55.04	0	7-8	7.59	55-57	55.16	0	7-8	7.87
N55-75C4	74-75	74.61	0	1-2	1.68	74-75	74.61	0	1-2	1.68
N55-75C5	74-75	74.68	0	9	9.00	74-75	74.68	0	9	9.00

The results of the filtering process are presented in Table 9.5. The average values in the table show the average number of pipes produced by the generator and the average number of pipes that do not exist in the template network, which is also reflected by the respective pipe ranges. The randomly generated layouts have not produced any matching layout, while the non-random generation had the matching layouts in six batches, displayed in the shaded cells of the table. Four additional batches had no layouts complying with the template, but the filtering tool indicated the difference in only 1 or 2 pipes. These are mostly the same pipes as can be seen in Table 9.6 showing the output layout of the network filtering module

The filtering results prove that the batches of 100 layouts are apparently insufficiently variable and too small for the limitations imposed by the number of links in the template. Consequently, the networks that fit to it can be generated only with the low level of complexity. Finally, the results in Table 9.5 show similarities for some categories, regardless the selected range of number of pipes, and/or maximum number of nodal connections, which again can be attributed to the restrictions caused by the interference between these parameters and the total number of nodes.

To get more insight into the pattern of network configurations, one random- and four of the non-random batches highlighted in Table 9.5 have been regenerated in the size of 1000 nets; these results are shown in Table 9.7. The codes in the brackets stand for unrestricted nodal connectivity (U), and the one maximised at 4 connections per node. Finally, the random batch R55-75(U) has been regenerated in the size of 5000 layouts, with the results shown in Table 9.8.

Table 9.6 Results of network filtering for NfullC4, unrestricted connectivity

Network filtering tool (Ver. 1.00 by N.Trifunovic)

Report file: g50t1FULL.cpx4cf.xls

Total number of links in the template file (g50t1.inp): 130

Number of networks to be filtered: 100

No	File name	NoLinks	Match	Deficit	PiDif1	PiDif2	Status	Comment
1	Input_001.inp	74	73	1	49		Rejected!	Match<Tem
2	Input_002.inp	74	73	1	49		Rejected!	Match<Tem
3	Input_003.inp	75	74	1	49		Rejected!	Match<Tem
4	Input_004.inp	74	73	1	49		Rejected!	Match<Tem
5	Input_005.inp	74	73	1	49		Rejected!	Match<Tem
6	Input_006.inp	75	74	1	49		Rejected!	Match<Tem
7	Input_007.inp	75	74	1	49		Rejected!	Match<Tem
8	Input_008.inp	75	74	1	49		Rejected!	Match<Tem
9	Input_009.inp	76	75	1	49		Rejected!	Match<Tem
10	Input_010.inp	74	72	2	49	68	Rejected!	Match<Tem
11	Input_011.inp	74	72	2	49	68	Rejected!	Match<Tem
12	Input_012.inp	75	73	2	49	68	Rejected!	Match<Tem
13	Input_013.inp	74	72	2	49	68	Rejected!	Match<Tem
14	Input_014.inp	74	72	2	49	68	Rejected!	Match<Tem
15	Input_015.inp	75	73	2	49	68	Rejected!	Match<Tem
16	Input_016.inp	75	73	2	49	68	Rejected!	Match<Tem
17	Input_017.inp	75	73	2	49	68	Rejected!	Match<Tem
18	Input_018.inp	76	74	2	49	68	Rejected!	Match<Tem
19	Input_019.inp	74	72	2	49	68	Rejected!	Match<Tem
20	Input_020.inp	74	72	2	49	68	Rejected!	Match<Tem
21	Input_021.inp	75	73	2	49	68	Rejected!	Match<Tem
22	Input_022.inp	74	72	2	49	68	Rejected!	Match<Tem
23	Input_023.inp	74	72	2	49	68	Rejected!	Match<Tem
24	Input_024.inp	75	73	2	49	68	Rejected!	Match<Tem
25	Input_025.inp	75	73	2	49	68	Rejected!	Match<Tem
26	Input_026.inp	75	73	2	49	68	Rejected!	Match<Tem
27	Input_027.inp	76	74	2	49	68	Rejected!	Match<Tem
28	Input_028.inp	75	73	2	49	69	Rejected!	Match<Tem
29	Input_029.inp	75	73	2	49	69	Rejected!	Match<Tem
30	Input_030.inp	76	74	2	49	69	Rejected!	Match<Tem
31	Input_031.inp	75	73	2	49	69	Rejected!	Match<Tem
32	Input_032.inp	75	73	2	49	69	Rejected!	Match<Tem
33	Input_033.inp	76	74	2	49	69	Rejected!	Match<Tem
34	Input_034.inp	76	74	2	49	69	Rejected!	Match<Tem
35	Input_035.inp	76	74	2	49	69	Rejected!	Match<Tem
36	Input_036.inp	77	75	2	49	69	Rejected!	Match<Tem
37	Input_037.inp	74	73	1	49		Rejected!	Match<Tem
38	Input_038.inp	74	73	1	49		Rejected!	Match<Tem
39	Input_039.inp	75	74	1	49		Rejected!	Match<Tem
40	Input_040.inp	74	73	1	49		Rejected!	Match<Tem
41	Input_041.inp	74	73	1	49		Rejected!	Match<Tem
42	Input_042.inp	75	74	1	49		Rejected!	Match<Tem
43	Input_043.inp	75	74	1	49		Rejected!	Match<Tem
44	Input_044.inp	75	74	1	49		Rejected!	Match<Tem
45	Input_045.inp	76	75	1	49		Rejected!	Match<Tem
46	Input_046.inp	74	72	2	49	68	Rejected!	Match<Tem
47	Input_047.inp	74	72	2	49	68	Rejected!	Match<Tem
48	Input_048.inp	75	73	2	49	68	Rejected!	Match<Tem
49	Input_049.inp	74	72	2	49	68	Rejected!	Match<Tem
50	Input_050.inp	74	72	2	49	68	Rejected!	Match<Tem
51	Input_051.inp	75	73	2	49	68	Rejected!	Match<Tem
52	Input_052.inp	75	73	2	49	68	Rejected!	Match<Tem
53	Input_053.inp	75	73	2	49	68	Rejected!	Match<Tem
54	Input_054.inp	76	74	2	49	68	Rejected!	Match<Tem
55	Input_055.inp	74	72	2	49	68	Rejected!	Match<Tem
56	Input_056.inp	74	72	2	49	68	Rejected!	Match<Tem
57	Input_057.inp	75	73	2	49	68	Rejected!	Match<Tem
58	Input_058.inp	74	72	2	49	68	Rejected!	Match<Tem
59	Input_059.inp	74	72	2	49	68	Rejected!	Match<Tem
60	Input_060.inp	75	73	2	49	68	Rejected!	Match<Tem
61	Input_061.inp	75	73	2	49	68	Rejected!	Match<Tem
62	Input_062.inp	75	73	2	49	68	Rejected!	Match<Tem
63	Input_063.inp	76	74	2	49	68	Rejected!	Match<Tem
64	Input_064.inp	75	73	2	49	69	Rejected!	Match<Tem
65	Input_065.inp	75	73	2	49	69	Rejected!	Match<Tem
66	Input_066.inp	76	74	2	49	69	Rejected!	Match<Tem
67	Input_067.inp	75	73	2	49	69	Rejected!	Match<Tem
68	Input_068.inp	75	73	2	49	69	Rejected!	Match<Tem
69	Input_069.inp	76	74	2	49	69	Rejected!	Match<Tem
70	Input_070.inp	76	74	2	49	69	Rejected!	Match<Tem
71	Input_071.inp	76	74	2	49	69	Rejected!	Match<Tem
72	Input_072.inp	77	75	2	49	69	Rejected!	Match<Tem
73	Input_073.inp	75	74	1	49		Rejected!	Match<Tem
74	Input_074.inp	75	74	1	49		Rejected!	Match<Tem
75	Input_075.inp	76	75	1	49		Rejected!	Match<Tem
76	Input_076.inp	75	74	1	49		Rejected!	Match<Tem
77	Input_077.inp	75	74	1	49		Rejected!	Match<Tem
78	Input_078.inp	76	75	1	49		Rejected!	Match<Tem
79	Input_079.inp	76	75	1	49		Rejected!	Match<Tem
80	Input_080.inp	76	75	1	49		Rejected!	Match<Tem
81	Input_081.inp	77	76	1	49		Rejected!	Match<Tem
82	Input_082.inp	75	73	2	49	69	Rejected!	Match<Tem
83	Input_083.inp	75	73	2	49	69	Rejected!	Match<Tem
84	Input_084.inp	76	74	2	49	69	Rejected!	Match<Tem
85	Input_085.inp	75	73	2	49	69	Rejected!	Match<Tem
86	Input_086.inp	75	73	2	49	69	Rejected!	Match<Tem
87	Input_087.inp	76	74	2	49	69	Rejected!	Match<Tem
88	Input_088.inp	76	74	2	49	69	Rejected!	Match<Tem
89	Input_089.inp	76	74	2	49	69	Rejected!	Match<Tem
90	Input_090.inp	77	75	2	49	69	Rejected!	Match<Tem
91	Input_091.inp	75	73	2	49	69	Rejected!	Match<Tem
92	Input_092.inp	75	73	2	49	69	Rejected!	Match<Tem
93	Input_093.inp	76	74	2	49	69	Rejected!	Match<Tem
94	Input_094.inp	75	73	2	49	69	Rejected!	Match<Tem
95	Input_095.inp	75	73	2	49	69	Rejected!	Match<Tem
96	Input_096.inp	76	74	2	49	69	Rejected!	Match<Tem
97	Input_097.inp	76	74	2	49	69	Rejected!	Match<Tem
98	Input_098.inp	76	74	2	49	69	Rejected!	Match<Tem
99	Input_099.inp	77	75	2	49	69	Rejected!	Match<Tem
100	Input_100.inp	76	74	2	49	70	Rejected!	Match<Tem

Filtered: 100   Rejected: 100

Table 9.7 Results of network filtering for five batches of 1000 nets

Batch Code	Time (sec)	Pipe range	Avg.	Net match	MisM range	Avg. miss
R55-75 (U)	1597	58-75	70.85	3	0-10	5.66
NfullC2(4)	1662	50-56	52.25	477	0-1	0.52
NfullC4(U)	1216	74-79	75.59	0	1-2	1.75
N55-75C2(4)	1695	55-57	55.22	375	0-1	0.63
N55-75C4(U)	1211	74-75	74.71	0	1-2	1.70

Table 9.8 Results of network filtering for random batch of 5000 nets

Batch code	Time (hrs)	Pipe range	Avg.	Net match	MisM range	Avg. miss
R55-75 (U)	Approx. 8	57-75	70.80	1	0-10	5.66

Two representative layouts of the batches from Table 9.7 have been shown in Figures 9.7a to 9.7e, the first one and the last one in the series of 1000. Although slightly better, the results for 1000 layouts do not show much different patterns than in case of 100 layouts per batch. Remarkably (or actually not), the random network generator has kept the average number of pipes not matching the template, at 5.66 in all three cases: 100, 1000 and 5000 layouts. Hence, the overall impression is that the layout of template can play essential role in the matching of the generated files. To document this, Table 9.9 presents the increase in number of matching layouts per maximum number of links in each of these, allowed not to exist in the template.

Table 9.9 Compliance with the template, for non-matching links

Batch Code	Net match for the max. number of non-matching pipes			
	0	1	2	3
R55-75(U)1000	3	7	21	93
R55-75 (U)5000	1	19	118	409
NfullC2(4)1000	477	1000	1000	1000
NfullC4(U)1000	0	252	1000	1000
N55-75C2(4)1000	375	1000	1000	1000
N55-75C4(U)1000	0	304	1000	1000

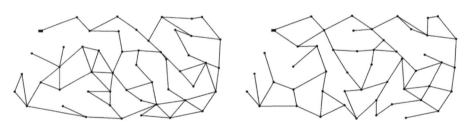

Figure 9.7a Sample of generated networks R55-75(U): layout nr.1 - left, layout nr.1000 - right

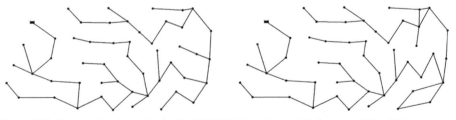

Figure 9.7b Sample of generated networks NfullC2(4): layout nr.1 - left, layout nr.1000 - right

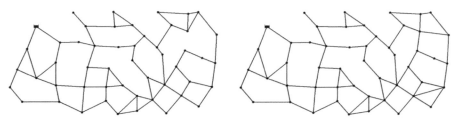

Figure 9.7c Sample of generated networks NfullC4(U): layout nr.1 - left, layout nr.1000 - right

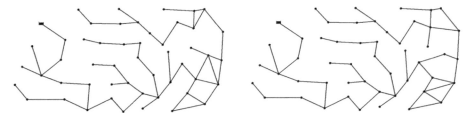

Figure 9.7d Sample of generated networks N55-75C2(4): layout nr.1 - left, layout nr.1000 - right

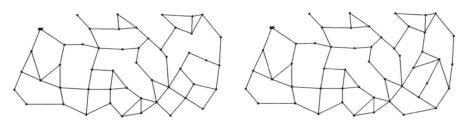

Figure 9.7e Sample of generated networks N55-75C4(U): layout nr.1 - left, layout nr.1000 - right

It shows that an adaptation of one or two connections can significantly increase the number of matching layouts. Furthermore, while the random number generator uses the same pattern regardless the selected number of networks, the total number of possibilities in non-random generation is huge. Consequently, the stricter the template is (i.e. with the lesser number of connections) the larger the sample of generated networks should be.

9.4.3 Network Initialisation and Optimisation

To provide a variety of layouts, the optimisation on the least-cost diameter has been conducted for all five categories of the filtered networks shown in Table 9.9. The default GA-settings shown in Tables 9.2 and 9.3 have been used with the following diameter range (total 15, in mm): 80, 100, 125, 150, 200, 250, 300, 350, 400, 450, 500, 600, 700, 800, and 1000. Total 1817 filtered networks ranging between 55 and 79 pipes have been optimised in approximately 400 minutes i.e. in less than seven hours, as shown in Table 9.10. The hydraulic simulations have been conducted with the nodal elevations and demands shown in

Figures 9.4, while the pipe lengths have been calculated from nodal coordinates. The uniform k-value of 0.5 mm was used in all the cases. These values have been set using the network initialisation module.

Table 9.10 Overview of GA optimised networks

Batch code	Total nets	Non-matching links	Optimisation time (min)
R55-75 (U)5000	409	3	97
NfullC2(4)1000	477	0	91
NfullC4(U)1000	252	1	64
N55-75C2(4)1000	375	0	75
N55-75C4(U)1000	304	1	77

9.5   RESULTS AND DISCUSSIONS

All the optimised networks have been further analysed in the network diagnostics module. Figures 9.8a to 9.8c show the costs against the ADF_{avg} (9.8a-left), NBI (9.8a-right), PBI (9.8b-left), I_n (9.8b-right), p_{min} (9.8c-left) and p_{max} (9.8c-right), respectively. The following observations can be made looking at these diagrams:

- The 409 randomly generated networks show better spread of the values than the rest. This is not a surprise, also given the larger number of generated networks in this sample (5000).
- The non-random networks show lesser variation, specifically the samples of 477 and 375 filtered networks of low complexity. This gives impression that the differences in configuration of these networks are relatively small i.e. many more networks should have been generated to arrive at better spread of values.
- The low-complexity samples (477 and 375) appear to be the most expensive in general, while the least reliable according to the values of ADF_{avg} and NBI (Figure 9.8a). This contradicts the results in Figure 9.8b where these networks, although more expensive, appear also to be more reliable. This originates from the higher pressures these networks have, confirming the head-driven nature of PBI and I_n, already discussed in Chapters 5 and 7. This also raises some concern about these parameters as 'reliable' reliability measures.
- The best 'value for money' is visible in the samples of 252 and 304 networks, originating from the same family of non-random generated networks of complexity category 4. These are in any case the cheaper networks and more reliable according to the values of ADF_{avg}.
- Somewhat surprising, the values of NBI in case of the samples of 252, 304 and 409 cover the wider range of values which gives impression that the GA-optimiser in this case has not produced clearly branched skeleton, as obviously is the case with the samples of 477 and 375 pipes. Repetitive simulations with modified GA-settings could likely throw more light on this situation. At least, more complex networks would require larger number of generations. The preliminary conclusion is that the level of complexity and (even) spatial distribution of demands can influence the branched structure of secondary mains in optimised looped networks.
- The second peculiarity regarding the optimisation runs is that the minimum pressures in the two categories of the networks of lower complexity (the samples of 477 and 375

pipes) have been lower than the threshold of 20 mwc, as low as 16-17 mwc. Again, the selection of GA-settings is the likely reason; repetitive simulations should have been done by increasing the penalty to arrive at better results. That step has been however skipped because the two categories are already more expensive and less reliable than the other three. The most favourable i.e. sufficiently looped samples, have been optimised on the minimum pressures in the range of 19.5 to 20.5 mwc, which has been considered as satisfactory.

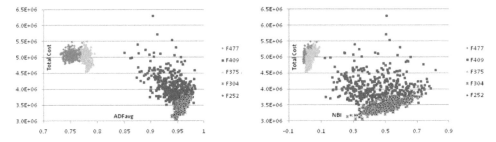

Figure 9.8a Diagnostics of networks from Table 9.10: ADF_{avg} - left, NBI - right

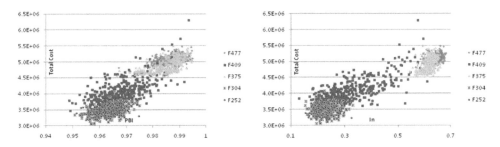

Figure 9.8b Diagnostics of networks from Table 9.10: PBI - left, I_n - right

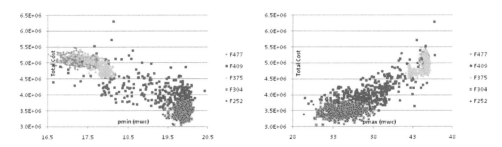

Figure 9.8c Diagnostics of networks from Table 9.10: p_{min} - left, p_{max} - right

Despite the contradiction between the values of ADF_{avg} and NBI on one side, and those of PBI and I_n on the other side, the focus in the individual assessment of the networks has been given to the samples of 409 networks (random) and those of 252 and 304 networks (non-random, complexity category 4). These three samples have been further zoomed-in within the price range from three to four million and ADF-range between 0.9 and 1, which can be seen in Figure 9.9.

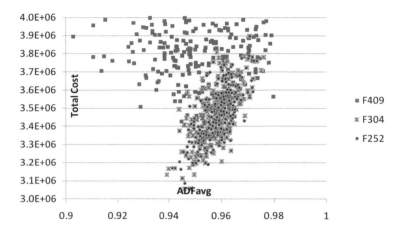

Figure 9.9 The ADF_{avg} values in the cost range between 3 and 4 million

9.5.1 Diagnostics of Single Networks

To illustrate the other features of NEDRA package, six networks have been picked from the diagram in Figure 9.8a. Two of them represent each of the samples: one as being the cheapest, and the other as the most reliable i.e. with the highest value of ADF_{avg}. These networks have been listed in Table 9.11 with the layouts showing the diameters and pressure distribution shown in Figures 9.10a to 9.10f. Two other layouts have been added to the table, to give impression about the entire price- and ADF_{avg} range.

Table 9.11 Overview of selected networks

Net nr.	Batch code	Property	Cost (10^6)	p_{min}-p_{max} (msl)	ADF_{avg}	NBI	PBI	I_n
1727	R55-75 (U)	Cheapest	3.46	19.86 - 36.69	0.945	0.362	0.962	0.237
4611	(F409)	Least reliable	5.13	18.58 - 44.08	0.850	0.241	0.987	0.564
3765		Most expensive	6.29	18.13 - 45.67	0.938	0.504	0.993	0.575
3307		Most reliable	4.00	20.01 - 38.28	0.983	0.785	0.962	0.363
691	NfullC4(U)	Cheapest	3.06	19.64 - 31.70	0.948	0.354	0.957	0.209
081	(F252)	Least reliable	3.17	20.03 - 34.74	0.941	0.281	0.955	0.196
508		Most expensive	3.90	19.85 - 32.24	0.971	0.673	0.964	0.192
757		Most reliable	3.76	20.06 - 35.85	0.974	0.696	0.970	0.262
353	N55-75C4(U)	Cheapest	3.06	19.83 - 30.90	0.946	0.312	0.958	0.194
760	(F304)	Least reliable	3.13	19.89 - 35.65	0.939	0.261	0.957	0.251
218		Most expensive	4.07	19.96 - 33.45	0.974	0.704	0.966	0.224
962		Most reliable	3.78	19.47 - 34.90	0.976	0.725	0.969	0.272

The shaded values show the absolute maximum/minimum values regardless the sample. The most of those originate from the random sample. The other two samples show mutually similar results given the similar pattern of non-random generation.

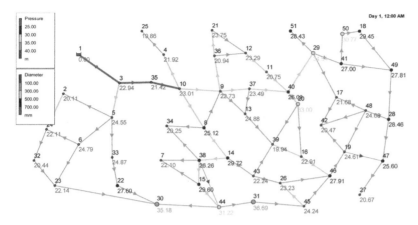

Figure 9.10a Sample R55-75(U): GA optimised layout nr.1727 (the cheapest)

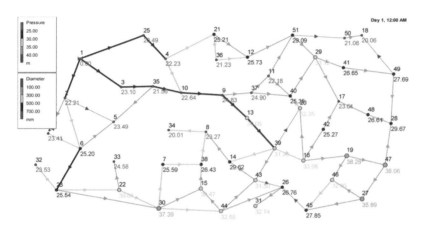

Figure 9.10b Sample R55-75(U): GA optimised layout nr.3307 (the most reliable)

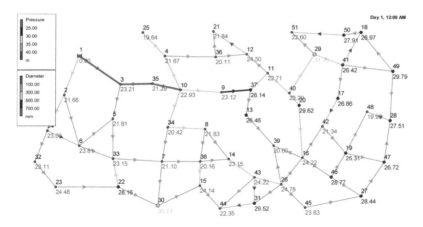

Figure 9.10c Sample NfullC4(U): optimised layout nr.691 (the cheapest)

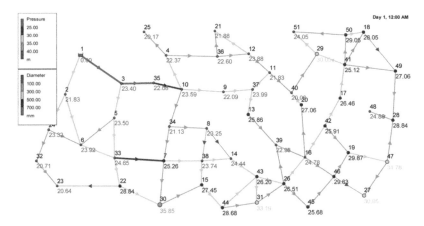

Figure 9.10d Sample NfullC4(U): optimised layout nr.757 (the most reliable)

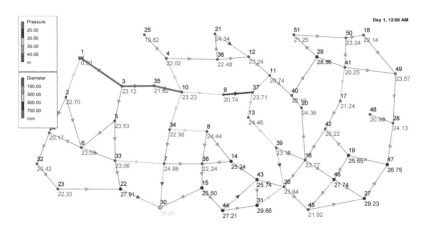

Figure 9.10e Sample N55-75C4(U): optimised layout nr.353 (the cheapest)

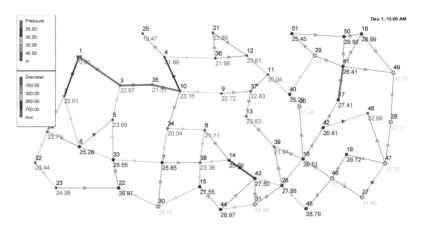

Figure 9.10f Sample N55-75C4(U): optimised layout nr.962 (the most reliable)

The hydraulic reliability diagrams have been further drawn for each of the six networks, which are shown in Figures 9.11a to 9.11c.

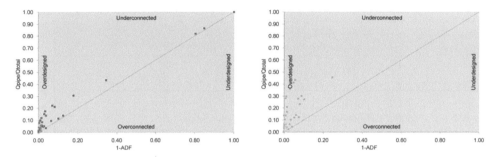

Figure 9.11a HRD: layout nr.1727 (the cheapest) - left, layout 3307 (the most reliable) - right

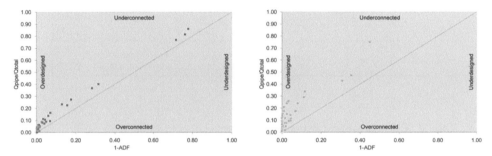

Figure 9.11b HRD: layout nr.691 (the cheapest) - left, layout 757 (the most reliable) - right

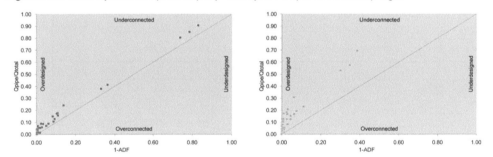

Figure 9.11c HRD: layout nr.353 (the cheapest) - left, layout 962 (the most reliable) - right

The diagrams depict clearly that the cheapest networks (on the left) are less reliable than those on the right. Amongst the three most reliable networks, the layout nr.3307 would possible qualify as the best overall, having the least number of pipes whose failure causes severe damage. The layout nr.962 could also qualify to win in case the problem of three distinct pipes can be solved. The failure of any of these pipes will be causing the loss of demand in the range of 30 to 40 %, which can be seen both in Figure 9.11c (right) and in the Tables 9.12 and 9.13. Table 9.12 shows the number of pipes causing the loss of demand in categories of 10%, while Table 9.13 shows the equivalent percentage. Both confirm the impression obtained from the HRDs that the layout nr.3307 is the most reliable one, with only two pipes causing the demand loss above 10%.

Table 9.12 Number of pipes causing loss of demand, per ADF range

Net nr.	0-10%	11-20%	21-30%	31-40%	41-50%	51-60%	61-70%	71-80%	81-90%	91-100%
1727	1	2					1		2	67
3307								1	1	72
691			3				1	1	3	67
757						1	2		2	70
353		1	2				2		4	66
962							3		1	71

Table 9.13 Percentage of pipes causing loss of demand, per ADF range

Net nr.	0-10%	11-20%	21-30%	31-40%	41-50%	51-60%	61-70%	71-80%	81-90%	91-100%
1727	1.37%	2.74%					1.37%		2.74%	91.78%
3307								1.35%	1.35%	97.30%
691			4.00%				1.33%	1.33%	4.00%	89.33%
757						1.33%	2.67%		2.67%	93.33%
353		1.33%	2.67%				2.67%		5.33%	88.00%
962							4.00%		1.33%	94.67%

9.5.2 Impact of Worst Case Failure

The network diagnostics module also categorises the loss of demand per node, giving impression about the coverage and severity of the worst case pipe failure, which is shown in Tables 9.14 and 9.15.

Table 9.14 Percentage of nodes with corresponding ADF after the worst pipe failure

Net nr.	0-10%	11-20%	21-30%	31-40%	41-50%	51-60%	61-70%	71-80%	81-90%	91-100%
1727	100%									
3307	2%		2%	2%		8%	12%	14%	16%	44%
691	22%		4%	6%	8%	16%	10%	20%	2%	12%
757	6%	2%	4%	6%	6%	8%	10%	28%	16%	14%
353	24%	4%	2%	4%	12%	16%	18%	18%		2%
962	4%	4%	2%	2%	6%	10%	8%	14%	22%	28%

Table 9.15 Percentage of original demand affected by the worst pipe failure (ADF range in %)

Net nr.	0-10%	11-20%	21-30%	31-40%	41-50%	51-60%	61-70%	71-80%	81-90%	91-100%
1727	100%									
3307	3.28%		3.28%	4.92%		14.75%	12.46%	8.85%	12.79%	39.67%
691	34.10%		6.89%	10.82%	9.84%	12.46%	6.89%	11.15%	1.97%	5.90%
757	8.20%	1.97%	8.52%	8.20%	8.52%	11.48%	17.05%	20.00%	7.87%	8.205
353	35.41%	7.54%	1.97%	4.59%	14.75%	15.41%	10.16%	8.52%		1.64%
962	5.90%	4.26%	4.26%	3.61%	10.82%	13.44%	13.11%	14.75%	14.10%	15.74%

The results in Table 9.14 show the percentages of the nodes and corresponding ADF category in which they will fall after the pipe failure causing the lowest available demand fraction, ADF_{min}. Clearly, the weakness of the layout nr.1727 is that the source has only one pipe connection and obviously all nodes will fall in the first category because their ADF after the failure of that pipe equals zero; thus, 100%. On contrary, the worst case failure in the layout nr.3307 will cause only 2% of the nodes (one single node in this case) reduce their demand for more than 90%, while the demand of 44% of the nodes (total 22 nodes) will be affected less than 10% (and some maybe not at all).

The results in Table 9.15 show the percentages of original demand affected by the worst case failure. Again, for the layout nr.3307, 3.28% of the original demand will be reduced between 90 and 100%, while 39.67% of the original demand will be affected between 0 and 10%. Comparing these results with those in Table 9.14, it can be concluded if the extreme pipe failure affects small/large number of nodes and/or small/large demand. In other words, an assumption can be made to which extent the distribution area and major consumers have been affected.

9.5.3 Economic Considerations

All cost calculations in this case study have been conducted applying the following default values in the formula $a + bX^n$, as shown in Table 9.16.

Table 9.16 Input parameters used for cost calculations

Factor/X	D (mm)	V (m³)	Q (l/s)	O&M costs %	
a	0	0	n/a	for pipes	0.5
b	1	222	n/a	for storage	0.8
n	1	1	n/a	for pumps	n/a
Loan repayment period				25 years	
Annual interest rate				8%	
Annual inflation rate				0%	
Energy cost per pumped kWh				n/a	

Due to relatively simple supplying (gravity) scheme, no sensitivity analyses have been done by modifying these parameters. The pipe investment costs have been calculated taking into considerations the pipe lengths calculated from nodal coordinates and the diameters as determined in GA optimisation. The storage volume has been assumed at 10,000 m³.

As already indicated in Table 9.11, the most expensive amongst the three most reliable layouts is nr.3307 with total annual cost of four million. Based on the input in Table 9.16, more detailed costs for the six layouts from Tables 9.12 to 9.15 are given in Table 9.17.

As can be expected in case of gravity networks i.e. in the absence of pump energy costs, the asset costs of the transmission part of the network become overwhelmingly dominant. This is additionally amplified by relatively large lengths of the pipes.

Table 9.17 Annual cost structure of selected networks (in millions)

Net nr.	Property	Pipes	%	Storage	%	O&M	%	Total
1727	Cheapest	3.04	88.68	0.208	6.07	0.180	5.25	3.46
3307	Most reliable	3.59	89.58	0.208	5.19	0.209	5.22	4.00
691	Cheapest	2.69	87.92	0.208	6.80	0.161	5.27	3.06
757	Most reliable	3.36	89.24	0.208	5.53	0.197	5.24	3.76
353	Cheapest	2.69	87.92	0.208	6.80	0.161	5.27	3.06
962	Most reliable	3.38	89.27	0.208	5.50	0.198	5.23	3.78

9.5.4 Future Demand Growth

The final calculation has been done for the layout nr.3307 and nr.962, by predicting the demand growth of 2% over 15 years, total 32%. The trends of I_n, NBI, ADF_{avg} and ADF_{min} are shown in Figures 9.12 and 9.13, respectively.

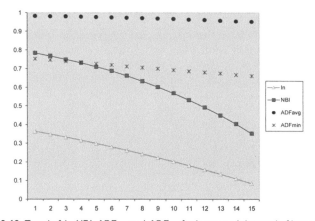

Figure 9.12 Trend of I_n, NBI, ADF_{avg} and ADF_{min} for increased demand of layout nr.3307

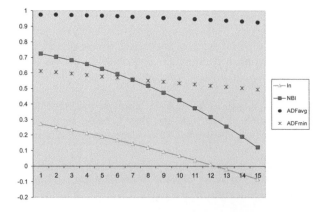

Figure 9.13 Trend of I_n, NBI, ADF_{avg} and ADF_{min} for increased demand of layout nr.962

Here as well, it shows that the layout nr.3307 performs better than nr.962. Next to the fact that the buffer of the latter network disappears faster than in the case of the first one, the impact of the worst case failure shall be more amplified with the growth of demand in this network. This can be observed in Tables 9.18 and 9.19, both showing the reduction in the percentage of original demand moderately affected by the failure, on the account of the increase of the percentage of significantly affected demand. This impact is further amplified by the actual growth of the demand of 32%, suggesting the network renovation in order to mitigate the problems with its performance, actually in both cases.

Table 9.18 Original demand affected by the worst pipe failure in layout nr.3307 (ADF range in %)

Q↑2%	0-10%	11-20%	21-30%	31-40%	41-50%	51-60%	61-70%	71-80%	81-90%	91-100%
1	3.28%		3.28%	4.92%	0.00%	14.75%	12.46%	8.85%	12.79%	39.67%
2	3.28%	3.28%		4.92%	2.30%	12.46%	12.46%	8.85%	13.77%	38.69%
3	3.28%	3.28%		4.92%	2.30%	13.77%	14.43%	6.89%	12.46%	38.69%
4	3.28%	3.28%		4.92%	2.30%	13.77%	14.43%	6.89%	12.46%	38.69%
5	6.56%			4.92%	2.30%	13.77%	14.43%	9.18%	11.15%	37.70%
6	6.56%			4.92%	2.30%	13.77%	16.07%	7.54%	11.15%	37.70%
7	6.56%			4.92%	2.30%	15.74%	14.10%	7.54%	11.15%	37.70%
8	6.56%			4.92%	2.30%	15.74%	14.10%	7.54%	11.15%	37.70%
9	6.56%		4.92%		6.56%	13.11%	12.46%	7.54%	11.15%	37.70%
10	6.56%		4.92%		8.85%	14.43%	10.49%	12.13%	6.23%	36.39%
11	6.56%		4.92%		8.85%	14.43%	12.13%	10.82%	8.20%	34.10%
12	6.56%		4.92%		8.85%	15.41%	11.15%	11.48%	7.54%	34.10%
13	6.56%		4.92%	2.30%	7.87%	14.10%	11.15%	11.48%	7.54%	34.10%
14	6.56%		4.92%	2.30%	7.87%	18.03%	7.87%	11.80%	6.56%	34.10%
15	6.56%		4.92%	2.30%	13.77%	12.13%	9.18%	11.48%	7.21%	32.46%

Table 9.19 Original demand affected by the worst pipe failure in layout nr.962 (ADF range in %)

Q↑2%	0-10%	11-20%	21-30%	31-40%	41-50%	51-60%	61-70%	71-80%	81-90%	91-100%
1	5.90%	4.26%	4.26%	3.61%	10.82%	13.44%	13.11%	14.75%	14.10%	15.74%
2	5.90%	4.26%	4.26%	3.61%	10.82%	13.44%	13.11%	14.75%	14.10%	15.74%
3	8.20%	1.97%	4.26%	3.61%	13.44%	10.82%	13.11%	14.75%	17.70%	12.13%
4	8.20%	1.97%	4.26%	6.89%	10.16%	10.82%	13.11%	17.70%	14.75%	12.13%
5	10.16%		4.26%	10.16%	6.89%	13.11%	10.82%	17.70%	14.75%	12.13%
6	10.16%		4.26%	14.43%	2.62%	13.11%	10.82%	20.33%	12.13%	12.13%
7	10.16%		7.87%	10.82%	2.62%	13.11%	15.41%	15.74%	13.11%	11.15%
8	10.16%		7.87%	10.82%	2.62%	13.11%	17.05%	14.10%	13.77%	10.49%
9	10.16%	4.26%	3.61%	10.82%	11.48%	4.26%	17.05%	16.39%	11.48%	10.49%
10	10.16%	4.26%	3.61%	10.82%	13.44%	2.30%	19.67%	14.75%	10.49%	10.49%
11	10.16%	4.26%	3.61%	10.82%	13.44%	8.20%	16.07%	14.43%	8.85%	10.16%
12	10.16%	4.26%	3.61%	10.82%	13.44%	10.82%	17.05%	10.82%	8.85%	10.16%
13	10.16%	4.26%	6.89%	7.54%	13.44%	10.82%	17.05%	12.79%	6.89%	10.16%
14	14.43%		6.89%	7.54%	13.44%	13.11%	14.75%	14.10%	6.89%	8.85%
15	14.43%		10.16%	4.26%	13.44%	13.11%	14.75%	14.10%	6.89%	8.85%

9.5.5 Final Choice

Leaving the analysis in this stage would likely lead to a conclusion that 220,000 per annum is worthwhile additional cost for the improved performance of the layout nr.3307, which then qualifies as the winner amongst 13,000 generated layouts. To apply a 'finishing touch', the two final layouts have been produced by modifying the layouts nr.3307 and nr.962 manually, which is shown in Figures 9.14 and 9.15, respectively.

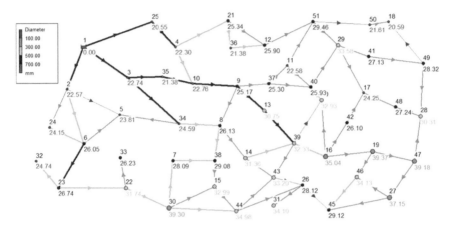

Figure 9.14 Modified layout nr. 3307

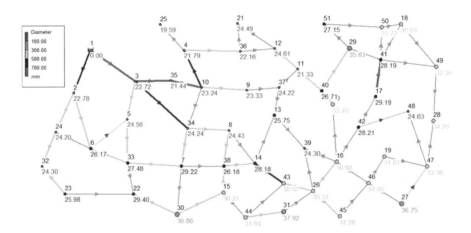

Figure 9.15 Modified layout nr. 962

The modification included elimination of the pipes that do not exist in the template network. There are two such pipes in the layout nr.3307: (1) pipe 5-35 in Figure 9.10b has been reconnected as 5-34, keeping the same diameter of 125mm, and (2) the peripheral pipe 23-30 has been reconnected as 8-9 with the same diameter of 250mm. The only pipe that differs from the template in the layout nr.962 is 31-43 which has been reconnected as 14-30 keeping the diameter of 150mm. Furthermore, to improve the hydraulic performance in both layouts, an additional pipe of 600mm has been connected between the nodes 3 and 34. In all these cases, the pipe lengths have been recalculated according to the coordinates of the newly connected nodes. Eventually, the hydraulic simulations have given the pressures shown in Figures 9.14 and 9.15 and two HRDs as in Figure 9.16.

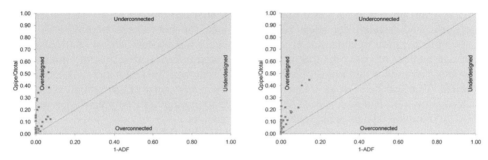

Figure 9.16 HRD: layout nr. 3307 - left, layout nr. 962 - right

The summary and comparison of the main results with the original layouts are shown in Table 9.20.

Table 9.20 Comparison of originally generated and modified networks

Net nr.	Property	Cost (10^6)	p_{min}-p_{max} (msl)	ADF_{avg}	NBI	PBI	I_n
3307	Original	4.00	20.01 - 38.28	0.983	0.785	0.962	0.363
M3307	Modified	4.11	20.55 - 39.37	0.993	0.891	0.965	0.426
962	Original	3.78	19.47 - 34.90	0.976	0.725	0.969	0.272
M962	Modified	3.98	19.60 - 37.92	0.986	0.818	0.974	0.372

Both the figure and the table show the improvement of reliability at relatively minor cost increase. Yet, the addition of the pipe 3-34 was more useful in case of the layout nr.3307 and to lesser extent in case of nr.962; further improvement of that layout would require change of more pipe routes/diameters. Interesting to observe is that the modified layout nr.962 costs approximately the same as the original layout nr.3307 and with relatively similar values of the parameters in the table, which can be considered as a coincidence.

9.6 CONCLUSIONS

This chapter has dealt with the decision support tool for network design and reliability assessment (NEDRA), compiled from the software applications specifically developed for the research presented in Chapters 5 to 8. The tool has been composed of the modules for network generation and filtering, initialisation, GA-optimisation on the least-cost diameter, diagnostics of network geometric-, hydraulic-, and economic parameters and finally the assessment of its resilience. Computations can be done for a single network or a sequence of networks of virtually unlimited size and number, mutually similar or different in properties. The programme has been illustrated on a design example of a gravity network of one source and 50 nodes. 13,000 network layouts have been generated and further filtered to arrive at

1817 layouts satisfying the configuration of the network template. Those have been then optimised and analysed on various reliability measures against the costs comprising the network investment, operation and maintenance. The least of conclusions from the programme performance and the results of the case study can be summarised in the following bullets:

- NEDRA has proven to be pretty robust package that can process a very large sample of networks within relatively short period of time. The bottle-neck time-wise is in the network generation module; not necessarily because this process is slow, but mostly due to huge number of networks that is to be generated to arrive at sufficient variety of the layouts. To produce thousands of layouts of moderate size, several hours will normally be needed. That can be seen as a long time, but the whole network generation is one-time activity, anyway. Running, say, a 48-hour calculation on a fast computer to arrive at e.g. 50-100 thousand layouts is not unthinkable assignment.

- Single design of a network would rarely deal with thousands or even hundreds of nodes/pipes. Yet, the number of combinations in even a moderate network size of a few dozen nodes can be enormously high, both in random and non-random generation process. In this generation approach, the large number of combinations will be regularly a must in order to arrive at sufficiently versatile layouts. This variety and the shorter time of generation can also be enhanced by careful selection of the network generation settings.

- The network filtering process has been very quick, but the number of layouts that comply with the network template can be surprisingly low. This will depend on the number of generated networks and complexity of the template. In general, the more complex the template and the larger number of generated layouts, the larger number of filtered layouts will be. To further increase this number, it may be sensible to also filter the layouts that have a few connections not matching the template, because in an otherwise promising layout, these links can be manually reconnected in the final stage.

- The selected EO-optimiser has proven to be pretty fast, but not arriving at quite logical sequence of diameters, typical for engineering practice. For instance, it has been not that uncommon to see a series of pipes where larger diameter pipe is inserted in between the two pipes of smaller diameters. Yet, even in this kind of layout, the optimiser has managed to find combinations of fairly satisfactory reliability because quite some layouts were optimised creating a buffer, which somehow defeats the assumption elaborated in previous chapters that GA makes distinctly branched skeleton of larger pipes within looped configurations. This phenomenon should therefore be investigated further. Secondly, the EO-optimiser has found some solutions resulting in pressures below the preferred threshold. Thorough assessment/selection of GA-optimiser and input settings has not been within the scope of this research and it is yet to be investigated to which extent the anomalies in the results can be eliminated by more appropriate selection of the settings, in the first place the number of generations and the penalty. All the GA-simulations have been run only once per batch of networks and with fixed settings; hence, no consistency could have been found to point the source of peculiar results. What could however have been observed is that the pressures below the threshold were more common for some classes of generated networks, e.g. those of lower complexity. Eventually, those have not qualified anyway, for the sake of their low connectivity i.e. low reliability at relatively high price. Reasonably looped layouts, such as those finally analysed, were optimised with the minimum pressure that has fallen regularly around the threshold of 20 mwc.

- Although not entirely justified, a (properly) GA-optimised design appears to be good starting point in reliability analysis; it avoids any bias possible by visual assessment of network configurations, which can be easily deceiving. GA provides quick reference design for large number of layouts that can be manually adapted in the final stage. Additional information obtained from the corresponding HRD is further instrumental in assessing and solving the bottlenecks After all, the finishing touch should always be done by humans; computers do only what they are programmed to do.

- In the analysis of large number of layouts, the network buffer index, *NBI*, and the corresponding HRD, have proven to be good indicators of network resilience, which can distinguish between the layouts that may look pretty similar. The other two measures, the pressure buffer index, *PBI*, and the network resilience, I_n, have shown less compliance with the loss of demand. Especially, the latter coefficient gives pretty conservative values, significantly lower than the other indices, which does not reflect the resilience as high as it really appears to be.

In the absence of sound theory emerging from the research on pattern recognition for reliability assessment of water distribution networks, NEDRA is the concrete outcome that can be effectively used in practice, to efficiently analyse numerous alternatives and scenarios. The real development and testing of the package has just started with this research. Throughout a huge number of simulations conducted, a couple of ideas have already been considered how to improve it further. Most of them deal with the network generation module that is presently the biggest bottle-neck of NEDRA. For instance:

- The list of input settings can be simplified. Some combinations of the settings yield pretty similar configurations.
- The existing network generation algorithm can be upgraded by introducing a condition that the node can be connected with another one only if that connection is available in the template.
- More radically, an alternative algorithm of network generation developed specifically for design purposes could be based on building the library of layouts applying gradual decomposition of the template.

Furthermore, in the process of GA-optimisation some diameters could be preset e.g. the skeleton of secondary mains, and kept fixed. This feature exists in some other GA-software but has not been built in the EO yet.

Last but not the least, NEDRA should pass through a much more rigorous testing before being integrated, made more user friendly, and possibly distributed for public use.

REFERENCES

1. Chase, D.V. (2000). *"Chapter 15: Operation of Water Distribution Systems."* Water distribution systems handbook, Mays,L.W., Editor-in-Chief, McGraw-Hill, New York, NY.
2. *Evolving Objects (EO)*, distributed under the GNU Lesser General Public License by SourceForge (http://eodev.sourceforge.net)
3. Keijzer, M., Merelo, J., Romero, G., and Schoenauer, M. (2002). Evolving Objects: A General Purpose Evolutionary Computation Library. In P. Collet, C. Fonlupt, J.-K. Hao, E. Lutton & M. Schoenauer (Eds.), *Artificial Evolution* (Vol. 2310, pp. 829-888): Springer Berlin / Heidelberg.
4. Kujundzić, B. (1996). *Veliki vodovodni sistemi*. Udruženje za tehnologiju vode i sanitarno inžinjerstvo, Beograd. (in Serbian Language)
5. OFWAT. (1996). Leakage of Water in England and Wales.
6. Rossman, A. L. (2000). *EPANET2 Users Manual*. Water Supply and Water Resources Division, National Risk Management Research Laboratory of Computer and System Sciences, Cincinnati OH 45268.
7. Tanyimboh, T. T., Burd, R., Burrows, R. and Tabesh, M. (1999). Modelling and Reliability Analysis of Water Distribution Systems, *Water Science and Technology*, IWA, Vol. 39, No.4, 249-255.
8. Trifunović, N. (2006). *Introduction to Urban Water Distribution*, Taylor & Francis Group, London, UK, 509 p.

CHAPTER 10

Conclusions

_____

10.1   RESEARCH SUMMARY

The research presented in this manuscript consists of theoretical part dealing with definition of network measures describing the node connectivity, the hydraulic performance and eventually the resilience, and the practical part comprising the development of decision support tool for reliability-based network design.

The pressure-driven demand algorithm presented in Chapter 3: 'Emitter Based Algorithm for Pressure-Driven Demand Calculations of Water Distribution Networks ', uses EPANET emitter coefficients that become active in all the nodes where the pressure drops below the set threshold value. The options to disconnect the pipes or set the demand at zero when the node pressure becomes negative, resulting from extreme topography, have also been considered. The threshold pressure is used to calculate the value of emitter coefficient from the baseline demand and fixed exponent.

The network generation tool presented in Chapter 4: ' Spatial Network Generation Tool for Performance Analysis of Water Distribution Networks ' has been developed based on the principles of graph theory. It uses the information from the set of junctions prepared in the EPANET input file and generates the selected number of networks either by random- or non-random generation. The generated networks receive the properties that are assigned in random or controlled way (fixed values, or calculated from coordinates, for pipe lengths, or in the process of GA-optimisation, for pipe diameters).

The research presented in Chapter 5: 'Hydraulic Reliability Diagram and Network Buffer Index as Indicators of Water Distribution Network Reliability' presents an alternative way of expressing the reliability index being derived from a diagram named the *hydraulic reliability diagram* (HRD), showing the correlation between the pipe flows under the regular supply conditions and the loss of demand caused by their failure. The reliability index derived from the position of the dots on the graph, adding proportional weighting to the pipes carrying more flow under normal condition, has been proposed as the *network buffer index* (NBI), which takes into consideration only the consequences of the failure and not the chance that it would happen.

The research from Chapter 6: 'Impacts of Node Connectivity on Reliability of Water Distribution Networks' analyses the impacts of node connectivity on network resilience. Two approaches have been compared to evaluate this connectivity: one which does it in relative terms and in simplified way, by taking into consideration the connectivity of the two most dominant node categories in three different ways (eventually comparing three different connectivity indices), and the other which considers the connectivity measures based on graph theory used in the NodeXL spreadsheet template available in public domain.

The research presented in Chapter 7: 'Diagnostics of Regular Hydraulic Performance of Water Distribution Networks and its Relation to the Network Reliability' assesses the potential of water distribution networks to sustain a certain level of failure based on the survey of their hydraulic parameters under regular supply conditions. Three different reliability measures have been analysed: the *network power index* (NPI), the *pressure buffer index* (PBI), and the *network residence time* (NRT). Furthermore, several statistical analyses have been conducted to check the correlations between the pipe properties, specifically the volumes and flows, as possible indicators of the loss of demand after the pipe failure.

The research discussed in Chapter 8: 'Economic Aspects of Decision Making in Reliability Assessment of Water Distribution Networks' compares the implications of increased investment and/or operational costs for the reliability of two network configurations with different supplying scheme. The economical aspects have been taken into consideration by applying the Present Worth method. The network layouts have been manipulated through different topographical patterns, altitude ranges and economic scenarios. The network reliability has been assessed for each of these combined scenarios, and additionally tested for 20 design scenarios comprising gradual increase of the pipe capacity and reduction of the pumping capacity, while targeting the similar threshold pressure.

And at the end, Chapter 9: 'Decision Support Tool for Design and Reliability Assessment of Water Distribution Networks' illustrates the network design tool named NEDRA, composed of the modules for network generation and filtering, initialisation and optimisation on the least-cost diameter, diagnostics of network geometric-, hydraulic-, and economic parameters and the assessment of its resilience. Computations using this software can be done for single network or a sequence of networks of virtually unlimited size and number, mutually similar or different in properties. The tool has been demonstrated on a design example of a gravity network of 50 nodes. 13,000 network layouts have been generated and further filtered to arrive at 1817 layouts satisfying the configuration of the network template. Those have been then optimised and analysed using various reliability measures against the costs comprising the network investment, operation and maintenance, eventually arriving at the most economically justified and reliable layout.

10.2   CONCLUSIONS IN RESPONSE TO RESEARCH QUESTIONS AND HYPOTHESES

The conclusions presented in Chapters 3 to 9 have been reformulated to directly address the research questions raised in Chapter 2.

Is the available demand fraction (ADF) true descriptor of network reliability? The ADF is certainly the most transparent index because this is the only value between 0 and 1 that can be directly converted into the loss of demand. Nevertheless, the average value of ADF may hide pipe bursts causing rather significant loss of demand. That is why the only complete picture about the network reliability can be obtained from the hydraulic reliability diagram (HRD).

Are the demand-driven based reliability measures, sufficiently accurate? The two reliability measures from the literature used for benchmarking, the resilience index of Todini (I_r), and the network resilience of Prasad and Park (I_n), have proven to follow the same trend as the indices developed in this research, in the first place the NBI. Nevertheless, the I_r and specifically the I_n, evaluated some networks with pretty low values, which has been in contradiction with relatively high value of ADF_{avg} for these networks. This leads towards a conclusion that two indices from the literature are, as their formulas also show, indeed mostly sensitive on the node heads and to a lesser degree on the network connectivity. The side effect here is that in the extreme events of pressure loss, both indices may yield arbitrarily high- either even negative value.

How the change in node connectivity affects the levels of service? The results showed that the connectivity measures alone can hardly be correlated to the reliability measures that are also based on pipe geometry and hydraulic performance. The approach based on graph theory has shown more consistent results of more simple measures such as the graph density (GD), which correlate better than the more complicated ones e.g. the betweenness centrality (BS). The difference between undirected and directed graphs, the latter coinciding with flow directions in the pipes, analysed in this research has been insignificant. Although used mostly for social networks, NodeXL has been user friendly add-on although lacking direct link with EPANET software. The range of features enabling good visual presentation of graphs gives impression that this MS Excel template can be used for further water distribution analysis, for instance for network zoning, district metering areas, etc.

What is the effect of selected supply schemes, with or without balancing tanks in the system? This question was only indirectly addressed within the economic analyses done for various scenarios i.e. reliability levels. As it could have been expected, the supplying schemes with more pumping have higher operational and lower investment costs. Those schemes with balancing tank appear to be more reliable and less costly than those with direct pumping, provided that the tank is properly designed and operated in combination with pumping station. Because all hydraulic simulations were snap-shots, the balancing volume was just roughly estimated, instead of being optimised. An extended period simulation would have yielded more useful results for this purpose.

What is the link between the pressure levels and the effects of potential failure in the system? The results point fairly good matching with the reliability i.e. resilience measures. The logical response of the indices occurs after the demand and diameter increase, or having various degree of network connectivity applied in the model. The pipe correlations of geometric and hydraulic parameters show less consistent patterns. Different network categories have been showing different level of correlations of particular parameters. GA-optimised networks show generally better correlations between the volumes and flows, while the networks with more buffers have better correlations of the hydraulic parameters involving friction losses. In most of the cases, the results have also shown the trend of improved correlations after gradually increasing the level of demand, but no firm conclusions could have been drawn.

Is there a link between the reliability and energy balance in the network? The network pressure index values (*NPI*) seem to be generally more susceptible to nodal elevations (i.e. the pressures) than are the pressure buffer index values (*PBI*) values. Oppositely, the *PBI* values are more susceptible to the demand growth (i.e. the friction losses) than are the *NPI* values. Additional buffer achieved by the increase of pipe diameters shows different effect on the improvement of network resilience. The different form of the curves for the same reliability index results from the presentation in different ranges of ADF_{avg}.

What are the differences between the most economic design and the most reliable design? Contrary to the assumption that 'the cheapest' means 'the least reliable', in the economic scenario of high O&M costs and unfavourable loans leading to substantial annuities, the least cost design scenario is likely to be more reliable than the one with increased costs for O&M. In addition, the reliability of networks laid in topographies which allow substantial use of gravity for water conveyance, appears to be more dependent on network capacity i.e. pipe volume.

Under what conditions is the increase of investment costs more affective for improvement of the reliability than the increase of operation and maintenance costs? Reliability analysed on the level of whole network, or significant part of it, will be much more efficiently maintained by investing into additional pipe capacity rather than by installing additional pumping capacity. The solutions that favour additional pumping on account of reduced pipe diameters hide calamities with potentially serious implications for service levels.

Furthermore, the following conclusions address the research hypothesis made.

Based on a given topography, network layout and demand scenario, there is a unique reliability/resilience footprint that can be described as a function of ADF. This footprint reflects the network buffer. Although, it could not have been mathematically proven that the footprint obtained by HRD is a unique one, this diagram has offered consistently logical and transparent results, leading to the wide range of NBI-values, easy to be interpreted. The overall feeling from numerous analyses is that the combination of HRD and NBI offers the most complete picture of the network reliability/resilience.

Reliability measures derived using demand-driven hydraulic models are less accurate than those derived by the use of pressure-driven demand models. As elaborated above, they are less accurate in general, not offering direct translation to the loss of demand, as well as they are less consistent in more extreme range of pressure drops.

Increasing the connectivity between the pipes improves network reliability in general. For a given supply scheme, there is an optimal network geometry that can be described by a 'shape index', which can be correlated to other reliability measures. This hypothesis could not have been proven. The connectivity measures proposed in this research showed less consistency than those of NodeXL. They appear to be rather coarse and reflect the difference in connectivity only in case of visibly different network layouts. The simpler network grid index (*NGI*) and the average value of network connectivity factor (*NCF$_{avg}$*) performed often better than the more complex network connectivity index (*NCIx*) or the network shape index (*NSIx*). Also, having the most of these values within the range between 0 and 1, regardless the number of nodes and links, gives an impression that all these measures would work better as parameters of more complex formula including the network resistance and supplying heads. Getting a lower value of the connectivity index for much larger network can be confusing at a time.

It is possible to make a quick reliability snapshot of a system by looking at typical hydraulic indicators. There is a clear implication from the interrelation between the pressures, flows/velocities and network resistance. The quick snapshot is obtained by the simulations done by NEDRA package but the implication of the interaction between the main hydraulic parameters is not always clear. The trends of single parameter change in the single network are mostly obvious but none of the indicators has shown 100% clear correlation with the ADF, regardless the layout.

The networks with generally higher pressures, despite potential for increased leakage, have more of a buffer to maintain the minimum service level during a single event of the component failure. This has been mostly indicated by the two resilience indices from the literature but not always reflected in the values of *ADF$_{avg}$* or *NBI*. Clearly, the network reliability is influenced by the combination of geometric and hydraulic properties, rather than by one of these groups alone.

The reliability is disproportional to the increase of the system resistance i.e. hydraulic loss. This has proven to be mostly true, but here as well, the correlation is not fully consistent.

The most economic design shall always be cheaper than the most reliable design. There is a threshold velocity and/or network resistance that can be taken as a border between the increase of investment- or operational costs, resulting in the most effective reliability improvement. This hypothesis has also not been proven. The cheapest, in this case the GA-optimised design, showed in quite some cases the higher resilience than the more expensive design comprising high operational- and low investment costs. Moreover, a unique threshold velocity or network resistance does not exists, but the diagrams comparing the investment and operational costs clearly showed a trend of reliability saturation, making further investment into the network renovation or operation from the perspective of reliability improvement, useless.

10.3 RECOMMENDATIONS FOR FURTHER RESEARCH

The consistency of some of the results in this study makes further research on reliability patterns justified. Nevertheless, this should be done with more substantial integration of hydraulic and geometrical parameters that influence network resilience. The statistical analyses done hint some patterns but mostly visible in extreme cases of optimised diameters or increased demand. Despite large number of simulations, the analysed samples could be considered too homogeneous and still small i.e. statistically insignificant. To continue with this approach, many more case networks of different characteristics will have to be analysed. In parallel, an attempt should be made to build more solid mathematical foundation into the concept, than was the case in this research. Secondly, the economic considerations of reliability stayed short of sensitivity analysis of the economic parameters. More time should be spent to verify the related conclusions, also adding the results of EPS hydraulic analysis to the considerations.

As for the tool, NEDRA package showed robustness that can with some further adaptations make it suitable for generation and processing of much larger number of network layouts, say, 100,000. Even in this version, an option of generating more versatile layouts within the same run is possible, by skipping some of the matrix columns in the process of non-random generation. Nevertheless, this feature has not been sufficiently tested in this research. Some degree of simplification of network generation settings is certainly possible and preferable as the current setup of menus is not quite user friendly. Nevertheless, this is just the first version of the software that has showed potential to be further improved and used in practice.

Regarding the network generation tool for the design purposes i.e. using the network template, a further improvement of the algorithm should consider an alternative of the template decomposition or instead, an option of filtering the layouts based on the nodal connectivity in the template, during the process of generation and not after it has been completed. Last but not the least, the further testing of GA-optimisation module is recommended, also by considering an options of multi-criteria optimisation. That aspect was not considered as an objective of this research, but some of the results ask for more explanation.

Finally, going back to the PDD algorithm being the heart of all reliability considerations, the room exists for further improvement if that one is to be applied in networks with more extreme topographies. The current algorithm has proven to be very effective in scenarios that avoid negative pressures. None of the two proposed measures, to set the negative node demands to zero or close the pipes with connected to the nodes with negative pressures, has been fully satisfactory, if the negative pressure would occur in the PDD mode of hydraulic calculation.

Name	Nemanja Trifunović	
Year of birth	1959	
Place of birth	Zagreb, Yugoslavia	
Nationality	Dutch/Serbian	
Present position	Associate Professor of Water Supply	
Years with firm	From 1990 – present	

EDUCATION

1984	-	Bachelor of Science in Civil Engineering, Specialisation in Hydrotechnics, University of Belgrade, Yugoslavia.
1990	-	Master of Science in Civil Engineering, Specialisation in Hydraulics, University of Belgrade, Yugoslavia.

TRAINING

2004	-	Certificate, Creative Learning Methods, Hohai University and National Hydraulic and Research Institute, Nanjing, China
2010	-	Certificate, University Teaching Qualification. 3TU Education, the Netherlands

EMPLOYMENT RECORD

1990 - present	-	Water Supply Expert, Department of Urban Water and Sanitation, UNESCO-IHE Delft, the Netherlands.
1988 - 1990	-	System Development Engineer, Municipal Water Supply and Sewerage Authority, Belgrade, Yugoslavia.
1987 - 1988	-	Compulsory military service.
1986 - 1987	-	Assistant Lecturer, Institute of Hydrotechnics, Faculty of Civil Engineering, University of Belgrade, Yugoslavia.

KEY QUALIFICATIONS

Mr. Nemanja Trifunović is specialist in the field of water distribution, in general, and in application of computer models in urban distribution networks, in particular. Apart from lecturing assignments, he is involved in research guidance as well as in organisation of various forms of training, including online learning modules. Lecturing and advisory missions conducted mostly in Africa and Asia included participation in educational and training programmes, and capacity-building projects. Next to his academic duties, Mr. Trifunović served as the programme coordinator of the Sanitary Engineering Masters programme and online course coordinator, and was the director of two large capacity-building projects conducted at Kwame Nkrumah University of Science and Technology (KNUST) in Kumasi, Ghana, in period 2001-2011. Currently, he is coordinator of the programme in water distribution at UNESCO-IHE, and the director of the capacity building project at Tshwane University of Technology in Pretoria, South Africa, implemented in consortium with Delft University of Technology, NHL University of Applied Science, SWO Dutch training institute for water management, and World Waternet.

PROFESSIONAL EXPERIENCE

1990 – present	Water Supply Expert: (1) Education and research in water transport and distribution: design and operation, use of computer models, reliability assessment, water demand management, non-revenue water and leakage. (2) Curriculum development for postgraduate- and tailor made programmes in the field of water transport and distribution; application of modern teaching methods and tools. (3) Identification, formulation and implementation of capacity building projects. (4) Design and management of post-graduate programmes in Water Supply Engineering.
2010 – present	Project Director, Capacity Building Project on Enhancing Institutional Capacity in Water and Waste Water Treatment, at Tshwane University of Technology, Pretoria, South Africa.
2007 – 2009	Coordinator of Online education at UNESCO-IHE; member of committee for educational innovation.

Update: December 2011

2005 – 2009	Project Director, Water Resources Management and Environmental Sanitation for Ghana and West Africa, Capacity Building Project at Kwame Nkrumah University of Science and Technology, Kumasi, Ghana.
2001 – 2004	Project Director, Water Supply and Environmental Sanitation for Ghana and West Africa, Capacity Building Project at Kwame Nkrumah University of Science and Technology, Kumasi, Ghana.
2001 – 2003	Programme Coordinator, The Masters Programme in Sanitary Engineering, UNESCO-IHE.
1999 - 2001	Resident Project Leader, Water Supply and Environmental Sanitation for Ghana and West Africa, Capacity Building Project at Kwame Nkrumah University of Science and Technology, Kumasi, Ghana.
1994 - 1997	Programme Coordinator, The Masters Programme in Sanitary Engineering, IHE.

PUBLICATIONS 2003-2010

BOOKS AND CHAPTERS

Odai, S.N., Trifunović, N., Eds. 2010. *Training Needs and Research Gaps in West Africa.* Proceedings of the Regional Workshop on Integrated Water Resources Management held in Accra, Ghana, February 2-3, 2009.

Trifunović, N. 2006. Reprint 2008. *Introduction to Urban Water Distribution*, 509 p., Taylor & Francis Group, UK.

Trifunović, N. 2002. *Small Community Water Supplies/IRC Technical Paper Series 40* edited by Smet J. and van Wijk C., Chapter 20: *Water Transmission,* Chapter 21: *Water Distribution,* International Water and Research Centre (IRC), 585 p., the Netherlands.

JOURNAL PAPERS

Ghebremichael, K., Gebremeskel, A., Trifunović, N., Amy, G. 2008. *Modelling Disinfection By-products: Coupling Hydraulic and Chemical Models*, Water Science & Technology – WSTWS 8.3/2008 p.289-295, IWA Publishing 2008.

Trifunović, N. 2006. *Water Demand: Quantities, Patterns and Trends*, Journal for Water and Sanitary Technology 5/2006, Association for Water Technology and Sanitary Engineering, Belgrade, Serbia, 21 p. (in Serbian/English).

CONFERENCE PAPERS

Trifunović, N., Sharma, S., Pathirana, A. 2009. *Modelling Leakage in Distribution System Using EPANET.* Manuscript presented at the IWA Water Loss 2009 Conference, Cape Town, South Africa, April 26 – 30, 2009.

Trifunović, N., Odai, S.N. 2010. *Capacity Building Experiences from Collaboration between KNUST and UNESCO-IHE.* Manuscript presented at the Regional Workshop on Integrated Water Resources Management, Accra, Ghana, February 2-3, 2009.

Trifunović, N., Vairavamoorthy, K. 2008. *Use of Demand-Driven Models for Reliability Assessment of Distribution Networks in Developing Countries.* Manuscript presented at the 10th International Water Distribution System Analysis Conference, Kruger National Park, South Africa, August 17 – 20, 2008.

Trifunović, N. 2008. *First Experiences from Running Distance Learning and Problem-Based Learning Programme in Water Distribution at UNESCO-IHE.* Manuscript presented at the International Research Symposium on Problem Based Learning, University of Aalborg, Denmark, June 30 – July 1, 2008.

Lee, K.W., Trifunović, N., Vairavamoorthy, K. 2007. *Application of Genetic Algorithms in Reliability Optimisation of Distribution Networks.* Manuscript presented at the International Conference on Computing and Control for the Water Industry, Leicester, UK, September 3-5, 2007. Supplementary Proceedings published under the title *Water Management Challenges in Global Change* by Ulanicki et al. (eds), De Montfort University, Leicester, UK, 2007.

Trifunović, N., Sharma, S., Figueres, C. 2007. *Urban Water Demand Management and Strategies in Continental Europe.* Manuscript prepared for the International Workshop on Water Demand Management in Urban Areas in Light of Tourism Development, Muscat, Sultanate of Oman, August 27 – 28, 2007.

Yoo, T.J., Trifunović, N., Vairavamoorthy, K. 2005. *Reliability Assessment of the Nonsan Distribution Network by the Method of Ozger.* Manuscript prepared for the 31st WEDC International Conference on Maximizing the Benefits from Water and Environmental Sanitation, in Kampala, Uganda, October 31 – November 4, 2005.

Totsuka, S.N., Trifunović, N., Vairavamoorthy, K. 2004. *Intermittent Urban Water Supply under Water Starving Situations.* Manuscript prepared for the 30th WEDC International Conference on People-Centered Approaches to Water and Environmental Sanitation, in Vientiane, Lao PDR, October 25-29, 2004.

Odai, S.N., Andam, K.A., Trifunović, N. 2004 *Strategic Partnerships for Sustainable Water Education and Research in Developing Countries.* Manuscript presented at the International Conference on Water Resources of Arid and Semi Arid Regions of Africa, in Gaborone, Botswana, August 3-6, 2004. Proceedings published under the title *Water Resources of Arid Areas* by Stephenson, Shemang and Chaoka, Balkema, 2004.

Trifunović, N., Umar, D. 2003. *Reliability Assessment of the Bekasi Distribution Network by the Method of Cullinane.* Manuscript presented at the International Conference on Computing and Control for the Water Industry in London, UK, September 15-17, 2003. Proceedings published under the title *Advances in Water Supply Management* by Maksimović, Butler and Ali Memon, Balkema, 2003.

PROFESSIONAL AFFILIATIONS

Member of International Water Association (IWA)
Member of American Waterworks Association (AWWA)
Member of Serbian Association for Water Technology and Sanitary Engineering

*For Product Safety Concerns and Information please contact
our EU representative GPSR@taylorandfrancis.com Taylor & Francis
Verlag GmbH, Kaufingerstraße 24, 80331 München, Germany*

T - #0107 - 230425 - C310 - 246/174/17 - PB - 9780415621168 - Gloss Lamination